D0462962

The Best American Science Writing 2001

The Best American Science Writing 2001

Editor: Timothy Ferris

Series Editor: Jesse Cohen

An Imprint of HarperCollins*Publishers*

(continued on page 334)

THE BEST AMERICAN SCIENCE WRITING 2001

HarperCollins books may be purchased for educational, business, or sales promotional use. For information please write: Special Markets Department, HarperCollins Publishers Inc. 10 East 53rd Street, New York, NY 10022.

FIRST EDITION

Designed by Cassandra J. Pappas

Library of Congress Cataloging-in-Publication Data has been applied for.

ISBN 0-06-621164-6 HARDCOVER

ISBN 0-06-093648-7 TRADE PAPERBACK

01 02 03 04 05 BVG/RRD 10 9 8 7 6 5 4 3 2 1

Contents

Introduction by Timothy Ferris vii

JOHN UPDIKE | *Transparent Stratagems* 1

MICHAEL S. TURNER | *More Than Meets the Eye* 3

NATALIE ANGIER | *In Mandrill Society, Life Is a Girl Thing* 13

JOEL ACHENBACH | *Life Beyond Earth* 19

ERIK ASPHAUG | *The Small Planets* 32

JOHN ARCHIBALD WHEELER | *How Come the Quantum?* 41

STEPHEN S. HALL | *The Recycled Generation* 44

RICHARD PRESTON | *The Genome Warrior* 61

Peter J. Boyer | *DNA on Trial* 90

John Terborgh | *In the Company of Humans* 112

James Schwartz | *Death of an Altruist* 120

Ernst Mayr | *Darwin's Influence on Modern Thought* 134

Greg Critser | *Let Them Eat Fat* 143

Andrew Sullivan | *The He Hormone* 154

Malcolm Gladwell | *John Rock's Error* 171

Helen Epstein | *The Mystery of AIDS in South Africa* 187

Debbie Bookchin and Jim Schumacher | *The Virus and the Vaccine* 212

Stephen Jay Gould | *Syphilis and the Shepherd of Atlantis* 232

Tracy Kidder | *The Good Doctor* 247

Jacques Leslie | *Running Dry* 275

Robert L. Park | *Welcome to Planet Earth* 302

Alan Lightman | *A Portrait of the Novelist as a Young Scientist* 310

Freeman J. Dyson | *Science, Guided by Ethics, Can Lift Up the Poor* 313

About the Contributors 317

Introduction by Timothy Ferris

WRITING ANYTHING that's any good is hard work, but science writers labor under a particular, and rather peculiar, set of constraints. Science is new—only about 400 years old, as a going concern—and prodigious, having transformed our conception of the universe and of our place in it. But precisely because its impact has been so rapid and so monumental, science has not yet been absorbed into our common consciousness. Readers come to the printed page already knowing something about crime and punishment, love and loss, triumph and tragedy—but not, necessarily, about the roles played by theory and observation in identifying a virus or tracing the curvature of intergalactic space. Hence science writers have to keep *explaining* things, from the significance of scientific facts to the methods by which they are adduced, while simultaneously holding the reader's attention and moving the story along. It's as if business reporters had to constantly explain what is meant by "turning a profit," or sportswriters by "scoring a touchdown."

Unsurprisingly, we science writers are often misunderstood. People tend to assume that we write computer software manuals or those buckram-bound engineering textbooks assigned to students in technical institutes. Fellow authors dismiss us as "translators," bringing to mind Robert Frost's quip that "poetry is what gets lost in translation." Editors may think us narrow. A quarter–century ago, when I was struggling to move away from writing about politics and rock music in order to concentrate on astronomy, the editor of a major magazine pressed me to do an article, called "The Bionic Man," on artificial body parts. When I declined, he became impatient.

"Well, what do you *want* to write about?" he asked, throwing up his hands, like a motorist cut off in traffic.

"Astronomy," I replied.

"You've already *written* about astronomy!"

"Yes, but I like it. It was my original interest in life."

"Aren't you afraid of becoming some sort of Johnny One Note?"

"Well, not really. You know, what's out there is nearly all of what there is—something like ninety-nine, then a decimal point, then twenty-eight more nines per cent of everything, by weight. Covering nearly everything doesn't seem all that limiting. And it leads to lots of other things."

I've been on the wrong side of arguments with editors more often than it is comfortable to recall, but on this occasion I turned out to have been right. Astronomy did lead to everything else. It led me into other sciences, of course—among them physics, chemistry, and biology—and also, by many winding paths, to poetry, literature, history, philosophy, art, music, and into conversation with some of the smartest and most creative people in the world.

Nor did it turn out to be all that much a case of taking the road less traveled by. True, science writers were relatively rare back when I was starting out. First-rate science journalists like Walter Sullivan and David Perlman were nearly as conspicuous for their scarcity as for their enviable talent, as were popularizing scientists like the physician Lewis Thomas, the physicist Steven Weinberg, and the young planetary astronomer Carl Sagan, who had recently been turning up on the Johnny Carson show. But as science has grown considerably in recent decades, so, fortunately, have the ranks of capable science writers, and the best of them are very good indeed.

They do not, in my experience, expect to get halo-effect credit for the scientific breakthroughs they write about, or to be granted allowances for the burdens they bear in gaining a sufficient command of difficult subjects to describe them in clear, simple terms. They ask only to play, on a level field, with all the other writers—including the "real" writers, a term which, in the shuttered world of the summer workshops at least, means those who dare to engage in what Norman Mailer calls the "existential" art of writing fiction. (My own admittedly biased opinion, for what it's worth, is that all fiction aspires to the state of nonfiction.)

The contents of this anthology demonstrate, I think, that they have earned that right. In putting it together I was gratified—if also a bit frustrated—to find that more excellent science writing was published in American periodicals in the year 2000 than would fit in this volume. How to proceed? One might, for instance, survey "the year in science," as the encyclopedia yearbooks put it, resolving to include at least one article on each major development, but that's what the scientific reviews are for. Instead, we elected to concentrate on science *writing*—on the best writing out there, pretty much regardless of its subject. This seems a defensible approach, although one must admit that in the prismatic world of science writing today, there is no legitimate single standard by which to rank works on a one-dimensional scale of "best" to "worst." In the end one can only plead the vagaries of personal taste, and apologize in advance to those unjustly excluded or overlooked.

As for what is here, I confess to feeling delighted, enlivened, and even (however perversely) a bit proud of it. What a banquet!

The book opens with a lovely poem, on the lucidity of undersea life, by John Updike, the only capable writer I know of who really does seem to dash these things off effortlessly and yet produce lasting results. His poem on neutrinos, "Cosmic Gall," may be the most widely quoted and reprinted verse on a scientific subject by any living writer. (Indeed, it was quoted on page one of James Gleick's introduction to *Best American Science Writing 2000.)*

Then we have "More Than Meets the Eye," by the University of Chicago cosmologist Michael S. Turner, a man of such seriousness of purpose that he retains his personal dignity and professional reputation despite a penchant for wearing baseball caps and Betty Boop T-shirts. Its subject, the mysterious matter (or energy) that exerts gravitational influence throughout the universe yet emits no light, has developed in recent years from a cosmological pain in the neck into what is beginning to look like the tip of a big, perhaps exotic, and certainly fascinating iceberg.

As in other fields, some science writers are traditionalists, doing well what their predecessors have done, while others are innovators. As her piece on mandrill society illustrates, Natalie Angier is an original. If she weren't writing about science (a dismal thought) I very much doubt that anyone else in the field would be capable of pulling off her idiosyncratic combination of wit, poise, and rock-solid command of the relevant science. One has to work hard to seem this effortless, and be quite serious to be this funny.

Speaking of which, Joel Achenbach is simultaneously serious and funny, too, and if he has ever written a boring paragraph I have yet to read it. His

"Life Beyond Earth" manages to convey quite a lot of science while remaining buoyant as a helium party balloon. It originally appeared in *National Geographic*, a magazine notorious for being more looked-at than read—it's said that even the readers of *Playboy* are less preoccupied by the pictures—so it is gratifying to present it here, where its unadorned literary merits may better be appreciated.

As I write this, NASA's NEAR ("Near Earth Asteroid Rendezvous") space probe is being gradually dropped into ever-lower orbits around the asteroid Eros, with the goal of concluding its year-long reconnaissance of the Manhattan-sized rock by deliberately bringing it down to the surface. Eros is big for an asteroid but still so low in mass, compared to the planets, that it is possible that the spacecraft, though not built to land, may survive its gentle "crash." In "The Small Planets," Erik Asphaug provides the sprightliest evocation that I have read of the strange conditions that prevail on these low-gravity worlds.

John Archibald Wheeler would, I think, be known as a leading philosopher were he not better known as a leading physicist. He has pondered the significance of quantum physics for nearly as long as the field has existed, and his splendid "How Come the Quantum?" explores the riddle that sits at the center of it all.

The next three pieces concern the biological sciences, which seem so far to be living up to the oft-quoted prediction that biology will be to the 21st century what physics was to the 20th—a fountainhead of eye-opening new facts, fresh ideas, and surprises both good and bad. Stephen S. Hall's "The Recycled Generation" brings us up to date on the hot—and politically heated—subject of stem cell research. Richard Preston's "The Genome Warrior" takes an admirably unvarnished look at the race to decode the human genome. And Peter J. Boyer's "DNA On Trial" sheds welcome light on the interface between science and the law, two powerful institutions with different and often inharmonious ways of thinking about things.

We then fan out to zoology and evolution theory, with John Terborgh's "In the Company of Humans," a warm yet unsentimental reflection on evidence that wild animals can get emotionally attached to people, James Schwartz's "Death of an Altruist," a moving account of the strange life and death of the sociobiologist George Price, and a lucid summation of Darwin's lasting influence by the eminent evolutionary thinker Ernst Mayr.

The medical sciences have long prompted science writing that is both nuanced and emotionally charged, as the seven medical pieces included here illustrate. Greg Critser's "Let Them Eat Fat" is both amusing and sufficiently alarming that I should think one could lose weight just by rereading it (although this hasn't yet quite worked for me). Andrew Sullivan's "The He Hormone" is an unforgettable first-person meditation on testosterone, the much-maligned hormone that, among other things, makes birds sing. In "John Rock's Error," Malcolm Gladwell considers what might have been a radically different career for The Pill. Helen Epstein's "The Mystery of AIDS in South Africa" amounts to a bit of scientific research in itself: It is self-doubting, open-ended, and admirably committed to the ongoing search for, if not truth, then what the musician Lou Reed (in response to a questionnaire asking celebrities what they regarded as "better than sex") called "back to back facts." In "The Virus and the Vaccine," Debbie Bookchin and Jim Schumacher examine the alarming possibility that millions of persons were contaminated, through vaccination of all things, with the lethal simian virus SV40. Rounding out our medical contributions, the inimitable Stephen Jay Gould weighs in with "Syphilis and the Shepherd of Atlantis," and the redoubtable Tracy Kidder, in "The Good Doctor," delivers a memorable portrait of the saintly physician Paul Farmer.

While looming large in the lives of those who practice it and write about it, science is only a part—albeit a powerful, promising, and in some ways threatening part—of the wider world of human life. So this anthology ends with four explorations of what might be called science in context. In "Running Dry," Jacques Leslie offers a penetrating analysis of Earth's rapidly diminishing supply of fresh water. Robert L. Park's "Welcome to Planet Earth" is a sprightly vision of the UFO controversy, graced by one of the most evocative and energetic first-person accounts of a flying saucer sighting to have been written by an empirically astute observer. The former astrophysicist Alan Lightman—who, unlike some scientists playing similar roles, is a real writer, with real things to say—reflects on his prior career in "A Portrait of the Novelist as a Young Scientist." We end with "Science, Guided by Ethics, Can Lift Up the Poor," by Freeman Dyson, a physicist who happens also to be a singularly penetrating thinker and essayist.

Oddly, reading over these pieces elicits in me a sense of almost paternalistic pride in my fellow science writers, overworked and underappreciated as they may be. They are not my children, nor I their parent. But I am honored to call them my brothers and sisters.

—T.F.

Rocky Hill Observatory, 2001

The Best American Science Writing 2001

JOHN UPDIKE

Transparent Stratagems

(BASED ON AN ARTICLE IN *SCIENTIFIC AMERICAN*,
"TRANSPARENT ANIMALS," BY SÖNKE JOHNSEN)

To be unseen: a key to sea survival,
within that boundless and unsolid mass
where up is slightly brighter and down is cobalt
deepening to purple, death everywhere.

Here in still silence evolution has
the scope of volume and the breadth of slaughter
it needs to be inventive. Venus's
girdles, so-called, millimeters thick

but six feet long, pass jellyfish whose maws
are filmy, four-cornered food-traps betrayed
by eight red gonads. Even retinas,
retaining light, are not, therefore, see-through,

and hence *Cystoma,* a kind of roach of glass,
back-stroking slyly by, has optic discs
both huge and thin, and a needle-slender gut,
since food digesting also is opaque—

some pinlike guts are always vertical,
no matter how the creature's body tilts,

to cast the smallest shadow. Protocols
of great discretion mark the watery feast.

where ambush shares the table with deceit.
Siphonophores have stinging organs shaped
like baby fish, and when a predator
approaches these, an unsuspected bulk

engulfs it, swallowing. Gelatinous
means near-invisible, but delicate;
a passing fin can shred a filmy beast,
and scientists destroy what they would study.

Down here, the very skin can hide—refraction
indices douse reflectivity
with furtive microscopic surface bumps
more minuscule than half of light's wavelength,

while body cells secrete their fat in droplets
scaled to be overlooked. Still, we are seen
and eaten. Death knows who is here, though you
avoid display, stay home, and think clear thoughts.

MICHAEL S. TURNER

More Than Meets the Eye

FROM *THE SCIENCES*

On a clear, moonless night, far from city lights, the sky provides an everyday illustration of what astronomers mean by the universe: stars are everywhere, and the closer you look the more stars you see. Even so, there is still more dark than light, more space between the stars than stars themselves. A powerful telescope reveals that the parts of the sky that seem dark to the eye are actually brimming with galaxies. In 1995 the Hubble Space Telescope pointed at the same patch of sky for ten days running to create the last word in starry nights; the image, known as the Hubble Deep Field, strongly suggested that every quarter square degree of the sky—an area about the same size as the disk of a full moon—encompasses millions of galaxies. Yet even in the Hubble Deep Field, which pictured every galaxy in that part of the sky, the space between galaxies appears empty.

But the space between galaxies is not empty at all. A much better picture in that regard was created in 1997 by a team of astronomers led by J. Anthony Tyson of Bell Labs in Murray Hill, New Jersey. Tyson and his colleagues began with an image of a single galaxy whose light, because of the position of the galaxy far behind a cluster of other galaxies, was bent by the gravity of the intervening cluster. In extreme cases of that phenomenon, called gravitational lensing, multiple images of a single distant galaxy can appear around the cluster; more typically, the image of the distant galaxy is sheared and distorted.

The image of the galaxy that Tyson and his colleagues studied appeared as several ghostly arcs scattered about the cluster. The distorted image enabled the workers to map the distribution of mass in the intervening cluster. Strangely, most of the light-distorting mass in the cluster was nowhere to be seen in the original image—not in the stars and not in the galaxies, either. More recently, another study of the heavens, the Sloan Digital Sky Survey at the Apache Point Observatory in Sunspot, New Mexico, has discovered through the same technique that the very idea of individual galaxies may be an illusion—the mass of every galaxy extends into the mass of its neighbors. Together, those studies illustrate how the universe is really put together: galaxies are like the bright lights on a dark Christmas tree, distant points of color that decorate a much larger, though unseen, web.

Astronomers can account for 100 billion galaxies—some 10^{22} stars—and yet the evidence from a variety of observations points to one conclusion: all the ordinary matter in the universe—the atoms in mountains and planets and stars and galaxies—makes up only about 5 percent of the universe. The overwhelming bulk of the cosmos is now thought to exist in two exotic forms. One form is a bath of subatomic particles, created in the aftermath of the big bang, that washes through ordinary matter and provides the infrastructure of the universe. The other is a mysterious kind of energy associated with the vacuum itself, energy that has enough power to determine both the shape of the universe and its fate. Only in the past few years have cosmologists been able to fully appreciate the dominance of unseen, or dark, matter and energy in the universe. Physicists and astronomers together are in the process of conducting experiments that should determine the precise identity of the dark-matter particles. And they have just begun to unravel the dark-energy mystery. In the next fifteen years—and I say this with some measure of optimism—astronomers will finally be able to identify the nature of the full inventory of ingredients that make up the universe.

Dark matter and dark energy have little effect on conventional matter over familiar distances. Instead, they make their presence known through their prodigious gravitational effects. In tracking them down, therefore, astronomers have had to study gigantic assemblages of matter, extending across spans of millions and billions of light-years. Perhaps the first to take that sweeping viewpoint was the Swiss-American astronomer Fritz Zwicky. In the 1930s Zwicky traced the motions of individual galaxies within great clusters of galaxies and made a remarkable discovery: the individual galaxies are moving too fast to be held together in a cluster by the force of gravity exerted by the starry matter visible within them. From his measurements, Zwicky concluded that the great clusters of galaxies must be held together by the gravitational effect of some unseen mass, which he dubbed "dark matter."

Zwicky's work, however, proved to be too far ahead of its time. Astronomers paid little attention to his result until the 1970s, when the astronomer Vera C. Rubin of the Carnegie Institution of Washington and others discovered a similar effect within individual galaxies. Our own Milky Way galaxy, in fact, would fly apart if its stars alone were exerting the only gravitational forces holding it together. Rather, the stars of the Milky Way are embedded in a great sphere of matter—a structure that is referred to as the galactic halo.

It was once thought that galactic halos and other manifestations of the so-called missing mass (a colossal misnomer, because it is the light, not the mass, that is missing) could be accounted for by various kinds of extremely dim stars or rarefied but extensive clouds of gas. Such stars and clouds have been observed, but not in masses great enough to account for the cohesion of gravitational clusters.

Could ordinary matter be hidden in some other way? A technique pioneered by my late colleague the astrophysicist David N. Schramm, and refined by me and others, including the astronomer Gary Steigman of Ohio State University in Columbus, promised to provide astronomers with a tally of all the ordinary matter ever created. In the earliest fractions of a second after the big bang, all ordinary matter existed in the form of quark soup—an intensely hot and dense mass of the elementary particles that, under less extreme conditions, comprise the neutrons and protons of ordinary atoms. As the universe expanded and cooled—though still within less than ten microseconds after the beginning of time—the quarks quickly agglomerated into neutrons and protons. In the next three minutes the neutrons joined up with protons to form nuclei of deuterium, a heavy form of hydrogen, as well as nuclei of helium and lithium. Schramm realized that the amount of deuterium formed was a highly sensitive indicator of the average density of ordinary matter present at the big bang: the more matter created, the more helium and the less deuterium would be present in today's universe. The key to making a census of the ordinary matter present in the universe was to measure the abundance of cosmic deuterium.

Measurements of cosmic deuterium began thirty years ago, when a detector was left on the moon by Apollo astronauts. But in 1998 the astronomer

David Tytler of the University of California at San Diego and his colleagues made the definitive observation of deuterium in primordial clouds of gas from a time that predates stars. Tytler's findings indicate that ordinary matter accounts for only one-seventh of the mass measured to exist in the universe. Whatever the source of the gravity that keeps galaxies and clusters of galaxies from flying apart, it cannot be clouds of gas, dim stars, extrasolar planets, black holes or anything else of that sort. The deuterium measurement, which marked the realization of Schramm's thirty-year dream, was announced just months before his death. He and I co-authored his last paper, which discussed those results.

Many exotic solutions have been proposed for the dark-matter problem, ranging from modifying the law of gravity to tucking the extra mass away into hidden dimensions of space-time. But the most promising candidates for dark matter have arisen from the study of particle physics. Neutrinos, for instance, were generated in copious abundance in the earliest moments of the big bang. They are so ubiquitous, in fact—about ten trillion (10^{13}) of them wash through every square centimeter of the universe each second—that their total mass could account for all the gravitational effects caused by dark matter, even if the mass of a single neutrino was quite small. But forty years of experimental searching failed to establish any solid evidence that neutrinos have any mass at all.

That changed in 1998. A large detector built in a zinc mine near Kamioka, Japan, had been measuring two kinds of neutrinos generated by cosmic-ray interactions with the atoms in the earth's atmosphere. Data from that experiment showed that at least one species of neutrino has a small mass. But the finding proved to be less significant for cosmology than theorists had expected. The measured neutrino mass turned out to be too small to account for more than a fraction of the dark matter. Although the experiment confirmed the principle that dark matter could consist of particles, it was obvious that the solution to the dark-matter puzzle itself would have to come from somewhere else.

The Kamioka result actually pleased most cosmologists. By 1990 neutrino matter had fallen out of favor as a probable bulk constituent of the dark matter. Because neutrinos travel nearly as fast as the speed of light, they conglomerate only on the grandest scale: the first structures that would form in a neutrino-dominated universe are enormous—larger than clusters of galaxies.

Those large structures would then fragment, forming the visible galaxies. But the Hubble Deep Field and other studies of the early universe clearly show that, instead, most galaxies were assembled in the first few billion years after the big bang, and only later did clusters and large structures begin to coalesce.

The observed structure and the natural history of the universe therefore suggest that dark matter must move relatively slowly—far slower than the speed of light. Cosmologists describe such slow-moving material as cold dark matter. And it turns out that among the particles posited by theorists in their attempts to formulate a unified account of the forces of nature, there are two that could have been spawned just after the big bang in the numbers and at the speeds necessary to create the cosmic structure that is visible to astronomers. Neither of the two particles has yet been observed, however, and the properties that have been attributed to them are strikingly divergent. Yet both are now considered the most promising answers to the question, What is the dark matter made of?

One of the candidate particles is the neutralino—a particle predicted by a theory called supersymmetry. Supersymmetry is a key test of the most promising theory that unifies gravity with the other forces of nature. The testable element of supersymmetry is its prediction that every known fundamental particle—the quarks, the electron, the photon and myriad others—has a doppelgänger particle called its superpartner. The neutralino, for instance, is the superpartner of the photon and of a particle known as the Higgs boson. Superpartners are much more massive than their ordinary-matter counterparts, and all but the lightest of them—the neutralino, whose mass is about a hundred times the mass of the proton—would long since have decayed into energy and less massive particles. My colleagues and I, as well as other investigators around the world, have calculated that the number of neutralinos left over from the earliest moments after the big bang should now be just about enough to account for the dark matter.

The second leading candidate for the dark-matter particle is the axion. The mass of an axion should be less than one million-millionth the mass of an electron, or less than 10^{-17} times the mass of the neutralino. The axion was postulated as part of a clever scheme to solve a subtle but nagging inconsistency within the standard model of particle physics. Axions would be generated in the early universe by a mechanism that is very different from the one that produced neutralinos. In spite of their low mass, the number of axions created during that early epoch could also be large enough to make up the entire complement of dark matter.

WHETHER THEY ARE heavy or light, particles of cold dark matter will be hard to detect. Much like the neutrino, the neutralino and the axion would have no charge and would interact with ordinary matter only occasionally. It is that great elusiveness that keeps dark matter "dark." But to smoke out the true nature of dark matter, cosmologists and particle physicists have worked to develop schemes to detect the neutralino and the axion, should they exist at all.

If neutralinos exist, they should fill the cosmos at a density of about one neutralino, on average, in the volume of a coffee cup. Each neutralino should be whizzing by at about one-thousandth the speed of light, so that millions of them would pass through the coffee cup per second. Nevertheless, only about one neutralino per month (or fewer) passing through that volume would leave any trace of its transit. It would do so by depositing a small amount of energy when it bumps into an atomic nucleus in the cup. By building ultra-sensitive detectors filled with many kilograms of matter for neutralinos to strike, and placing them deep underground, dark-matter detectives hope to find evidence of neutralinos. More than six such experiments are in progress around the world. This past February two of them, the Cryogenic Dark Matter Search at Stanford University in Palo Alto, California, and the Dark Matter Experiment at the Gran Sasso Underground Laboratory in Italy, began operating at a sensitivity capable of unambiguously detecting neutralinos.

Particle physicists, who are perhaps more impatient than cosmologists, have their own preferred way of searching for neutralinos: they hope to conjure them from scratch in the aftermath of highly energetic collisions between subatomic particles. At the Tevatron accelerator at Fermilab in Batavia, Illinois, and, within a few years, at the Large Hadron Collider at CERN, the European laboratory for particle physics located outside Geneva, Switzerland, physicists are trying to create the neutralino in the laboratory by generating temperatures and densities of energy similar to that which existed in the first billionth of a second of cosmic history. The signature of the creation of a neutralino would be the disappearance of huge amounts of energy, as that energy was converted into the mass of the neutralino (in accordance with Einstein's equation, $E=mc^2$, equating energy and mass).

But what if dark matter is instead made up of the featherweight axions? Detecting *them* requires an apparatus that is completely different from a neutralino detector. They interact with matter even more weakly than neutrinos do, making them almost impossible to detect.

Nevertheless, in the presence of an intense magnetic field, axions would have a slight tendency to convert into pure energy. Given their mass, such a decay would transform them into microwave photons at frequencies of around one gigahertz—about the same as those of a cellular phone. Because each cubic centimeter of the universe could be jammed with as many as 10 trillion axions, even such a rare event as axion decay should generate a few microwave photons per second—enough to be readily detected. An axion detector, therefore, is, in essence, a microwave receiver immersed in a strong magnetic field. Two such detectors, one in Japan and one in the United States, are already listening for the axion's call.

THE AMOUNT OF MATTER in the universe is bound up with other fundamental cosmological questions: the overall shape of the cosmos, the rate at which the universe is expanding, and whether that expansion will slow, halt or even reverse. According to Einstein's theories, the shape of the universe and the average density of matter in it are directly connected. Too much mass, and the universe would curve back on itself, as if space acted in a three-dimensional way like the two-dimensional skin of a balloon. Too little mass, and the universe would curve "away from" itself at every point, as if space acted like the two-dimensional surface of a saddle. For one specific density of matter, the space and time of the universe would have no curvature at all; instead, it would have a flat, Euclidean geometry.

In spite of the multitude of galaxies visible in the Hubble Deep Field image, astrophysicists have determined that the cumulative mass of all the stars accounts for a humbling 0.5 percent—one two-hundredth—of the mass needed to make a flat universe. The total amount of ordinary matter, as derived from Tytler's measurements of primordial deuterium and Schramm's calculations, is ten times as great, but even that makes up only 5 percent of the necessary mass. And the inventory of mass—both normal and dark—required to hold together galaxies and galactic clusters brings the count up to only 35 percent. Such a universe would be shaped like a saddle.

Many theorists, myself included, were disappointed with that answer. We believed that a theory first proposed by the particle physicist Alan H. Guth of the Massachusetts Institute of Technology in Cambridge—that the universe underwent an incredibly rapid expansion in the first moments after the big bang and then settled down to a much slower rate—explained many observations better than competing ideas did. Such a period of cosmic inflation, however, would leave the geometry of the universe flat, not curved. Supporting evidence for that geometry, we hoped, would come from another kind of cosmic bookkeeping, one that took a top-down approach.

The most promising such bookkeeping relied on measurements of the variations in intensity of the cosmic microwave background, the faint glow left over from the big bang itself. Such variations provide a map of how primordial matter was distributed before the stars and the galaxies were formed. The sizes of the hot and cold spots on the map depend on the shape of the universe: they are bigger in a heavy universe and smaller in a light one. This past April investigators operating Boomerang, a microwave-background-mapping experiment aboard a balloon that circumnavigated the South Pole, announced that their data confirmed that the sizes of the spots were consistent with a flat universe.

Such results offer some comfort to advocates of cosmic inflation, but they still leave cosmology in a quandary. How can a flat universe be reconciled with the finding that matter—both ordinary and dark—accounts for only 35 percent of the mass necessary for a flat universe? Was there some other factor at work? In 1995 several theorists, including Lawrence M. Krauss of Case Western Reserve University in Cleveland, Ohio, and me, suggested that the way out of the quandary was to revive an idea first proposed by Einstein—and later repudiated by him as his greatest blunder. A layer of energy spread smoothly across the universe, what Einstein called a "cosmological constant," could flatten the geometry of the universe. Although such an energy layer would not clump the way matter does, and therefore would not affect the formation of galaxies and clusters of galaxies, its presence could be measured in other ways.

ASTRONOMERS HAVE TRIED to weigh the universe much as Edwin P. Hubble first did in the 1920s, by measuring the rate at which its expansion has slowed over time. That rate of change is directly related to its total mass: the more matter there is, the more its mutual gravitational attraction can counteract the momentum of the expanding universe, and thus the greater the slowdown. Too much mass—and too great a deceleration—and the universal expansion would someday reverse itself and collapse in a "big crunch." Too slight a slowdown, and the universe would expand forever. If the deceleration was "just right," the universe would expand forever, to be sure, but more and more slowly toward a definite, asymptotic limit on its eventual size. Measuring the slowing of the universe was tantamount to predicting its fate.

To measure that effect, astronomers look for objects of comparable intrinsic brightness that are scattered throughout the visible universe. The currently favored "standard candles" are certain types of supernovas, which are bright enough to be seen billions of light-years away. Astronomers compare the brightness of the supernova, which provides a gauge of its distance, and its spectrum, which reflects its velocity away from the earth, to calculate how the cosmic expansion has changed over time. By 1998 the astronomer Saul Perlmutter of the University of California at Berkeley and his colleagues had measured enough of those distant supernovas to calculate a definitive result. At almost the same time, another team, led by the astronomer Brian P. Schmidt of the Mount Stromlo Observatory in Australia, concluded a similar survey and found the same result. To their surprise, both Perlmutter and Schmidt discovered that the universe was not slowing down at all. It was expanding at an ever-increasing rate.

A universe that is flying apart at an increasingly faster pace seems to be at odds with what physicists understand about gravity. But that kind of antigravitational drive is exactly what one would expect to see if a cosmological constant is filling in about 65 percent of the mass-energy needed to flatten the cosmic geometry.

THE REVIVAL OF the cosmological constant seemed to solve everything at once. What's more, though Einstein proposed the constant as a fudge factor, it is now known that such a term corresponds to an actual physical quantity, the energy of the quantum vacuum. According to quantum mechanics, the vacuum is not empty; it is filled with particles living on borrowed time and borrowed energy. Particles and antiparticles pop into existence for brief instants and then disappear, reverting to pure energy. The existence of such a simmering sea of particles has been confirmed by numerous experiments, and ratified by the awarding of two Nobel prizes.

Although the existence of the quantum-vacuum energy has been known for some time, physicists still have no clear idea how large it is. Estimates range from the absurdly large to the simply insignificant. But even if physicists cannot be certain that quantum-vacuum energy is the source of the acceleration of cosmic expansion, it seems clear that some form of energy is driving the universe apart.

For want of a better term, and with a nod to Zwicky, I call the mystery ingredient dark energy. Dark matter and dark energy are the yin and yang of the universe. Dark matter, like all matter, draws mass toward itself to form material aggregates on every scale. That property enabled dark matter to create the cosmic structure visible throughout the universe. Dark energy, in contrast, is repulsive, and it is distributed smoothly throughout the cosmos. Those properties make dark energy a negligible factor in understanding the way that a sparrow flies or a galaxy spins. They do suggest, however, that gravity works in previously unsuspected ways on the largest scales. Indeed, the ultimate identity of dark energy may provide a critical clue to apprehending the quantum nature of gravity.

The presence of dark energy breaks the direct link between cosmic geometry and destiny. In order to determine the ultimate fate of the universe, physicists must understand not only the shape of the universe but also the nature of dark energy.

FOR ALL THE MYSTERIES that remain, it is remarkable that one of the age-old questions of physics has now been provisionally answered: What is the universe made of? The recent results from Boomerang and the discovery of the acceleration of cosmic expansion seem to pull everything together. If they are correct, dark matter and dark energy account for the vast preponderance of the cosmos: about 95 percent of the universe is made of stuff that workers have yet to fully identify. Another 4 percent is made up of clouds of extremely hot hydrogen and helium gas—ordinary matter, but generally invisible. Nearly half of 1 percent is made up of neutrinos left over from the big bang. That leaves just 0.5 percent for the material locked up in stars—all the stars, in all the galaxies. And the atoms that have been formed since the big bang in the hearts of stars and the aftermath of supernovas, atoms of carbon and oxygen and iron and everything else in the periodic table save hydrogen and helium, all that accounts for just two parts in 10,000 of the bulk of the universe.

It is humbling to discover that the chemical elements that we are made of account for so little of the composition of the universe. Even the stars themselves pale in significance beside the dark matter and dark energy that make up the "real" universe. In fact, it would not be far-fetched to say that the billions upon billions of stars and the galaxies that populate what is usually thought of as the universe are insubstantial. On the other hand, perhaps their jewel-like rarity is one more reason to regard them as special.

Natalie Angier

In Mandrill Society, Life Is a Girl Thing

FROM THE *NEW YORK TIMES*

It was a moment when the sublime met the ridiculous, Jane Goodall by way of the Marx Brothers.

Kate Abernethy and her fellow primatologists were driving through the Lope Reserve of Gabon, in Central Africa, a gorgeous checkerboard mosaic of open savanna grassland and dense, galleried tropical rainforest. Suddenly they spied a group of mandrills crossing the road just a few yards in front of them, the distinctively colorful primates loping along in snappy single file.

In a flash, the scientists realized that the great monkey god had smiled on them, and that the military procession gave them a chance to get an accurate count of a mandrill group, something that, although the baboon-like creatures have been known about and admired for centuries, nobody had ever managed to do. Yanking out a hand-held video camera, the researchers started shooting, literally, from the hip. As the large, raucous battalion passed in front of her, Dr. Abernethy, of the International Center for Medical Research in Gabon, tried to quickly describe on the videotape the sex of each individual, and whether it was adult or juvenile.

"Female, female, female," she muttered, keeping her voice low so as not to frighten the monkeys. "Juvenile, female, female, big male, big male, female."

The mandrills kept crossing.

"Female, female, female, big male, female, female . . ."

Remember the classic scene in "Night at the Opera," when people keep piling into Groucho's room, and just when you think it's stuffed beyond capacity, another guest pushes through the door?

"Female, female, female, juvenile, big male, female . . ."

Finally, Dr. Abernethy's immortal utterance (as heard on the video, which can be seen at the new Congo exhibit at the Bronx Zoo): "Oh, forget it!" And still the mandrills kept on coming . . .

"It was one of the most exciting experiences I'd ever had with primates in the wild," said Elizabeth Rogers of the University of Edinburgh in Scotland, who has worked at the Lope Reserve periodically since 1984. "The size of that group was really over the top, and there we were, filming it."

The scientists' excitement only mounted when they replayed the tape and got the census they had sought: 604 mandrills, the largest aggregation of nonhuman primates ever observed. Nor was that particular mandrill group the most populous the scientists would encounter. In five years of rigorous study at Lope, the scientists have tallied hordes of 700, 800, up to 1,350.

Importantly, the primatologists have determined that the mandrill groups are extremely cohesive. Baboons are known to congregate while feeding in groups of maybe 150 individuals, but such crowds are fleeting, and will fission into their core social subunits of 30 or 40 animals once an area has been picked clean. By contrast, a group of 800 mandrills is a tightknit and stable society, traveling, foraging, breeding, playing, nuzzling and bickering together year after year. (The average mandrill lives for 15 years in the wild.)

It's like a movable version of Winesburg, Ohio—with one big difference. In Sherwood Anderson's fictional community, as in nearly all primate societies, males are outstanding, if not always upstanding, members of the community.

But as Dr. Abernethy and her co-authors E. Jean Wickings and Lee J. T. White of the Wildlife Conservation Society describe in a new report that was presented last month at an ecology meeting in Orlando, and that will be published this year, the enduring mandrill social group consists almost entirely of females and their dependent offspring.

As astonished as the researchers were by the magnitude of a mandrill society, they have been equally surprised to discover that adult male mandrills interact with the group only when the females are in estrus, from June through early November. Once the breeding season is through, the males disappear, and spend the rest of the year in distinctly unsimian solitude. They don't even bother to form itinerant bachelor bands, as many male primates do. And should a male happen to spy a passing flock of females in the off-season, Curious George he ain't.

"He'll simply ignore them," Dr. Abernethy said in a telephone interview from Gabon, adding, "Male mandrills don't invest in political relationships, as chimpanzees will. They don't establish long-term bonds with females. When they move into a group, at the beginning of a breeding season, they have to establish everything from scratch."

Of the approximately 235 known species of primates, only the orangutan matches the male mandrill in its taste for the monkish life. The new findings overturn almost everything that has been said about the behavior and social life of Mandrillus sphinx to date, and also call into question existing models of why primates form social groups.

"Everybody knew that mandrills were amazing animals, the type that gets featured by Disney," said Dr. John Oates, a zoologist and conservationist at Hunter College, referring to the mandrills' appearance in the animated blockbuster, *The Lion King*. But now mandrills turn out to be "so amazing," he said, "that all we can do is throw up our hands and speculate wildly about how they got to be the way they are."

On a more somber note, the discoveries raise questions about the species' long-term prospects as a free-ranging creature. Conservationists are alarmed by how the recent rise in the so-called bush meat trade—the hunting of wild animals in general and primates in particular—might affect mandrill populations.

Because the new studies show that mandrills travel over unusually long distances in spectacularly large numbers, conservationists are concerned that the monkeys will be especially vulnerable to hunters using the ever-expanding network of logging roads to penetrate into forests that were once inaccessible to hunting. Some conservation organizations now rank the bush meat trade as the single greatest threat to primates and other wildlife, outstripping in severity even the familiar eco-villain, habitat loss.

"In case of species like mandrills, we have a real problem from hunting," said Dr. John G. Robinson, a vice president of the Wildlife Conservation Society in the Bronx, which finances the mandrill project in Gabon. "People like to eat them, they're very tasty, and they come in the right size. You can smoke them and transport them very easily. And because mandrills move in very large groups, they're pretty easy to find and to intercept, and to hunt in batches of 50, 100 or even more."

At the moment, mandrills are classified as "near-threatened," rather than "endangered," but Dr. Robinson and his colleagues are taking steps while the numbers still offer hope for a creature that loves a crowd.

Among other things, the scientists are negotiating to expand the list of criteria that a logging company must meet if its wood is to be sanctioned as "sustainable." By current practice, wood is certified as sustainable if a logging company cuts down only so many trees a year, avoids clear cutting and replants a certain percentage of what it takes.

The market for such environmentally correct wood is growing faster than you can say "chain saw." Home Depot, the chain of home-improvement stores, has announced that it will buy sustainable wood whenever possible. Conservationists would like to append "sensitivity to wildlife" to the certification process, a stipulation that would impel logging companies to keep commercial hunters off their roads and away from the primate throngs.

That mandrills take group size to extremes is in keeping with the flamboyant creatures, which boast Guinness-type records in an array of categories. As Charles Darwin observed in 1871, the mandrill is the most vividly colored specimen in mammaldom, its lozenge-shaped muzzle of red and blue more like what you would expect on a bird than on the furred. The male gets gaudier still at sexual maturity, when his blue and red markings turn neon-bright and are further highlighted by the growth of dazzling white cheek ridges and a thick golden beard; and, just to be sure he is seen coming and going, his blue buttocks brighten as well.

At three feet in height and 90 pounds in weight, and with canine teeth the size of human thumbs, the adult male is the largest and most formidable of any monkey. Moreover, the sexual dimorphism, or size difference, between the male and the 25-pound female is among the most extreme in the primate order. Yet, though mandrills have been known about for centuries, prized as curiosities by potentates and long a favorite attraction in zoos, they have been little studied in the wilderness.

"They're extremely difficult to track when they're in the forest," Dr. Rogers explained.

Because mandrills look like Crayola versions of baboons, they were thought until recently to be close cousins of the Papio family, even members of it. Thus, mandrill society was usually described as baboonish in structure, composed of an alpha male and his dogged coterie of females. Some researchers suggested that the male mandrill's bright colors made it easier for him to lead "his" females through the forest.

Last year, however, a DNA study published in *The Proceedings of the Na-*

tional Academy of Sciences revealed that mandrills are neither baboons nor their next of kin. Instead, mandrills turn out to be most closely related to mangebays, short-legged tree-dwelling monkeys. Now the research at Lope is showing that, behaviorally, mandrills are a class unto themselves.

In addition to using video cameras to record troops as they traverse openings in the forest, the researchers have also managed to put radio collars on a number of adult mandrills, an approach that revealed the loneliness of the off-season male.

By collecting thousands of scat samples, the scientists have learned much about mandrill diet. It turns out that mandrills are finicky omnivores. They eat a little bit of a lot of things—nuts, leaves, insects, fruit, grass, fungi, small vertebrates. They pick and choose and throw a lot away in disgust. A species of nut that pleases them in one spot may repel them in the next. They travel constantly as they forage, covering three or five miles a day, far more than most primates travel. Their home range is accordingly huge, 115 square miles or so, the largest such range known for any primate.

The researchers have also learned how violent a male's life can be.

"Of the six males that we collared, only one had all its canines intact," Dr. Abernethy said. "All the rest had broken teeth, scars on the face, ripped nostrils up to the eye. One had such severe fresh puncture wounds that we could see the cartilage. We couldn't collar him, and I don't know if he survived after we let him go." The males get these wounds in violent battles with one another during mating season.

As it turns out, the males do not have much time to make their reproductive mark. They do not have harems, after all, and they cannot very well monopolize a female society numbering 800 strong.

Once females start coming into estrus, males from all around descend on the group. The males start grunting, panting, heaving noisily from the chest hour upon hour. They are big, colorful, loud and impossible to ignore. They fight, yes, and they suffer from the lacerating canines of their competitors, but the primatologists propose that the males would suffer even greater damage if not for their dramatic appearance.

"Having seen the risks they take during actual fights," Dr. Abernethy said, "I'd suspect that the incredible investment they make in color and auditory signaling is an attempt to minimize that risk." In other words, a male in his fully pigmented and baritone glory is trying to get the attention of at least some females as quickly as possible, before a rival suitor decides to go for his throat.

The researchers do not yet understand why mandrill societies have evolved as they have. They do not know why so many females stick together, and they are just starting to do the studies to see whether all the females in a given troop are related.

Does large group size protect the female, and if so, from whom? Perhaps from predators, or perhaps from mandrill males. Whereas female baboons often display the wounds of the alpha males that seek to control them, female mandrills do not seem to be the recipients of mate abuse, despite the male's vastly superior size.

Can mandrills find safety in numbers should human hunters come to call? Don't bet a buffalo nickel on it.

JOEL ACHENBACH

Life Beyond Earth

FROM *NATIONAL GEOGRAPHIC*

Something astonishing has happened in the universe. There has arisen a thing called life—a flamboyant, rambunctious, gregarious form of matter, qualitatively different from rocks, gas, and dust, yet made of the same stuff, the same humdrum elements lying around everywhere.

Life has a way of being obvious—it literally scampers by, or growls, or curls up on the windowsill—and yet it's notoriously difficult to define in absolute terms. We say that life replicates. Life uses energy. Life adapts. Some forms of life have developed large central processing networks. In at least one instance, life has become profoundly self-aware.

And that kind of life has a big question: What else is alive out there?

There may be no scientific mystery so tantalizing at the brink of the new millennium and yet so resistant to an answer. Extraterrestrial life represents an enormous gap in our knowledge of nature. With instruments such as the Hubble Space Telescope, scientists have discovered a bewildering amount of cosmic turf, and yet they still know of only a single inhabited world.

We all have our suppositions, our scenarios. The late astronomer Carl Sagan estimated that there are a million technological civilizations in our galaxy alone. His more conservative colleague Frank Drake offers the number 10,000. John Oro, a pioneering comet researcher, calculates that the Milky

Way is sprinkled with a hundred civilizations. And finally there are skeptics like Ben Zuckerman, an astronomer at UCLA, who thinks we may well be alone in this galaxy if not in the universe.

All the estimates are highly speculative. The fact is that there is no conclusive evidence of any life beyond Earth. Absence of evidence is not evidence of absence, as various pundits have wisely noted. But still we don't have any solid knowledge about a single alien microbe, a solitary spore, much less the hubcap from a passing alien starship.

Our ideas about extraterrestrial life are what Sagan called "plausibility arguments," usually shot through with unknowns, hunches, ideologies, and random ought-to-bes. Even if we convince ourselves that there must be life out there, we confront a second problem, which is that we don't know anything *about* that life. We don't know how truly alien it is. We don't know if it's built on a foundation of carbon atoms. We don't know if it requires a liquid-water medium, if it swims or flies or burrows.

Despite the enveloping nebula of uncertainties, extraterrestrial life has become an increasingly exciting area of scientific inquiry. The field is called exobiology or astrobiology or bioastronomy—every few years it seems as though the name has been changed to protect the ignorant.

Whatever it's called, this is a science infused with optimism. We now know that the universe may be aswarm with planets. Since 1995 astronomers have detected at least 22 planets orbiting other stars. NASA hopes to build a telescope called the Terrestrial Planet Finder to search for Earth-like planets, examining them for the atmospheric signatures of a living world. In the past decade organisms have been found thriving on our own planet in bizarre, hostile environments. If microbes can live in the pores of rock deep beneath the earth or at the rim of a scalding Yellowstone spring, then they might find a place like Mars not so shabby.

Mars is in the midst of a full-scale invasion from Earth, from polar landers to global surveyors to rovers looking for fossils. A canister of Mars rocks will be rocketed back to Earth in the year 2008, parachuting into the Utah desert for scrutiny by scientists in a carefully sealed lab. In the coming years probes will also go around and, at some point, into Jupiter's moon Europa. That icy world shows numerous signs of having a subsurface ocean—and could conceivably harbor a dark, cold biosphere.

The quest for an alien microbe is supplemented by a continuing effort to find something large, intelligent, and communicative. SETI—the Search for Extraterrestrial Intelligence—has not yielded a confirmed signal from an alien civilization in 40 years of experiments, but the signal-processing technology grows more sophisticated each year. The optimists figure it's only a matter of time before we tune in the right channel.

No one knows when—or if—one of these investigations might make a breakthrough. There's a fair bit of boosterism surrounding the entire field, but I'd bet the breakthrough is many years, if not decades, away. The simple truth: Extraterrestrial life, by definition, is not conveniently located.

But there are other truths that sustain the search for alien organisms. One is that, roughly speaking, the universe looks habitable. Another is that life radiates information about itself—that, if nothing else, it usually leaves a residue, an imprint, an echo. If the universe contains an abundance of life, that life is not likely to remain forever in the realm of the unknown.

Contact with an alien civilization would be an epochal and culturally challenging event, but exobiologists would settle gladly for the discovery of a tiny fossil, a mere remnant of extraterrestrial biochemistry. One example. One data point to add to the one we have—Earth life. That's what we need to begin the long process of putting human existence in its true cosmic context.

EXOBIOLOGISTS GO to the worst places on Earth, or at least the most extreme—the driest, coldest, most Mars-like or Europa-like environments they can find.

If you want to track down the exobiologist Jack Farmer from Arizona State University, you look for him in Death Valley, on the shores of nearby Mono Lake, or swimming beneath the ice shelf in Antarctica. If seeking Chris McKay, you might check out the Atacama Desert of Chile or some island north of the Arctic Circle.

The place to find Penny Boston is in the nastiest cave imaginable. I tagged along with Boston on one of her trips to a wet, bat-ridden cave in southern Mexico called Villa Luz. Boston has been studying the microbes that thrive there—in environments where a human being not wearing a gas mask would perish.

"All my life I've wanted to cross the cosmos, go to other planets," says Boston. "This is probably as close as I'll get at my age."

Boston and her friend Diana Northup, a librarian and cave biologist in New Mexico, are undeterred by the face-smashing gas masks they must wear or by the constant wetness, the darkness, the bats, or the slight possibility that a belch of carbon monoxide would kill everyone. Nor are they overly concerned about the various threats of malaria and dengue fever and whatever other exotic diseases they might pick up here. Before we entered Villa Luz, I asked if there was any danger of encountering an unknown, Ebola-like pathogen. "We think it's moderately unlikely," Boston said.

The cave floor was covered with water of varying depths and no transparency, and we walked gingerly so as to avoid discovering unmapped deep water. By caving standards, though, this was a walk in the park—no ropes required, just some crawling and scrambling through low-ceilinged passages.

Eventually we reached the deepest, largest chamber, known as the Great Hall. Midges flitted, spiders spun webs, bats zagged and zigged just over our heads, emitting their high-pitched sonar. Red rock walls were covered with green slime, black muck, gooey white gypsum paste, and limestone in the process of being dissolved by sulfuric acid.

Just as I was thinking how much this cave resembled the human nasal cavity, we came to the snottites (Boston is lobbying to have the word recognized as a scientific term). Snottites are gelatinous structures formed by microbial wastes. They dangle from the ceiling. Boston and her team have been measuring their growth, trying to understand the metabolism of the microbes and their long-term effect on the geology of the cave. Dry weather since her last visit seemed to have inhibited the growth of the structures.

Mike Spilde, another member of the team, splashed over to where I'd been inspecting a water bug whose shell was covered with eggs. He reached into a spring burbling from under a rock and pulled out some gray wads the consistency of cooked cabbage. These are known, in keeping with the theme of the place, as phlegm balls. They are vibrant microbial communities, not clinging to life in a narrow niche but proliferating in it, replicating up a storm.

Taking a break back on the surface, Boston placed some of her cave work in context.

"We have discovered"—she means scientists in general—"organisms thriving in environments harsh to us but essential to them. It broadens your perspective. We all suffer to some extent from 'expertitis' in science. It's good for your soul, and good for your intellect, and good for your work to have your imagination stretched, to be open to the possibilities."

The most tantalizing possibility is that the universe hums with life and that in the coming centuries we will find it. An exobiologist's abiding optimism is fired by the knowledge that living things are primarily constructed of hydrogen, nitrogen, carbon, and oxygen—the four most common chemically active elements in the universe. And life is inextricably interwoven with nonlife; not even the sharpest razor can perfectly slice them apart.

We also know that a functioning ecosystem does not require sunlight or photosynthesis. In the early 1990s researchers found that the basaltic rock deep beneath Washington State contains an abundance of microbes totally cut off from the photosynthetic world. Even more complex life can adapt to hostile places. When scientists in the deep-sea submersible *Alvin* went tooling around the mid-ocean ridges, they found hot vents covered with shrimp and mouthless tube worms.

What remains unknown is whether life can survive over time in narrow ecological niches on largely barren worlds. Could life survive in aquifers far below the harsh surface of Mars? What could endure the cold, dark environment of Europa's hypothesized ocean? Can an alien world have just a little bit of life, or are biospheres an all-or-nothing proposition?

The cave at Villa Luz, as remote as it is, does not exist in isolation. It is a small, connected piece of a world that riots with life.

AS SCIENTISTS STRUGGLE to find a trace of life somewhere else in the universe, there exists for many people a more dramatic situation, one in which extraterrestrial life isn't microbial and slimy but rather intelligent, technological, and lurking in our midst. The believers in these aliens are not likely to be convinced that ETs are a bogus phenomenon. An ability to elude detection and confirmation, particularly by mainstream thinkers, is a presumed characteristic of the Visitors.

Having dropped in on a couple of UFO conventions and visited Roswell, New Mexico, and its UFO museum, I've come to the conclusion that it's not possible to win an argument about space aliens. True believers and skeptics rarely go over to the other side. I think it's fair to say, however, that flying-saucer aliens lack scientific stature. If they insist on being so jumpy, if they insist on abducting people in the middle of the night when no one else can verify their presence, then they have no right to enter a reputable natural history museum.

But neither are people who believe in the UFO narrative—which generally is dated to the 1947 sighting of some flying "disks" near Mount Rainier in Washington State—necessarily irrational, much less crazy, as they are sometimes depicted. Most people operate from the same instinct, which is to know the truth about the universe. That so many people would adopt a theory of aliens utterly contrary to that of mainstream science (and that of, among agencies, the U.S. Air Force, which spent 22 years investigating UFO reports) is a reminder of the special attraction of the idea of extraterrestrial life.

As many writers have noted, aliens are, for some people, the secular equivalent of angels and demons and ghostly spirits. The aliens are an extrapolation of modern astronomy and engineering (big universe, fast rocket ships), but they also possess some ancient urge to come to Earth and mess with human beings. What makes them so intriguing is that even scientists will concede that alien beings could very well be out there somewhere. Therefore the scenario in which they come to Earth requires only some imagination about transportation.

Many scientists don't wonder why aliens are buzzing the Earth in flying saucers—they wonder why they aren't. In 1950 Enrico Fermi, a physicist, asked some of his colleagues a question that would become famous: Where is everybody? Humans could theoretically colonize the galaxy in a million years or so, and if they could, astronauts from older civilizations could do the same. So why haven't they come to Earth? This is known as the Fermi paradox.

Could it be that they're observing us but not interfering? (The zoo hypothesis.) Did they come and leave artifacts and get bored and go away? (This is the "ancient astronauts" idea that posits the aliens as builders of pyramids and so forth.) Or could it be that for all intelligent species, interstellar travel is too expensive and time-consuming? (It's just less than 25 trillion miles from Earth to the nearest star beyond the sun.)

Or could it be possible that, at least in our part of the galaxy, the most technologically advanced species is the one right here on Earth?

Our contemporary culture did not invent this idea of life beyond Earth. The alien is a Hollywood stock character but not a Hollywood creation. More than 2,000 years ago the Greek philosopher Metrodorus of Chios wrote, "It is unnatural in a large field to have only one shaft of wheat, and in the infinite Universe only one living world." Four centuries ago Giordano Bruno was burned at the stake in part because he believed that there were inhabited worlds throughout the cosmos. Astronomers like Christian Huygens supplemented their purely scientific work with treatises on the characteristics of life beyond Earth. Huygens felt, for example, that aliens would probably have hands, like humans.

Missing from the debate, typically, was the one ingredient of a truly persuasive argument: Evidence. That seemed to change with the apparent discovery of the Martian canals. In 1877 Giovanni Schiaparelli, an Italian astronomer, found what he called *canali*, or channels, on the surface of the planet. The American astronomer Percival Lowell and a few colleagues took the idea from there.

In the final years of the 19th century, Lowell, using a new telescope he built near Flagstaff, Arizona, revealed the discovery of hundreds of canals and argued that these were the artificial creations of an intelligent Martian civilization. In fact, he wrote, the Martians would certainly have to be superior to us. He reasoned that their globe-spanning engineering projects were far beyond our own capabilities and that the ability of a race of creatures to live in harmony over the whole of a planet showed them to be of a more advanced character than our own squabbling selves. H. G. Wells tweaked the idea just a bit in his novel *The War of the Worlds,* in which the Martians come to Earth with deadly heat rays and dreams of conquest.

The Martians, alas, were doomed, except as cultural artifacts. When astronomers looked at Mars with more powerful telescopes, there were no canals anywhere. Lowell's canals were created in his mind's eye—a classic example of the saying "Believing is seeing." But there remained, into the 1960s, a fascination with waves of seasonal darkening on the surface. Could this be vegetation? The Martian prairies and forests were conclusively eradicated in 1965, when the Mariner 4 probe took 22 pictures of the surface. Mars was a cratered wasteland, reminiscent of the moon.

When the Viking landers descended to the Martian surface in 1976, they found no compelling sign of life and indeed discovered that the surface contains no trace of organic molecules. Though the mission was a fantastic triumph of science and technology, the absence of detectable life on Mars put exobiology into a two-decade funk.

The mood changed in the 1990s. Biologists were detecting organisms in such exotic environments on Earth that they were inspired to look anew at the rest of the solar system as potentially habitable. They also discovered signs that life appeared early in the Earth's history. Intriguingly, at about the time life arose on Earth, Mars was a much more hospitable planet than it is today. Images of the Martian surface indicate that the planet once had flowing rivers and perhaps an ocean. Life could even have started on Mars and spread to Earth aboard a meteorite.

Which brings up the most famous Martian meteorite: ALH84001. In 1996 a team of three NASA scientists based in Houston announced that this potato-size rock, found in Antarctica, contained what appeared to be Martian fossils.

The discovery was proclaimed at an unforgettable NASA press conference in Washington, D.C., on August 7, 1996. Everyone realized the historical glory of being right about these purported microfossils—and the reciprocal tarnish of being wrong. Dan Goldin, NASA Administrator, cautioned that the results were not definitive, but he said, "We may see the first evidence that life might have existed beyond the confines of this planet, the third rock from the sun."

The NASA team made a dramatic presentation, complete with graphics and the first, startling images of the microfossils, one of which looked like a worm (others a bit like Cheetos). But then came a dissenter, UCLA's J. William Schopf, who said that on a scale of one to ten of increasing probability of biological origin, he could only grant the alleged Martian fossils a two. So began, that day, an enduringly divisive scientific debate.

The NASA scientists had to admit that their four primary lines of evidence could each be explained nonbiologically. They had found, for example, PAHs, polycyclic aromatic hydrocarbons, which sometimes are associated with living things but which also can be found in car exhaust. They found grains of magnetite, which might have been produced inside microbes or might not have. In a sense the research raised the question of whether a series of possibilities add up to a probability. At the least it runs headlong into a Sagan dictum, which is that extraordinary claims require extraordinary evidence.

The NASA team saw its conclusions vigorously attacked. One damaging study showed that some of the microbe-like structures were merely flakes of the rock rendered more biological in appearance by the coating process used in the preparation of slides. Researchers also found contaminants from Earth inside the meteorite.

The team fought these challenges point by point, but after three years critics felt they'd pretty much killed off the Mars rock. Luann Becker, a geochemist at the University of Hawaii, told me, "I think we're beating a dead horse."

But Everett Gibson, part of the NASA meteorite team, sees this as a typical scientific resistance to a revolutionary idea. "Science," he said, "doesn't accept radical ideas quickly."

There was a time when scientists didn't believe that meteorites could possibly fall from the sky. There was a time when plate tectonics—the movement and collision and subduction of vast slabs of the Earth's crust—was deemed a very strange idea. Are the Mars rock fossils in the same category? Or are they more like those canals?

IF LIFE SPRANG UP through natural processes on the Earth, then the same thing could presumably happen on other worlds. And yet when we look at outer space, we do not see an environment teeming with life. We see planets and moons where no life as we know it could possibly survive. In fact we see all sorts of wildly different planets and moons—hot places, murky places, ice worlds, gas worlds—and it seems that there are far more ways to be a dead world than a live one.

Within our solar system the Earth may be in a fairly narrow habitable zone, not too hot and not too cold, just the right distance from the sun that water can splash around on the surface in a liquid state. And there may be many other things that make life on Earth possible. The tectonic activity recycles the planet's carbon. Mars has no such mechanism, and this seemingly minor deficiency may be the reason Mars lost most of its atmosphere.

The search for extraterrestrial life is in some ways a search for constraints, for the things that limit the emergence of life or the evolution of complex organisms. For calculating the number of technological, communicative civilizations, the most popular theoretical tool is the Drake equation.

In 1960 an American astronomer named Frank Drake became the first person to conduct a sensitive radio search for signals from extraterrestrial civilizations. He aimed an 85-foot radio telescope at two nearby stars and, after one false alarm, found no intentional signals. The next year, preparing for a meeting of visionary thinkers (including the young Sagan), he made an outline for how to discuss the probability of detecting intelligent life, starting with the rate of star formation and the typical number of planets and working through to the longevity of civilizations. "I thought it was just a gimmick. It's amazing to me now that it's in the astronomy textbooks," he told me.

Going through the factors from left to right—$N=R_{*}f_{p}n_{e}f_{l}\,f_{i}f_{c}L$—you don't get very far before you hit some serious unknowns. Jill Tarter, who has dedicated her career to SETI, says, "The Drake equation is a wonderful way to organize our ignorance."

The only factor well understood, R_{*}, tells us the number of stars. Suffice it to say that there are lots of them, more than a hundred billion in our galaxy alone, maybe as many as 400 billion (and that doesn't count, of course, the billions of other galaxies). The second factor, f_{p}, the fraction of stars with planets, is rapidly coming into focus. There are still uncertainties, since the detection equipment can find only extremely massive planets. These behemoths aren't like the Earth. Many of the extrasolar planets discovered so far may have migrated toward the parent star over time, destroying any rocky, Earth-like planets along the way.

Eventually the Terrestrial Planet Finder (TPF) could help solve the next factor in the equation, n_e—the number of planets with habitable environments—and may even be able to glean some evidence of the following factor, f_l—the fraction on which life has originated. TPF, still many years from construction, would capture the feeble reflected light from a distant rocky planet, while nulling the far more brilliant light of the parent star. This remnant of light might amount to only a single pixel of data. The light could then be examined for the spectral signatures of, for example, oxygen, methane, ozone, or some indicator of a planet with biological processes. Thrilling as such a discovery would be, it's easy to imagine how it would echo the situation with the Mars rock. There would likely be no "proof" of life, merely an interpretation subject to much second-guessing.

Even on Earth the origin of life is a stubbornly enduring mystery. "How can a collection of chemicals form themselves into a living thing without any interference from outside?" asks Paul Davies, a physicist and writer. "On the face of it, life is an exceedingly unlikely event," he argues. "There is no known principle of matter that says it has to organize itself into life. I'm very happy to believe in my head that we live in a biofriendly universe, because in my heart I find that very congenial. But we have not yet discovered the Life Principle."

No one is even sure that life requires liquid water, though that seems a reasonable bet and is surely the case on Earth. Liquid water may be fairly scarce in the universe—Europa may help solve that issue—but another presumed ingredient of life, organic molecules, those made up primarily of carbon, are commonplace. That's why Jeffrey Bada, a pretty hardnosed researcher, thinks the universe is full of living things. "I don't see any way to avoid that," he said, sounding almost apologetic.

So let's assume that life can spring up in many places. Now comes f_i, another giant unknown in the Drake equation: How often does life evolve to a condition of intelligence?

There are those, like Ernst Mayr, one of the great biologists of the 20th century, who argue that high intelligence has occurred only once on Earth, among something like a billion species. Hence it is a billion-to-one long shot. But Paul Horowitz, a Harvard physicist, argues that the same data can be looked at the opposite way: That on the only planet we know of that has life, intelligence appeared. That's a one-for-one proposition.

I've never met anyone who thinks that if you rewound the tape of terrestrial evolution (to use Stephen Jay Gould's metaphor) and played it again, you'd wind up with a genetically identical human being the second time around. But there are those who say that an intelligent being is more likely under certain initial conditions. The paleobiologist Andy Knoll argues that intelligence is rooted in the emergence of structures that allow simple animals to sense their environment and seek food. "If we get to creepy crawlies that look for food, then at some point intelligent life may emerge," he says.

There are those who argue passionately that alien life would be nothing like us—in Fred Hoyle's novel *The Black Cloud* the alien is a gaseous cloud that decides to feed on our sun—and there are others who say the biology of the Earth is probably a pretty good example of what's out there.

Finding life somewhere else, even a single alien amoeba, might clarify the extent to which life evolves along parallel tracks—and whether it typically arrives at certain useful structures, such as eyeballs, wings, and large brains. Human beings have, by far, the biggest brains on Earth in ratio to body size. Did we get these things in our skulls through a random, improbable evolutionary quirk?

Lori Marino, a psychobiologist at Emory University, points out that dolphins appear to have undergone a dramatic increase in brain size in the past 35 million years, which may have a parallel in the quadrupling of brain size among hominids in the past few million years. By her reckoning, huge leaps in intelligence may be found among creatures on worlds everywhere else in the universe.

But it's also true that the data are scarce, and this is still a territory for, among others, philosophers and theologians. What does it mean to be "intelligent"? When we "think" or "feel" or "love," what is it that we are doing? When we ask if we are "alone," we really want to know if there are others out there in the universe who are, in key aspects, very much like ourselves. We seek the communicators—Drake's f_c, creatures who have the technology to send signals—storytellers, ideally.

EVERY THREE YEARS a bioastronomy meeting gathers many of the leading thinkers in the field. I went to the 1999 assemblage in August on the Big Island of Hawaii, and at the opening reception around a hotel pool a University of Toronto sociologist named Allen Tough offered a provocative theory:

"I think a probe is already here. It's probably been here a long time."

He didn't mean flying saucers. His alien probes would be much smaller—"nanoprobes," tiny robotic exploratory craft sent to Earth from advanced civilizations. The alien probes may, at some point, let themselves be known to human civilization. How? Where? "I think it will happen on the World Wide Web," said Tough.

Tough and about a dozen other visionaries had a pre-conference meeting to discuss what to do if human civilization receives a "high-content" message from extraterrestrials. There was much uncertainty about how well prepared humankind is for such an event. We might have trouble crafting a response. Should we be forthcoming about the flaws of our species? If we acknowledge our history of wars and slavery, could that be misinterpreted as a threat? What if, even as an international committee of well-meaning thinkers tried to put together a message, some guerilla radio broadcaster or "shock jock" beat everyone to it?

Bioastronomy also has its more down-to-Earth side. The meeting reminded me how much there is still to learn about our own little solar system. Exobiologist Jack Farmer made a simple yet stunning point one morning when he noted that neither the Viking landers in 1976 nor the Pathfinder spacecraft in 1997 carried to Mars the tool so vital to a geologist: a magnifying lens. Nor would the polar lander scheduled for a December 1999 landing carry such an instrument. Farmer's comment remained in my mind when Cindy Lee Van Dover, an oceanographer, noted that no one has ever made a dive in a deep-sea submersible to an active hot vent in the Indian Ocean to see what might be alive down there.

So before we worry about our dealings with the Galactic Empire, we have some serious fieldwork to do closer to home.

FREEMAN DYSON, a physicist, has argued that humans may engineer new forms of life that will be adapted to living in the vacuum of space or on the surface of frozen moons and comets and asteroids. In Dyson's universe, life is mobile, and planets are gravitational traps inhibiting free movement.

"Perhaps our destiny is to be the midwives, to help the living universe to be born," he said recently. "Once life escapes from this little planet, there'll be no stopping it."

But life must first survive this planet. The longevity of civilizations is the final factor in the Drake equation, the haunting letter L. Humans in their modern anatomy have been around only 125,000 years or so. It is not clear yet that a brain like ours is necessarily a long-term advantage. We make mistakes. We build bombs. We ravage our world, poison its water, foul its air. Our first order of business, as a species, is to make L as long an interval as possible.

I would hope that anyone who investigates this issue will come away with a renewed appreciation of what and who we are. In a universe of empty space and stellar furnaces and ice worlds, it is good to be alive. And we should remember that even if we find intelligent life beyond Earth, it may not be what we expect or even what we were searching for.

The alien may not speak to that part of our consciousness that we deem most important—our spirit, if you will. It may have little to teach us. The great moment of contact may simply remind us that what we most want is to find a better version of ourselves—a creature we will probably have to make, from our own raw elements, here on Earth.

ERIK ASPHAUG

The Small Planets

FROM *SCIENTIFIC AMERICAN*

Growing up in the Space Age, my friends and I would sometimes play the gravity game. One of us would shout, "Pretend you're on the moon!" and we'd all take the exaggerated slow strides we'd seen on television. "Pretend you're on Jupiter!" another would say, and we'd crawl on our hands and knees. But no one ever shouted, "Pretend you're on an asteroid!" In that pre-*Armageddon* era, who knew what "asteroid" meant? Now a grown-up who studies asteroids for a living, I still don't know how to respond.

Although we haven't seen any of the largest asteroids up close, they probably resemble shrunken, battered versions of the moon. In their weaker gravity, visiting astronauts would simply take longer strides. But below a few dozen kilometers in diameter, gravity is too feeble to press these so-called minor planets into even an approximately round shape. The smallest worlds instead take on a carnival of forms, resembling lizard heads, kidney beans, molars, peanuts and skulls. Because of their irregularity, gravity often tugs away from the center of mass; when added to the centrifugal forces induced by rotation, the result can seem absurd. Down might not be down. You could fall up a mountain. You could jump too high, never to return, or launch yourself into a chaotic (though majestically slow) orbit for days before landing at an unpredictable location. A pebble thrown forward might strike you on the head.

A gentle vertical hop might land you 100 meters to your left or even shift the structure of the asteroid underfoot. Even the most catlike visitor would leave dust floating everywhere, a debris "atmosphere" remaining aloft for days or weeks.

These aspects of asteroid physics are no longer only theoretical curiosities or a game for children. Space missions such as the Near Earth Asteroid Rendezvous (NEAR), the first probe to go into orbit around a minor planet, are dramatitally modernizing our perception of these baffling objects. But in spite of careful observations and the occasional proximity of these bodies to Earth, we know less about asteroids (and their relatives, the comets) than we knew about the moon at the dawn of space exploration. Minor planets exhibit a delicate interplay of minor forces, none of which can be readily ignored and none of which can be easily simulated in a laboratory on Earth. Are they solid inside, or aggregate assemblages? What minerals are they composed of? How do they survive collisions with other small bodies? Could a lander or astronaut negotiate an asteroid's weird surface?

Half-Baked Planets

My graduate studies began during the [first] Bush administration, when asteroids were mere dots—a thousand points of light known to orbit primarily in a belt between Mars and Jupiter. A few lesser populations were known to swoop closer to Earth, and then there were comets in the Great Beyond. From periodic variations in color and brightness, asteroids were inferred to be irregular bodies ranging in size from a house to a country, rotating every several hours or days. More detailed properties were largely the stuff of scientific imagination.

Asteroids residing closer to Mars and Earth commonly have the spectra of rocky minerals mixed with iron, whereas asteroids on the Jupiter side are generally dark and red, suggesting a primitive composition only coarsely differentiated from that of the primordial nebula out of which the planets began to coalesce 4.56 billion years ago. This timing is precisely determined from analysis of lead isotopes—the products of the radioactive decay of uranium—in the oldest grains of the most primitive meteorites. In fact, meteorites have long been suspected to derive from asteroids. The spectra of certain meteorites nearly match the spectra of certain classes of asteroids. We therefore have pieces of asteroids in our possession.

Many astronomers used to think that telescope observations, combined with meteorite analysis, could substitute for spacecraft exploration of asteroids. Although the puzzles proved more stubborn than expected, researchers have been able to piece together a tentative outline of solar system history. For the planets to have accreted from a nebula of dust and gas, there had to be an initial stage in which the first tiny grains coagulated into growing bodies known as planetesimals. These became the building blocks of planets. But in the zone beyond Mars, gravitational resonances with massive Jupiter stirred the cauldron and prevented any body from growing larger than 1,000 kilometers across—leaving unaccreted remnants to become the present asteroids.

The largest of these would-be planets nonetheless accumulated enough internal heat to differentiate: their dense metals percolated inward, pooling and perhaps forming cores, leaving behind lighter rocky residues in their outer layers. Igneous activity further metamorphosed their rock types, and volcanoes erupted on some. Although none grew large enough to hold on to an atmosphere, hydrated minerals found in some meteorites reveal that liquid water was often present.

Encounters among the planetesimals became increasingly violent as Jupiter randomized the orientation and ellipticity of their orbits. Instead of continuing to grow, the would-be planets were chiseled or blasted apart by mutual collisions. Their pieces often continued to orbit the sun in families with common orbital characteristics and related spectra. Many asteroids and meteorites are the rock- or metal-rich debris of these differentiated protoplanets. Other asteroids (and most comets) are more primitive bodies that for various reasons never differentiated. They are relics from the ur-time before planets existed.

The Sky Is Falling

A DECADE AGO no asteroid had been imaged in any useful detail, and many astronomers had trouble taking them seriously. The first asteroids, discovered in the early 1800s, were named in the grand mythological manner. But with the tenth, the hundredth and the thousandth, asteroids began taking on the names of their discoverers, and then of discoverers' spouses, benefactors, colleagues and dogs. Now, after a century of near-neglect, serious interest in asteroids is waxing as new observations transform them from dim twinkles in the sky into mind-boggling landforms. For this, asteroid scientists can thank National Aeronautics and Space Administration administrator Daniel S. Goldin and the dinosaurs.

Goldin's "faster, better, cheaper" mantra has been a boon to asteroid science, because a visit to a tiny neighbor is both faster and cheaper than a mission to a major planet. The specter of fiery death from above has also focused minds. The discovery of the Chicxulub crater in the Yucatan vindicated the idea that the impact of an asteroid or comet 65 million years ago extinguished well over half the species on Earth.

A repeat is only a matter of time, but when? Until we completely catalogue all significant near-Earth asteroids—a job we have just begun—poker analogies must suffice. (We will never completely catalogue the comet hazard, because each comet visits the inner solar system so rarely.) The chance of a global calamity in any year is about the same as drawing a royal flush; your annual chance of dying by other means is about the same as drawing three of a kind. None of us is remotely likely to die by asteroid impact, yet even scientists are drawn to the excitement of apocalypse, perhaps too often characterizing asteroids by their potential explosive yield in megatons instead of by diameter. Our professional dilemma is akin to notoriety in art: we want asteroids to be appreciated for higher reasons, but notoriety pays the bills.

Egged on by this nervous curiosity, we are entering the golden age of comet and asteroid exploration. Over a dozen have been imaged and each new member of the menagerie is welcomed with delight and perplexity. They are not what we expected, to say the least. Small asteroids were predicted to be hard and rocky, as any loose surface material (called regolith) generated by impacts was expected to escape their weak gravity. Aggregate small bodies were not thought to exist, because the slightest sustained relative motion would cause them to separate.

Reduced to Rubble

BUT OBSERVATIONS and modeling are proving otherwise. Most asteroids larger than a kilometer are now believed to be composites of smaller pieces. Those imaged at high resolution show evidence for copious regolith despite the weak gravity. Most of them have one or more extraordinarily large craters, some of which are wider than the mean radius of the whole body. Such colossal impacts would not just gouge out a crater—they would break any monolithic body into pieces. Evidence of fragmentation also comes from the available measurements for asteroid bulk density. The values are improbably low, indicating that these bodies are threaded with voids of unknown size.

In short, asteroids larger than a kilometer across may look like nuggets of hard rock but are more likely to be aggregate assemblages—or even piles of loose rubble so pervasively fragmented that no solid bedrock is left. This rubble-pile hypothesis was first proposed two decades ago by Don Davis and Clark Chapman, both then at the Planetary Science Institute in Tucson, but they did not suspect that it would apply to such small diameters.

Shortly after the NEAR spacecraft flew by asteroid Mathilde three years ago on its way to Eros, the late planetologist Eugene M. Shoemaker (for whom NEAR has been renamed) realized that the huge craters on this asteroid and its very low density could only make sense together: a porous body such as a rubble pile can withstand a battering much better than an integral object. It will absorb and dissipate a large fraction of the energy of an impact; the far side might hardly feel a thing. A fair analogy is a bullet hitting a sandbag, as opposed to a crystal vase.

What about the jagged shapes of most asteroids? Intuition tells us that dramatic topography implies solidity. But first glances can deceive. When measured relative to the fun-house gravity, no regional slope on any imaged asteroid or comet exceeds a typical angle of repose (about 45 degrees), the incline at which loose debris tumbles down. In the steepest regions, we do see debris slides. In other words, small bodies could as well be made of boulders or even sand and still hold their shape. Dunes, after all, have distinct ridges yet are hardly monolithic. Rapid rotation would contribute to an elongated, lumpy appearance for a rubble pile.

Direct support for the rubble-pile hypothesis emerged in 1992, when comet Shoemaker-Levy 9 strayed too close to Jupiter and was torn into two dozen pieces. Two years later this "string of pearls" collided with the giant planet. According to a model I developed with Willy Benz of the University of Bern, the comet could have disassembled as it did only if it consisted of hundreds of loose grains in a slow cosmic landslide. As the comet was stretched by Jupiter's tides, the grains gravitated into clumps much like water beading in a fountain. From this breakup we proposed that comets are likely to be granular structures with a density around two thirds that of water ice. What applies to comets might apply to asteroids as well.

When Nothing Matters, Everything Matters

YET THE RUBBLE-PILE HYPOTHESIS is conceptually troublesome. The material strength of an asteroid is nearly zero, and gravity is so low you are tempted to neglect that, too. What's left? The truth is that neither strength nor gravity can be ignored. Paltry though it may be, gravity binds a rubble pile together. And anyone who builds sand castles knows that even loose debris can cohere. Oft-ignored details of motion begin to matter: sliding friction, chemical bonding, damping of kinetic energy, electrostatic attraction and so on. (In fact, charged particles from the sun can cause dust at the surface to levitate.) We are just beginning to fathom the subtle interplay of these minuscule forces.

The size of an asteroid should determine which force dominates. One indication is the observed pattern of asteroidal rotation rates. Some collisions cause an asteroid to spin faster; others slow it down. If asteroids are monolithic rocks undergoing random collisions, a graph of their rotation rates should show a bell-shaped distribution with a statistical "tail" of very fast rotators. If nearly all asteroids are rubble piles, however, this tail would be missing, because any rubble pile spinning faster than once every two or three hours (depending on its bulk density) would fly apart. Alan Harris of the Jet Propulsion Laboratory in Pasadena, Calif., Petr Pravec of the Academy of Sciences of the Czech Republic in Prague and their colleagues have discovered that all but five observed asteroids obey a strict rotation limit. The exceptions are all smaller than about 150 meters in diameter, with an abrupt cutoff for asteroids larger than about 200 meters.

The evident conclusion—that asteroids larger than 200 meters across are multicomponent structures or rubble piles—agrees with recent computer modeling of collisions, which also finds a transition at that diameter. A collision can blast a large asteroid to bits, but those bits will usually be moving slower than their mutual escape velocity (which, as a rule of thumb, is about one meter per second, per kilometer of radius). Over several hours, gravity will reassemble all but the fastest pieces into a rubble pile. Because collisions among asteroids are relatively frequent, most large bodies have already suffered this fate. Conversely, most small asteroids should be monolithic, because impact fragments easily escape their feeble gravity.

Qualitatively, a "small" asteroid sustains dramatic topography, and its impact craters do not retain the debris they eject. It looks like a battered bunker in a war movie. A "large" asteroid is an assemblage of smaller pieces that gravity and random collisions might nudge into a rounded or, if spinning, an elongated shape. Its craters will have raised rims and ejects deposits, and its surface will be covered in regolith. But this size distinction is not straightforward. Asteroid Mathilde could be considered small, as it has no visible rims or ejects deposited around its enormous craters, or large, as it is approximately spheroidal. Tiny Dactyl could seem large, also being spheroidal and sustaining such well-developed craters. The ambiguity is a sign that the underlying science is uncertain.

Shock Value

Given that geophysicists are still figuring out how sand behaves on Earth and how landslides flow, we must be humble in trying to understand conglomerate asteroids. Two approaches are making inroads into one of their key attributes: how they respond to collisions.

Derek Richardson and his colleagues at the University of Washington simulate asteroids as piles of discrete spheres. Like cosmic billiards on a warped pool table—the warp being gravity—these spheres can hit one another, rebound and slow down because of friction and other forms of energy dissipation. If balls have enough collisional energy, they disperse; more commonly, some or all pile back together. Richardson's model is particularly useful for studying the gentle accretionary encounters in the early solar system, before relative velocities started to increase under the gravitational influence of nascent Jupiter. It turns out to be surprisingly difficult for planetesimals to accrete mass during even the most gentle collisions.

High-speed collisions, more typical of the past four billion years, are more complicated because they involve the minutiae of material characteristics such as strength, brittle fracture, phase transformations, and the generation and propagation of shock waves. Benz and I have developed new computational techniques to deal with this case. Rather than divide a target asteroid into discrete spheres, we treat it as a continuous body, albeit with layers, cracks, or networks of voids.

In one sample simulation, we watch a 6,000-ton impactor hit billion-ton Castalia at five kilometers per second. This collision releases 17 kilotons of energy, the equivalent of the Hiroshima explosion—and enough to break up Castalia. We simulate Castalia as a two-piece object held together by gravity. The projectile and an equal mass of Castalia are vaporized in milliseconds, and a powerful stress wave is spawned. Because the shock wave cannot propagate through vacuum, it rebounds off surfaces, including the fracture between the two pieces of the asteroid. Consequently, the far piece avoids damage. The near piece cracks into dozens of major fragments, which take hours to disperse; the largest ones eventually reassemble. This outcome is very sensitive to what we start with. Other initial configurations and material parameters (which are largely unknown) lead to vastly different outcomes. Asteroids that start off as rubble piles, for example, are hard to blast apart.

Rendezvous with Eros

We can also work backward, inferring the rock properties of an asteroid by trying out different initial guesses and comparing the simulations with observations. As an example, I have worked with Peter Thomas of Cornell University to re-create the largest crater on Mathilde as precisely as possible: its diameter and shape (easy enough), its lack of fracture grooves or damage to existing craters (somewhat harder) and the absence of crater ejecta deposits (very hard).

If we assume that Mathilde was originally solid and monolithic, our model can reproduce the crater but predicts that the asteroid would have cracked into dozens of pieces, contrary to observations. If we assume that Mathilde was originally a rubble pile, as Shoemaker suggested, then our impact model easily matches the observations. Kevin Housen of the Boeing Shock Physics Lab and his colleagues have also argued that Mathilde is a rubble pile, although they regard the craters as compaction pitslike dents in a beanbag—rather than excavated features.

Understanding asteroid structure will be crucial for future missions. A rubble pile will not respond like a chunk of rock if we hope to gather material for a sample return to Earth or in the more distant future, construct remote telescopes, conduct mining operations or attempt to divert a doomsday asteroid headed for Earth. The irregular gravity is also a problem; spacecraft orbits around comets and asteroids can be chaotic, making it difficult to avoid crashing into the surface, let alone point cameras and instruments. NEAR is therefore conducting most of its science a hundred kilometers or more away from Eros. At this distance the irregular, rapidly rotating potato exerts almost the same gravity that a sphere would. The spacecraft's deviation from a standard elliptical orbit will enable NEAR scientists to measure the density distribution within Eros.

Orbiting Eros at the speed of a casual bicyclist (corresponding to the low gravity), NEAR is beaming a torrent of data toward Earth. The primary objective is to clarify the link between asteroids and meteorites. Cameras are mapping the body to a few meters' resolution, spectrometers are analyzing the mineral composition, and a magnetometer is searching for a native magnetic field and for interactions with the solar field. Upcoming missions will probe asteroids and comets in ever greater detail, using a broader range of instruments such as landers, penetrators and sample returns.

These discoveries will help plug a vast conceptual hole in astronomy. We simply don't understand small planetary bodies, where gravity and strength compete on sometimes equal footing. Asteroids are a balancing act, as serene as the moon yet of cataclysmic potential, large enough to hang onto their pieces yet too small to lose their exotic shape. Neither rocks nor planets, they are something of Earth and Heaven.

JOHN ARCHIBALD WHEELER

How Come the Quantum?

FROM THE *NEW YORK TIMES*

What is the greatest mystery in physics today? Different physicists have different answers. My candidate for greatest mystery is a question now a century old, "How come the quantum?"

What is this thing, the "quantum"? It's a bundle of energy, an indivisible unit that can be sliced no more. Max Planck showed us a hundred years ago that light is emitted not in a smooth, steady flow, but in quanta. Then physicists found quantum jumps of energy, the quantum of electric charge and more. In the small-scale world, everything is lumpy.

And more than just lumpy. When events are examined closely enough, uncertainty prevails; cause and effect become disconnected. Change occurs in little explosions in which matter is created and destroyed, in which chance guides what happens, in which waves are particles and particles are waves.

Despite all this uncertainty, quantum physics is both a practical tool and the basis of our understanding of much of the physical world. It has explained the structure of atoms and molecules, the thermonuclear burning that lights the stars, the behavior of semiconductors and superconductors, the radioactivity that heats the earth, and the comings and goings of particles from neutrinos to quarks.

Successful, yes, but mysterious, too. Balancing the glory of quantum achievements, we have the shame of not knowing "how come." Why does the quantum exist?

My mentor, the Danish physicist Niels Bohr, made his peace with the quantum. His "Copenhagen interpretation" promulgated in 1927 bridged the gap between the strangeness of the quantum world and the ordinariness of the world around us. It is the act of measurement, said Bohr, that transforms the indefiniteness of quantum events into the definiteness of everyday experience. And what one can measure, he said, is necessarily limited. According to his principle of complementarity, you can look at something in one way or in another way, but not in both ways at once. It may be, as one French physicist put it, "the fog from the north," but the Copenhagen interpretation remains the best interpretation of the quantum that we have.

Albert Einstein, for one, could never accept this world view. In on-again, off-again debates over more than a dozen years, Bohr and Einstein argued the issues—always in a spirit of great mutual admiration and respect. I made my own effort to convince Einstein, but without success. Once, around 1942, I went around to his house in Princeton to tell him of a new way of looking at the quantum world developed by my student Richard Feynman.

Feynman pictured an electron getting from point A to point B not by one or another possible path, but by taking all possible paths at once. Einstein, after listening patiently, said, as he had on other occasions, "I still cannot believe God plays dice." Then he added, "But maybe I have earned the right to make my mistakes."

Feynman's superposed paths are eerie enough. In the 1970s, I got interested in another way to reveal the strangeness of the quantum world. I called it "delayed choice." You send a quantum of light (a photon) into an apparatus that offers the photon two paths. If you measure the photon that leaves the apparatus in one way, you can tell which path it took.

If you measure the departing photon in a different way (a complementary way), you can tell if it took both paths at once. You can't make both kinds of measurements on the same photon, but you can decide, after the photon has entered the apparatus, which kind of measurement you want to make.

Is the photon already wending its way through the apparatus along the first path? Too bad. You decide to look to see if it took both paths at once, and you find that it did. Or is it progressing along both paths at once? Too bad. You decide to find out if it took just one path, and it did.

At the University of Maryland, Carroll Alley, with Oleg Jakubowicz and William Wickes, took up the challenge I offered them and confirmed that the outcome could be affected by delaying the choice of measurement technique—the choice of question asked—until the photon was well on its way. I like to think that we may one day conduct a delayed-choice experiment not just in a laboratory, but in the cosmos.

One hundred years is, after all, not so long a time for the underpinning of a wonderfully successful theory to remain murky. Consider gravity. Isaac Newton, when he published his monumental work on gravitation in the 17th century, knew he could not answer the question, "How come gravity?" He was wise enough not to try. "I frame no hypotheses," he said.

It was 228 years later when Einstein, in his theory of general relativity, attributed gravity to the curvature of spacetime. The essence of Einstein's lesson can be summed up with the aphorism, "Mass tells spacetime how to curve, and spacetime tells mass how to move." Even that may not be the final answer. After all, gravity and the quantum have yet to be joined harmoniously.

On the windowsill of my home on an island in Maine I keep a rock from the garden of Academe, a rock that heard the words of Plato and Aristotle as they walked and talked. Will there someday arise an equivalent to that garden where a few thoughtful colleagues will see how to put it all together and save us from the shame of not knowing "how come the quantum"? Of course, in this century, that garden will be as large as the earth itself, a "virtual" garden where the members of my imagined academy will stroll and converse electronically.

Here, a hundred years after Planck, is quantum physics, the intellectual foundation for all of chemistry, for biology, for computer technology, for astronomy and cosmology. Yet, proud foundation for so much, it does not yet know the foundation for its own teachings. One can believe, and I do believe, that the answer to the question, "How come the quantum?" will prove to be also the answer to another question, "How come existence?"

STEPHEN S. HALL

The Recycled Generation

FROM *THE NEW YORK TIMES MAGAZINE*

Almost every weekday morning, usually before 10:30, an overnight delivery truck with an unusual cargo negotiates the hilly streets on the outskirts of Worcester, Mass., and comes to a halt in front of a brick-and-tinted-glass building called Biotech Three. The courier disappears into the building with one or two large gray containers and drops them off at a small company called Advanced Cell Technology. The gray cases look like toolboxes, but they are actually sophisticated shipping containers, commonly used for transporting materials used in animal-breeding work, and they contain the starting material for a series of experiments that may completely rewrite the tables of human longevity. Or they may be remembered only for being among the most ethically troubling scientific endeavors of our times.

Inside the gray cases are hundreds of cow eggs—big, plump and beautifully rotund oocytes, as they're technically known—each one painstakingly plucked the day before from the ovaries of cows slaughtered in Iowa. They are doused with a marinade of enzymes that prime them for imminent fertilization and then sealed in small plastic test tubes before being shipped overnight to Worcester. On the drizzly, overcast day in mid-December when I visited the laboratories at Advanced Cell, the shipment of eggs arrived a little before me. In all the time that cows have roamed the planet, their oocytes have never encountered the insults they were about to face that day. The eggs would be stripped of their DNA, deliberately fused with human cells and fooled into thinking that fertilization had occurred, in the fervent commercial hope that some sort of an embryo might result.

Experimenting with embryos created from two different species—to say nothing of engaging in a form of human cloning—is an enterprise fraught with scientific and social uncertainty; indeed, Congress forbids the National Institutes of Health to finance any research involving human embryos. So it's natural to wonder why a small, largely unknown and understaffed biotech company would risk public scorn and ethical outrage to perform such research. One possible reason is publicity, which a number of critics have been eager to suggest. But the real answer, insists Michael D. West, the dreamy-eyed 46-year-old entrepreneur who heads A.C.T., is the scientific chase for what he calls "the mother of all cells—the embryonic stem cell."

The embryonic stem cell is an almost mythically powerful and versatile human cell, fleetingly present during the earliest days of embryonal development. This one cell has the potential—the genetic blueprint and the biological know-how—to become any cell, any tissue, any organ in the human body. With the proper biochemical coaxing, for example, it can turn into heart muscle, which could replace tissue damaged by a heart attack. Or into brain cells, which could be used to treat Parkinson's disease. Or into retinal cells, which could be used to restore failing vision.

Imagine, in short, a cell so protean and potent that it could theoretically generate an infinite supply of replaceable body parts—organ and skin, sinew and bone, blood and brain—to knit the tatters of disease, injury or old age. Imagine further that, with the use of controversial technologies like cloning, you might one day donate a snippet of your own skin, allowing scientists to harvest stem cells that theoretically would become a self-generated and limitless supply of transplant tissue—tissue that would make a perfect immunological match with you because, after all, it is you. These ideas have not only been imagined; patents and licensing agreements are already in place.

"These are incredible cells," gushes Thomas B. Okarma of Geron Corporation, the California biotechnology company that controls many of the patents and licenses on stem cells. "I've been in biology since high school, and this is still a chilling technology when you see these things and realize what they do. The number of applications is mind-boggling." Last month, *Science* magazine dubbed stem-cell technology the "breakthrough of the year."

In spite of this promise, the scientific, commercial and ethical landscapes that intersect with these fascinating cells have been shaped, even turned upside down, by the right-to-life debate, because in order to obtain human embryonic stem cells by currently available technology, you must destroy a human embryo. Because of the Congressional ban on financing of this research, many scientists now find themselves in the awkward position of possibly losing their jobs if they try to use human stem cells to devise new treatments for many common diseases. "With stem cells," says Ronald McKay, a researcher at the National Institute of Neurological Disorders and Stroke in Bethesda, "you tickle them and they jump through hoops for you." McKay's group published a highly regarded paper in *Science* last summer showing how rat embryonic-stem cells could be used to treat a version of multiple sclerosis in rats, but he doesn't dare extend the research to test the method's effectiveness in humans. "I would get fired if I did that," McKay says, fairly squirming in his seat to make this seem like the most reasonable thing in the world.

The ban on N.I.H. financing has also had the collateral effect of relegating the technology to the private sector, where embryo research can proceed unencumbered. But the fact that so much of this controversial research now occurs in the private sector means that public discussion has become constrained. And complicating this commercial landscape is the fact that two of the companies most avidly competing over the research, Geron and Advanced Cell Technology, have a tangled corporate history dating back to business opportunities created by the federal ban.

The common figure wandering through all these unsettled landscapes, popping up again and again like some white-coated Zelig, is Michael West. West's insistence on pursuing controversial experiments in which cow eggs are used as cellular incubators in an attempt to create humanlike embryos comes at a particularly sensitive moment. Emboldened by a legal interpretation that seemed to show a way around the Congressional ban, the N.I.H. on Dec. 2 issued long-awaited guidelines explaining the rules by which the agency would finance human-stem-cell research. The 60-day comment period ends on January 31, and the agency hopes to begin soliciting grant proposals from university researchers sometime during the summer of 2000.

However, Representative Jay Dickey, a Republican from Arkansas who has blocked N.I.H. financing for embryo research since 1995, has already vowed to outlaw federally financed stem-cell research this spring, either through legislation or legal action. Senator Arlen Specter, meanwhile, who is entranced by the medical possibilities, has promised to introduce a bill explicitly allowing federal financing for the research.

West, whose laid-back, soft-spoken demeanor belies the soul of a headstrong provocateur, promises to be an impassioned voice in this debate, too. "I think a lot of the problem we have in trying to develop these new technologies for medicine is people's knee-jerk reaction to words like 'fetal' and 'embryo,' " he says. "You know, you use the word 'fetal' and people just go completely irrational."

UNTIL WE LEARN how to direct the fate of human embryonic stem cells—learn how to tell them, for example, that we'd like them to become a new liver or glistening white bone—their enormous promise still remains under Nature's lock and key. Many scientists believe it will take at least 10 years before we learn how to program the development of cells that could be transplanted into humans, and probably many more before we learn how to coax stem cells into creating something as grand and complex as a liver or a kidney.

But let's say some version of the science will be in practice by the year 2011, when the first of the baby boomers turn 65. And let's say the technology of stem cells will be complemented by promissory notes redeemed in related fields of gerontological science. What might the menu of spare parts look like?

The first step in creating those spare parts (which could be used not only for the aged but also for anyone in need of intervention) might well be the donation by the patient of a biopsy sample, which could be quickly cloned, thus creating an early embryo, which would produce stem cells in a week or so. Every cell and every tissue derived from those stem cells would be a perfect immunological match, which would immediately circumvent the big stumbling block of current transplant medicine: matching tissue.

At that point, it simply becomes a matter of matching the ailment to the right cell type. Researchers speak of creating nerve cells to treat spinal-cord injuries, stroke and Alzheimer's disease; glial cells for multiple sclerosis; pancreatic islet cells to treat diabetes; muscle cells for muscular dystrophy; chondrocytes for arthritis; hepatocytes for cholesterol metabolism; endothelial vessels to grow new blood vessels to replace vessels clogged by fat and plaque—there are more than 200 different cell types in the body, and stem cells can theoretically be nudged to form each one. If these cells are souped up, as has been proposed, with an enzyme that maintains its cellular youthfulness, we're talking not only about replacement parts, but also about parts that never grow old.

What might the world of stem-cell medicine look like? For one thing, every place in the developed world might look a lot more like Florida. Although the maximum attainable human life span is now approximately 120 years, only about 65,000 Americans have currently reached the age of 100. But a century from now, with new medical technologies in place, the Census Bureau predicts there will be 5.3 million people living to the age of 100 and perhaps much longer. To hear the optimists talk, there'll be much longer waits for tennis courts and tee times.

This is the kind of speculation that keeps ethicists and philosophers busy, but there are a lot of touchier issues on the immediate horizon—creating embryos for research purposes, combining biological material from different species and performing nuclear-transfer experiments involving adult human cells (also known as cloning). All three are being pursued in a single line of research at Advanced Cell Technology.

THE EXPERIMENTS at A.C.T. take place in a long, dark, narrow room in the middle of the lab. The lights are kept low to allow technicians to view the cells through microscopes while poking, turning and prodding them with micromanipulation devices. K. C. Cunniff, her blond hair spilling out over a white lab coat, begins by vacuuming all of the cow's genetic material from the egg cells, one after another, for more than an hour.

As I watch a magnified view of the proceedings on a television monitor, it's hard not to be impressed by the nimble piecework of the technicians. Cunniff, 26, deftly maneuvers one pipette to hold a round, plump egg cell in place. (They prefer cow eggs because they are inexpensive and easier to use than either pig or primate oocytes). She turns the egg, which had previously been stained with a fluorescing dye, as if rotating a ball in her hand, until she sees a telltale blue-green speck of DNA. Using her other hand, she nudges a sharp, hollow needle up against the surface of the egg. With a precise thrust, she enters the egg, sucks out the DNA and withdraws the needle, all in a matter of seconds. "Very quick," she says matter-of-factly. In, suction, out. Next egg. In, suction, out.

It's so matter-of-fact that you'd never guess this type of experiment was the focus of wildly fearful predictions only a few years ago. In 1997, when scientists at the Roslin Institute in Scotland announced the cloning of Dolly the sheep, there was a lot of fevered speculation about the possibility of cloning human beings. It was generally dismissed as both a distant and undesirable proposition, but the sudden appeal of embryonic stem cells has provided a surprisingly urgent medical justification for moving the technology of human cloning closer to reality.

In the case of A.C.T.'s experiments, the cow egg is being tested as a way to "reprogram," or reset, the adult DNA back to a pristine state resembling the moment of fertilization. In these experiments, the egg comes from a cow but the adult DNA comes from human donors; it's the human DNA, West believes, that orchestrates subsequent embryonic development. The technical term for this type of experiment is "nuclear transfer," but West prefers a more transparent phrase—"human therapeutic cloning." That modifier "therapeutic" signals that A.C.T. is not in the business of making babies, just embryos to get stem cells. And they're not alone in pursuing the nuclear-transfer approach. In May of 1999, to remarkably little public comment, Geron acquired Roslin Bio-Med, the Scottish company in charge of commercializing the research that produced Dolly.

As Cunniff moves another cow egg into position, Nancy Sawyer gazes at the TV monitor and makes a sound like "Unk!" as the needle punctures the egg. "That's so cool," she whispers of the image. Seen on the monitor, the image has a primal iconic beauty, like a dim grainy photograph of a distant planet. On each moonlike oocyte, one isolated lunar mare glows with an eerie greenish blue, betraying the location of genetic material that will soon be suctioned away.

At a nearby microscope, Sawyer, 24, performs the next step. Human skin cells, known as fibroblasts, have been provided by an anonymous donor. The human cells dot the screen like so many lumpy, transparent potatoes, much smaller than the egg cells. Sawyer loads them into a hollow needle, immobilizes a cow egg with suction and, with gentle pressure, inserts the needle tip just under the rind of the egg. With a squeeze of her hand, a single human skin cell, with its unique payload of human DNA, plops into the egg of a creature that moos. Sawyer even manages to maneuver her needle tip to literally tuck the hole closed.

After stuffing every cow egg with its little spud of human DNA, Sawyer prepares the next step. She gives the cells a zap of 120 volts. The jolt of electricity effectively fuses man and beast into a single biological fate. After one final step, this . . . this *thing* will believe it has been fertilized and, if all goes well, begin cleaving, or dividing, in the bubbling, momentous arithmetic of life lifting off the pad: 2 cells, 4 cells, 8 cells, 16 cells, 32 cells—blastocyst!

The odds are extremely long that any of the 100 or so cell fusions will result in a blastocyst, the hollow ball of cells that appear about a week after fertilization and, in a normal embryo, are destined to become the placenta. But if that ball of cells should form, a separate group of cells known as the inner cell mass assembles against one inside wall of the blastocyst, like bats huddling on the ceiling of a cave, and this is where the embryonic stem cells will appear. These stem cells exist only briefly before they begin to differentiate into more specialized tissues—there one moment and gone the next, transients vanishing into the great biology of becoming. "If I'm lucky," says Jose Cibelli, vice president of research at A.C.T., "I'll get one blastocyst. That's how low the efficiency is." And getting the blastocyst is just the first hurdle—can they isolate the stem cells and keep them alive in a test tube? And will they truly act like human embryonic stem cells?

Before leaving Worcester that day, I ask Cibelli when he will know if any blastocysts have formed from the experiment I've observed. He says he wouldn't expect to see anything before 9 or 10 days. I do the math in my head and make a mental note to get in touch either on December 24 or Christmas Day.

IN MANY WAYS Michael West is the shadow impresario of the field. As founder of Geron Corporation, one of this decade's most closely watched biotechnology companies, and now as president and C.E.O. of Advanced Cell Technology, West has achieved remarkable success as a kind of merchant of immortality, selling the idea that stem cells and related technologies might someday completely revise the tables of average human life span. And he is so convinced that the promise of stem cells justifies a controversial strategy like cow-human nuclear transfer that he is happy to foster, if not force, a national discussion of this technology.

In a recent issue of the journal Nature Medicine, for example, West, Cibelli and their colleague Robert Lanza argued the case for human therapeutic cloning because stem-cell research is so promising. "Does a blastocyst," they wrote, "warrant the same rights and reverence as that accorded a living soul—a parent, a child or a partner—who might die because we failed to move the moral line?" And that is what Mike West is trying to do at this touchy juncture of the stem-cell wars—with this company, in these experiments, in an ethical debate that has seemed too arcane and complicated to attract much public attention to date. He is putting his shoulder to the moral line that forbids embryo research and is trying to force some sort of social reckoning. The degree to which he and others succeed or fail may well determine if stem cells have a chance to live up to their promise as medicine or remain too hot to handle in the current political climate.

Stem cells burst into public consciousness when, in November 1998, James Thomson of the University of Wisconsin reported the creation of human embryonic stem (or E.S.) cell lines. Using leftover frozen embryos from in-vitro fertilization clinics in Wisconsin and Israel, Thomson and colleagues isolated stem cells and have shown that they can be maintained indefinitely in the lab—can be grown, frozen and then thawed, and still retain their power to develop into, say, heart-muscle cells or brain cells. Michael Shamblott and John Gearhart, at the Johns Hopkins School of Medicine, headed a separate effort to cultivate something called "embryonic germline" (E.G.) cells, which are harvested from a tiny speck of fetal tissue from an aborted fetus and then grown in the lab. Because of the Congressional ban on federal financing for human-embryo experimentation, both teams conducted the research with financing from Geron.

There are many technical hurdles to overcome. But the sheer power of the approach makes it clear that, if properly harnessed, stem cells could serve as a warehouse of spare human parts. This teasing hint of immortality is the cultural subtext that runs beneath the public fascination with the science, and no one has done more to promote that connection than West. The son of a truck mechanic, West is a onetime creationist and a self-styled truth seeker, and his entrepreneurial interest in the biology of aging derives from an obsessive, almost morbid fascination with death. "All I think about, all day long, every day, is human mortality and our own aging," he says.

The first time I visited Advanced Cell Technology, West showed up two hours late for our appointment, apologized with sheepish charm for his tardiness and began to spin out the kind of polished futurism he regularly conveys to scientists, investors and lay people. "I thought I'd show you some pictures," he said, and then proceeded to project slides on a screen in the company's conference room, delivering a lecture on the biology of aging to an audience of one. With his gently soothing Midwestern voice and relentlessly upbeat brand of biological positivism, he manages to make science sound almost like a cult. Former colleagues concede that he does not possess a crisp management style (punctuality, for example, being a continuing challenge), but even his critics admit he has a knack for looking beyond the horizon and dreaming deep. He'll talk for hours about saving endangered species through cloning, or the possibility of cloning pets, or why it makes more economic and ethical sense to pay $1 for a cow egg than $2,000 to surgically obtain a single human egg. "He's a very visionary guy," says James Thomson, a University of Wisconsin biologist, "but he's also a very good salesman."

Even as a high-school student growing up in Niles, Michigan, West was fascinated with aging and rejuvenation. He immersed himself in religion and philosophy. He learned Hebrew and Greek, he says, to read ancient texts in the original. He went on to study psychology at Rensselaer Polytechnic Institute, obtained a Masters of Science degree at Andrews University, a Seventh-Day Adventist school, and even studied creationism for a time in San Diego, he says, before convincing himself of the truth of Darwinian evolution. Following the death of his father in 1980, West worked in the family's truck-leasing business and then belatedly embarked on a career in science. He received his Ph.D. in cell biology from the Baylor College of Medicine in 1989 and started medical school at the University of Texas Southwestern Medical Center in Dallas.

That same year, West showed up unannounced and began to hang out at the University of Texas lab of Woodring Wright and Jerry Shay, who were researching the molecular biology of aging. Four years earlier, scientists had discovered a critically important enzyme called telomerase, which acts on telomeres, the little caps of DNA at the end of chromosomes; telomeres ordinarily grow shorter each time a cell divides, until the cells stop dividing altogether. It turned out that a handful of human cells—germ cells, cancer cells and embryonic stem cells—use telomerase to circumvent that shortening process, and thus also circumvent the aging process and achieve a cellular version of immortality.

As West learned more about telomeres, he eventually came to view the enzyme telomerase as a molecular version of the fountain of youth; it looked as if it might bestow immortality on normal cells and, conversely, could have the beneficial effect of pulling the plug on immortality in cancer cells if blocked. While still technically enrolled in medical school, West moved out to California and banged on doors in search of seed money for a biotech startup. Thus, in November 1990, he founded a company dedicated to the molecular causes of aging. He named it Geron—Greek for "old person." He eventually captured the interest of the most prestigious venture-capital firm on the West Coast, Kleiner Perkins Caufield & Byers, which along with other firms invested $7.6 million in Geron in 1992.

For a company that has lost tens of millions of dollars and is in no danger of curing aging anytime soon, Geron has managed to cast a spell on investors, the media and the lay public. In both technical articles and news releases, it has retailed a scientific vision (and vocabulary) that clearly push the right zeitgeist buttons. West and Geron spoke tirelessly of "immortalizing enzymes" and the "life extension" of cells; Geron is almost universally recognized as an "anti-aging" company. And in 1997, after winning a highly competitive race to clone (and patent) the human gene for telomerase, Geron actually had a real molecule around which to develop clinical products. The company currently has a number of promising directions for telomerase-based products, including potential anticancer applications, but the mythology is so firmly established that even though company officials insist Geron is no longer an "anti-aging company," this is still how it is inevitably portrayed.

From the very beginning, however, West had another big idea he wanted to pursue. "You need replaceable cells and tissues for the problems of aging as well," he said. "And it seemed to me that the ideal source for an aging population is to go back to the beginning of life." To the embryo, that is, and stem cells. And so, as early as 1992, he paid a visit to Roger Pedersen, a professor of obstetrics, gynecology and reproductive sciences at the University of California at San Francisco, and a leading expert on embryonic stem cells in mice. They both agreed that the time had arrived to explore the vast potential of such embryonic stem cells in human medicine, and West inquired into whether Pedersen would accept financing from Geron to do research on human stem cells.

Pedersen flatly refused: "This area of investigation is something that is at the headwaters, and it's not appropriate for private investors to control the headwaters of a stream of research."

Several years later, however, Pedersen got back in touch with West. Circumstances had changed, he said, and he was ready to deal.

THE CIRCUMSTANCES WERE political.

In 1975, federal regulations stipulated that any government support for in-vitro-fertilization research required the approval of a federal ethics advisory board. After a short and turbulent history, this board was disbanded in 1980 without a single research project having received government funds. One cynical stratagem of the Reagan and Bush administrations, according to Pedersen, was to block all attempts to reconstitute the panel, effectively thwarting such research throughout the 1980s.

Under the new Clinton administration, Congress did away with the phantom federal ethics review and set up a special N.I.H. committee to establish guidelines for human-embryo research. Everything seemed back on track when, early in 1995, Jay Dickey, the Republican congressman from Arkansas, successfully inserted a rider into the budget bill for the Department of Health and Human Services (which includes the N.I.H.) banning federal funds for human embryo research. Just as right-to-life politics had forced in-vitro fertilization and reproductive biology into the private sector, where lack of oversight and regulation has led to a series of well-documented scandals, stem-cell research seemed headed for similar privatization.

Those were the circumstances that had changed Roger Pedersen's mind. Geron reached an agreement with the University of California to finance Pedersen's work on embryonic stem cells. During his discussions with West, Pedersen mentioned that a scientist at the University of Wisconsin, James Thomson, was about to publish a paper announcing another breakthrough: the isolation of embryonic stem cells from rhesus monkeys. West was in Wisconsin the next day, and Geron signed up Thomson too.

"I would have been much happier with public support," Thomson admits. "But given the constraints, I welcomed the funding I got." Getting access to stem cells from a primate, an animal biologically close to humans, West realized, promised tremendous intellectual-property dividends. "We could just learn how to work with them, and file patents," he said. "But we'd have this head start on the whole world." When West later learned that another university researcher, John Gearhart at Johns Hopkins, had made significant progress isolating cells very similar to embryonic stem cells, he headed straight for Baltimore. "He just showed up on my doorstep one day," Gearhart recalled with a laugh.

With Pedersen acting as talent scout, West chased down and signed up three of the leading stem-cell researchers in the world. Geron began to assemble a staggering intellectual-property portfolio in a field with almost limitless medical potential. And because the Congressional right-to-life advocates had effectively tied the N.I.H.'s hands in terms of financing, there wasn't any competing research in university labs. The investment paid off spectacularly in November 1998, when both Thomson and Gearhart announced they had isolated human stem cells. The universities where the work was done retained the patents on the research, but Geron received exclusive rights to many applications.

Unfortunately, Mike West enjoyed this moment of triumph only vicariously, because by then he had left the company, unhappily. In 1997, according to several sources, Geron planned to spin off its entire stem-cell program into a separate company, which would include West, when the plan, in the words of a scientific board member of the company, "got clobbered by the company leadership." Thomas Okarma, who joined Geron in December 1997 and is now president, offers a different interpretation. "I was hired explicitly to run that program because it really wasn't moving," he said. In any event, the stem-cell research stayed at Geron, and West says he increasingly felt he could do more outside the confines of the company than inside.

The fact that West's current company, Advanced Cell Technology, is now competing against the company he founded may go a long way toward explaining why Mike West is so determined to find an alternative way of obtaining human stem cells, one that doesn't rely on existing human embryos in clinics or fetal material from abortions—the methods that Wisconsin and Johns Hopkins licensed to Geron. And it certainly explains why he perked up when, a few months after leaving Geron, he learned of an unusual experiment by Jose Cibelli, an Argentinian scientist working at Advanced Cell Technology. The good news was that Cibelli had tried to isolate human stem cells using a method that seemed to offer an alternative to Geron's approach. The bad news was that the strategy involved human cloning. And cows. "And I knew," West says, "that was going to be a problem."

IN THE SUMMER OF 1996, Cibelli was a graduate student at the University of Massachusetts at Amherst, working in the laboratory of James Robl, a respected developmental biologist. Cibelli had the radical idea of fusing some of his own cells with cow egg cells, in effect cloning himself—not to make a copy, of course, but as a way to get human stem cells. Interestingly, this wasn't the first time Robl had been confronted with such an idea. Several years earlier, a student in the lab, unbeknown to Robl, had fused human cells with the egg cells of a rabbit, and the cells had begun to divide. This time, Robl went to university officials and received institutional approval to proceed.

During July and August, Cibelli rinsed out some of the cells that lined the inside of his cheeks and tried fusing them with cow egg cells. "One day, Jose was about to go on vacation, and he was about to throw out the dish," Robl recalls. "I don't generally look at these things, but I did that day. And there was a blastocyst." In other words, the embryolike thing had moved beyond mere cell division and graduated to the stage where embryonic stem cells begin to form. As is routine when the experiment reaches this stage, Cibelli placed the fragile blastocyst on a bed of fetal mouse cells, which nourished its further development. "We watched it for about two more weeks," Robl says. "It looked, to my eye, not like a cow blastocyst. The morphology of the cells was different."

Cibelli and Robl did not publicly discuss the experiment at the time, nor did they prepare a scientific report for peer review. "We never considered a publication," Robl explains, "because there was not nearly sufficient data." But the University of Massachusetts did consider the experiment sufficiently novel to file a patent application. The United States patent was issued, virtually unnoticed, in August 1999.

Cibelli recounted this remarkable story to West in the spring of 1998, when West happened to be visiting A.C.T. "I was just flabbergasted," West recalls. "I mean, he showed me human embryos that had been made by cloning. And I had no idea—no one in the world had any idea—that it had been done. I thought, 'Oh my gosh, this is exactly what I want to be doing for the next 10 or 20 years of my life.' " More to the point, it was exactly the kind of technology that would allow him to get back into the stem-cell game.

West officially joined A.C.T. in October 1998 and immediately presided over an episode that was, for him, uncharacteristic—a major public-relations fiasco. As soon as West joined the company, Cibelli lobbied to resume his cloning experiments. West agreed, but wanted to disclose details of the 1996 experiment and gauge public reaction to the technology before starting it up again. "I didn't want to be accused of doing this in secret," he says. So he invited a film crew from the CBS newsmagazine "48 Hours" to film the work in progress.

Then, one week before the scheduled broadcast, on November 6, West got blindsided by his previous life. Geron announced Thomson and Gearhart's successes isolating stem cells. "I had no idea it was coming," West admitted. It wasn't just that the research made front-page headlines and drove Geron stock up 74 percent in one day. By the time CBS broadcast the show on Nov. 12, along with a news account of the experiments that appeared in that morning's *New York Times,* West's professed desire for openness looked like something entirely different: a bid to leverage "me too" publicity for his otherwise unknown company.

For an experiment that never received formal peer review, Cibelli's cow-human nuclear-transfer work got plenty of unofficial feedback, beginning with the White House. Clinton called it "deeply troubling." Thomas Murray, president of the Hastings Center and a leading bioethicist, wondered "if the timing of the announcement had to do with scientific competition, personal competition or positioning for funding from investors." Roger Pedersen was quoted as saying, "I smell a sham." (He claims he wasn't told who performed the experiment when he was asked to comment upon it.) Thomson regarded the whole affair as "unfortunate." Right-to-life pickets showed up outside Biotech Three in Worcester, marching around in cow masks.

Two days after news of the cow-human experiments broke, President Clinton asked his National Bioethics Advisory Commission to prepare a report on stem-cell research in general. The commission hastily convened hearings in January 1999 in Washington. And who should turn up, uninvited, to make an unscheduled presentation before the panel but Michael West.

"I READ AN EDITORIAL by an individual who wrote that science should stop so that ethics can catch up," West told the national commission that day. His ambition, he said, was "to communicate to people in public policy and in biomedical ethics, so that, simply, ethics can walk hand in hand with science." Ethics and modern biology, however, have rarely walked hand in hand, and stem-cell research has added new difficulties to the relationship.

In 1999, two high-powered advisory committees have, with one significant difference, endorsed the general idea of embryonic-stem-cell research. A committee established by the American Association for the Advancement of Science recommended in August that researchers be able to receive public funds for experiments on embryonic stem cells, but only using cell lines already created by researchers in the private sector. The National Bioethics Advisory Commission, by contrast, recommended in September that researchers financed by the N.I.H. should be allowed to create stem cells using human embryos already in existence (thousands of such embryos, frozen and destined to be discarded, exist at in-vitro fertilization clinics).

But no oversight or guidelines exist for private industry, and in February 1999, well before either committee delivered its opinion, Advanced Cell Technology quietly decided to resume its cow-human nuclear-transfer experiments. It seemed like an important decision for such socially sensitive research, so I asked West if it was something that went to the board or an ethics advisory panel for approval. West said it was strictly his decision. And there's the paradox. Now that abortion politics has forced so much of the research into the private sector, the transparency of the ethical conversation about it has become more obscured.

Everyone, including Mike West, insists on an open national debate about stem-cell research. But I began to notice that whenever I asked one question too many about exactly what work was being done, or even contemplated, the conversations became elliptical and vague. I was having lunch with West one day, for example, when I asked if the use of human donor eggs for cloning experiments was under consideration. Cibelli had told me he thought such a development was very likely. "I have to confer on this issue," West replied apologetically, leaning over to huddle with A.C.T.'s public-relations adviser. Then, after a pause but without directly answering the question, he expressed concern that such a program could exploit women.

For all their good intentions, ethicists may have allowed themselves to be placed in a difficult, possibly untenable position. At Geron, for instance, the company's ethics advisory board seems to have the ear of management. But, says Karen Lebacqz, who heads the Geron ethics advisory board, "they are perfectly at liberty to ignore all our advice." Further, as Lebacqz points out, she is not free to discuss certain aspects of the research. "Early last summer, they brought us a piece of research that they were going to fund. Several members of the board raised objections, so they decided not to pursue that particular line of research." What was the research under discussion? "I'm sorry, I really can't," she said.

The ethicists have become proxies for all of us, precisely because so much of this technology, for political reasons, is unfolding in the private sector. Yet they have limited power. West, for example, told me that when Geron first considered establishing an ethics board, the company determined that giving such a board the right to veto research projects would undermine its fiduciary responsibilities to shareholders. The ethicists have what West calls "the power of the pen," but what they can report back to us is constrained.

Their mere participation in the process, however, creates the appearance of oversight and ethical responsibility, and that is precisely what bothers David Cox, vice chairman of the genetics department at Stanford University and a member of the national bioethics commission. Cox says ethics advisory boards at biotech companies are "a joke": "They're supposedly doing ethical review, but the process by which they're working is backwards."

It's very hard to have a national debate on issues as socially and ethically important as cloning and the creation of embryonic stem cells when every conversation may ultimately bump up against corporate confidentiality. The problem of openness is compounded by the editorial policy at a number of leading scientific journals, which refuse to publish research if the results have previously been disclosed in public. That almost guarantees that breakthroughs in a controversial and competitive field of research like stem cells will land in the public's lap as scientific *faits accomplis,* just as Dolly did.

And it leaves us in the same scientifically uncertain and ethically queasy place we were more than a year ago. At that time, Thomas Murray of the national bioethics commission asked West if he thought the cow-human experiments resulted in human embryos that were potentially "viable"—in other words, embryos that, if implanted in a woman, could result in a live human being. West didn't answer, and A.C.T. still isn't answering. I asked Jose Cibelli, for example, if the A.C.T. scientists had made any progress overcoming the problem of biological incompatibility between a cow egg and a human cell, an issue involving small cellular organs called mitochondria; it is an obstacle repeatedly raised by the many scientists who remain skeptical about the approach and are dubious that a blastocyst would be created, especially since A.C.T. still hasn't published a sprig of data on it in more than a year.

"We think," Cibelli began to say, "but this is very preliminary. . . ." He shrugged and smiled. "You can say that we've had good progress," he continued with a little laugh. "We've had good progress, and we expect to have something to report in the near future. But I guess I need to be protective of the data for publication's sake."

Michael West sounded a similar theme when I told him how many complaints I'd heard about A.C.T.'s unpublished experiments. "We could publish now, the data we have," West assured me, well aware of the exasperation of the research community. "But," he continued, "we're trying to generate a real killer paper here. We're going to do a paper we're proud of."

On Christmas Eve, I sent an e-mail to Jose Cibelli, asking about the status of the cells I'd seen fused. The day after Christmas he wrote back to report that the nuclear-transfer units had "developed at the 'predictable' rate," while declining to specify exactly what that was. As soon as new techniques were in place, he continued, Advanced Cell Technology would report the results in a peer-reviewed journal.

AS CONGRESS PREPARES to debate the merits of stem-cell research in the coming months, we will undoubtedly hear rosy visions of the future of medicine with stem cells (as well as the contorted political logic that suggests that research on human embryos and cross-species nuclear-transfer experiments are permissible in the private sector, but morally indefensible in the public sector). But it's also worth thinking through the implications of immortalizing medicines, which I had the opportunity to do with Leonard Hayflick, the elder statesman (and elder contrarian) of the field.

Hayflick, now 71, is a well-known cell biologist; his discovery in 1961 that normal human cells grown in a test tube simply stop dividing after a specific number of cell divisions, known as the Hayflick limit, in effect introduced the notion of mortality into the biology of aging. We arranged to meet at the Union Club in Manhattan in early December, when Hayflick showed up to harangue fellow board members of the American Federation for Aging Research about their financing priorities.

Hayflick doubts we'll ever have a quick fix to arrest aging anytime soon, but he has lots of reasons to think it would be a very bad idea. Like a number of bioethicists, he believes the first line of division is economic. Access to the regenerative medicine of stem-cell and immortalizing enzymes is most likely to be a phenomenon available only to affluent segments of the population in the developed world.

In his book *How and Why We Age* and other writings, Hayflick has even gone to the trouble of imagining "bizarre situations" that might unfold if scientists were ultimately able to create a medication that would, from the moment treatment began, essentially freeze the process of aging at a certain point. He has imagined "children becoming biologically older than their parents" if a parent chose to stop aging at age 45, for example, while a child did nothing. How would you even know, he asks, the right age to stop at?

Karen Lebacqz has also pondered this distant future. "If we are successful with the use of stem cells or in the reprogramming of cells," she says, "it will mean that people are no longer dying of the things we are dying of today. What do we do with all of ourselves if we don't die?" We'll squander even more of the world's resources, she continues, and put even more pressure on the developing world.

But perhaps the ultimate argument against the implicit promise of immortality has to do with a simple biological fact: if we were to rejuvenate our brains, Hayflick argues, we might lose the most precious thing we have: our sense of self. "Given the possibility that we could replace all our parts, including our brain, then you lose your self-identity, your self-recognition. You lose who you are! You are who you are because of your memory."

There is a lot of scientific research and a lot of heated political debate to come before we arrive at any of those distant quandaries. And how we resolve the ethical conversations we're in the midst of having over stem cells will have a lot to say about whether we'll have a chance of reaching that future at all.

Richard Preston

The Genome Warrior

FROM *THE NEW YORKER*

Craig Venter is an asshole. He's an idiot. He is a thorn in people's sides and an egomaniac," a senior scientist in the Human Genome Project said to me recently. The Human Genome Project is a nonprofit international research consortium that since the late nineteen-eighties has been working to decipher the complete sequence of nucleotides in human DNA. The human genome is the total amount of DNA that is spooled into a set of twenty-three chromosomes in the nucleus of every typical human cell. It is often referred to as the book of human life, and most scientists agree that deciphering it will be one of the great achievements of our time. The stakes, in money and glory, to say nothing of the future of medicine, are huge.

In the United States, most of the funds for the Human Genome Project come from the National Institutes of Health, and it is often referred to, in a kind of shorthand, as the "public project," to distinguish it from for-profit enterprises like the Celera Genomics Group, of which Craig Venter is the president and chief scientific officer. "In my perception," said the scientist who was giving me the dour view of Venter, "Craig has a personal vendetta against the National Institutes of Health. I look at Craig as being an extremely shallow person who is only interested in Craig Venter and in making money. Only God knows what those people at Celera are doing."

What Venter and his colleagues are doing is preparing to announce, in the next few days or weeks, that they have placed in the proper order something like ninety-five per cent of the readable letters in the human genetic code. They refer to this milestone as First Assembly. They have already started selling information about the genome to subscribers. The Human Genome Project is also on the verge of announcing a milestone: what it calls a "working draft" of the genome, which is more than ninety-per-cent complete and is available to anyone, free of charge, on a Web site called Gen-Bank. It contains a large number of fragments that have not yet been placed in order, but scientists in the public project are scrambling to get a more complete assembly. Both images of the human genome—Celera's and the public project's—are becoming clearer and clearer. The human book of life is opening, and we hold it in our hands.

A human DNA molecule is about a metre long and a twenty-millionth of a metre wide—the width of twenty hydrogen atoms. It is shaped like a twisted ladder, and each rung of the ladder is made up of four nucleotides—adenine, thymine, cytosine, and guanine. The DNA code is expressed in combinations of the letters A, T, C, and G, the first letters of the names of the nucleotides. The human genome contains at least 3.2 billion letters of genetic code, about the number of letters in two thousand copies of *Moby-Dick.*

Perhaps three per cent of the human code consists of genes, which hold recipes for making proteins. Human genes are stretches of between a thousand and fifteen hundred letters of code, often broken into pieces and separated by long passages of DNA that don't code for protein. It is believed that there are somewhere between thirty thousand and possibly more than a hundred thousand genes in the human genome (there's great puzzlement about the number). Much of the rest of the genome consists of blocks of seemingly meaningless letters, gobbledygook. These sections are referred to as junk DNA, although it may be that we just don't understand the function of the apparent junk.

The conventional route for announcing scientific breakthroughs is publication in a scientific journal, and both Celera and the public project plan to publish annotated versions of the human genome later in 2000, perhaps in *Science.* It is even possible that they will announce a collaboration and publish together. Although right now the two sides look like armies maneuvering for advantage, the leaders of the Human Genome Project have consistently denied that they are involved in some kind of competition.

"They're trying to say it's not a race, right?" Craig Venter said to me recently, in a shrugging sort of way. "But if two sailboats are sailing near each other, then by definition it's a race. If one boat wins, then the winner says, 'We smoked them,' and the loser says, 'We weren't racing—we were just cruising.' "

I FIRST MET Craig Venter on a windy day in summer nearly a year ago, at Celera's headquarters in Rockville, Maryland, a half-hour drive northwest of Washington, D.C. The company's offices and laboratories occupy a pair of five-story white buildings with mirrored windows, surrounded by beautiful groves of red oaks and tulip-poplar trees. One of the buildings contains rooms packed with row after row of DNA-sequencing machines of a type known as the ABI Prism 3700. The other building holds what is said to be the most powerful civilian computer array in the world; it is surpassed only, perhaps, by that of the Los Alamos National Laboratory, which is used for simulating nuclear bomb explosions. This second building also contains the Command Center, a room stuffed with control consoles and computer screens. People in the Command Center monitor the flow of DNA inside Celera. The DNA flows through the Prism machines twenty-four hours a day, seven days a week.

That day last summer, Venter moved restlessly around his office. There had been a spate of newspaper stories about the race to decode the complete genome, and about the pressure Celera was putting on its competitors. "We're scaring the shit out of everybody, including ourselves," he said to me. Venter is fifty-three years old, and he has an active, cherubic face on which a smile often flickers and plays. He is bald, with a fuzz of short hair at the temples, and his head is usually sunburned. He has bright-blue eyes and a soft voice. He was wearing khaki slacks and a blue shirt, New Balance running shoes, a preppy tie with small turtles on it, and a Rolex watch. Venter's office looks into the trees, and that day leaves were spinning on branches outside the windows, flashing their white undersides and promising rain. Beyond the trees, a chronic traffic jam was occurring on the Rockville Pike. Celera is in an area along a stretch of Interstate 270 known as the Biotechnology Corridor, which is dense with companies specializing in the life sciences.

Celera Genomics is a part of the P.E. Corporation, which was called Perkin-Elmer before the company's chief executive, Tony L. White, split the business into two parts: P.E. Biosystems, which makes the Prism machine, and Celera. Venter owns five per cent of Celera's stock, which trades, often violently, on the New York Stock Exchange. In recent months, the stock has been tossed by waves of panic selling and panic buying. Currently, the company is valued at three billion dollars, more or less. At times, Craig Venter's net worth has slopped around by a hundred million dollars a day, like water going back and forth in a bathtub.

"Our fundamental business model is like Bloomberg's," Venter said. "We're selling information about the vast universe of molecular medicine." Venter believes, for example, that one day Celera will help analyze the genomes of millions of people as a regular part of its business—this will be done over the Internet, he says—and the company will then help design or select drugs tailored to patients' particular needs. Genomics is moving so fast that it is possible to think that in perhaps fifteen years you will be able to walk into a doctor's office and have your own genome interpreted. It could be stored in a smart card. (You would want to keep the card in your wallet, in case you landed in an emergency room.) Doctors would read the smart card, and it would show a patient's total biological-software code. They could see the bugs in the code, the genes that make you vulnerable to certain diseases. Everyone has bugs in his code, and knowing what they are will become a key to diagnosis and treatment. If you became sick, doctors could watch the activity of your genes, using so-called gene chips, which are small pieces of glass containing detectors for every gene. Doctors could track how the body was responding to treatment. All your genes could be observed, operating in an immense symphony.

Venter stopped moving briefly, and sat down in front of a screen and tapped a keyboard. A Yahoo! quote came up. "Hey, we're over twenty today," he said. (Celera's stock has since split. Adjusted for today's prices, it was trading at ten dollars a share; at the time of this writing, it was trading at around seventy dollars a share.) I was standing in front of a large model of Venter's yacht, the Sorcerer, in which he won the 1997 Trans-Atlantic Race in an upset victory—it was the only major ocean race that Venter had ever entered. "I got the boat for a bargain from the guy who founded Land's End," Venter said. "I like to buy castoff things on the cheap from ultra-rich people."

Venter went into the hallway, and I followed him. Celera was renovating its space, and tiles were hanging from the ceiling. Some had fallen to the floor. Black stains dripped out of air-conditioning vents, and sheets of plywood were lying around. Workmen were Sheetrocking walls, ripping up carpet, and installing light fixtures, and a smell of paint and spackle drifted in the air. We took the stairs to the basement and entered a room that held about fifty ABI Prism 3700 machines. Each Prism was the size of a small refrigerator and had cost three hundred thousand dollars. Prisms are the fastest DNA sequencers on earth. At the moment, they were reading the DNA of the fruit fly. This was a pilot project for the human genome. The machines contained lasers. Heat from the lasers seemed to ripple from the machines, even though they were being cooled by a circulation system that drew air through them. The lasers were shining light on tiny tubes through which strands of fruit-fly DNA were moving, and the light was passing through the DNA, and sensors were reading the letters of the code. Each machine had a computer screen on which blocks of numbers and letters were scrolling past. It was fly code.

"You're looking at the third-largest DNA-sequencing facility in the world," Venter said. "We also have the second-largest and the largest."

We got into an elevator. The walls of the elevator were dented and bashed. Venter led me into a vast, low-ceilinged room that looked out into the trees. This was the largest DNA-decoding factory on earth. The room contained a hundred and fifty Prisms—forty-five million dollars' worth of the machines—and more Prisms were due to be installed any day. Air ducts dangled on straps from the ceiling, and one wall consisted of gypsum board.

Venter moved restlessly through the unfinished space. "You know, this is the most futuristic manufacturing plant on the planet right now," he said. Outdoors, the rain came, splattering on the windows, and the poplar leaves shivered. We stopped and looked over a sea of machines. "You're seeing Henry Ford's first assembly plant," he said. "What don't you see? People, right? There are three people working in this room. A year ago, this work would have taken one thousand to two thousand scientists. With this technology, we are literally coming out of the dark ages of biology. As a civilization, we know far less than one per cent of what will be known about biology, human physiology, and medicine. My view of biology is 'We don't know shit.' "

Celera's business model provokes some interesting questions, and some observers believe the company could fail. For instance, it appears to be burning through at least a hundred and fifty million dollars a year. But who will want to buy the information the company is generating, and how much will they pay for it? "There will be an incredible demand for genomic information," Venter assured me. "When the first electric-power companies strung up wires on power poles, there were a lot of skeptics. They said, 'Who's going to buy all that electricity?' We already have more than a hundred million dollars in committed subscription revenues over five years from companies that are buying genomic information from us—Amgen, Novartis, Pharmacia & Upjohn, and others. After we finish the human genome, we'll do the mouse, rice, rat, dog, cow, corn, maybe apple trees, maybe clover. We'll do the chimpanzee."

ONE DAY AT Celera's headquarters, I was talking with a molecular biologist named Hamilton O. Smith, who won a Nobel Prize in 1978 as the co-discoverer of restriction enzymes, which are used to cut DNA in specific places. Scientists use the enzymes like scissors, chopping up pieces of DNA so that they can be studied or recombined with the DNA of other organisms. Without the means to do this, there would be no such thing as genetic engineering.

Ham Smith is in his late sixties. He is six feet five inches tall, with a shock of stiff white hair and a modest manner. "Have you ever seen human DNA?" he asked me, as he poked around his lab.

"No."

"It's beautiful stuff."

A box that held four small plastic tubes, each the size of a pencil stub, sat on a countertop. "These four tubes hold enough human DNA to do the entire human-genome project," Smith said. "There's a couple of drops of liquid in each tube."

He held up one of the tubes and turned it over in the light to show me. A droplet of clear liquid moved back and forth. It was the size of a dewdrop. Then he held up a glass vial, and rocked it back and forth, and a crystal-clear, syrupy liquid oozed around in it. He explained that this was DNA he'd extracted from human blood—from white cells. "That's long, unbroken DNA. This liquid looks glassy and clear, but it's snotty. It's like sugar syrup. It really is a sugar syrup, because there are sugars in the backbone of the DNA molecule."

Smith picked up a pipet—a hand-held device with a hollow plastic needle in it, which is used for moving tiny quantities of liquid from one place to another. His hands are large, but they moved with precision. Holding the pipet, he sucked up a droplet of DNA mixed with a type of purified salt water called buffer. He held the drop in the pipet for a moment, then let it go. The droplet drooled. It reminded me of a spider dropping down a silk thread.

"There the DNA goes, it's stringing," he said. "The pure stuff is gorgeous." The molecules were sliding along one another, like spaghetti falling out of a pot, causing the water to string out. "It's absolutely glassy clear, without color," he said. "Sometimes it pulls back into the tube and won't come out. I guess that's like snot, too, and then you have to almost cut it with scissors. The molecule is actually quite stiff. It's like a plumber's snake. It bends, but only so much, and then it breaks. It's brittle. You can break it just by stirring it."

The samples of DNA that Celera is using are kept in a freezer near Smith's office. When he wants to get some human DNA, he removes a vial of frozen white blood cells or sperm from the freezer. The vials have coded labels. He thaws the sample of cells or sperm, then mixes the material with salt water, along with a little bit of detergent. A typical human cell looks like a fried egg, and the nucleus of the cell resembles the yolk. The detergent pops the eggs and the yolks, and strands of DNA spill out in the salt water. The debris falls to the bottom, leaving tangles of DNA suspended in the liquid.

One of Smith's research associates, a woman named Cindi Pfannkoch, showed me what shattered DNA was like. Using a pipet, she drew a tiny amount of liquid from a tube and let a drop go on a sheet of wax, where it beaded up like a tiny jewel, the size of the dot over this "i." An ant could have drunk it in full.

"There are two hundred million fragments of human DNA in this drop," she said. "We call that a DNA library."

She opened a plastic bottle, revealing a white fluff. "Here's some dried DNA." She took up a pair of tweezers and dragged out some of the fluff. It was a wad of dried DNA from the thymus gland of a calf; the wad was about the size of a cotton ball, and it contained several million miles of DNA.

"In theory," Ham Smith said, "you could rebuild the entire calf from any bit of that fluff."

I placed some of the DNA on the ends of my fingers and rubbed them together. The stuff was sticky. It began to dissolve on my skin. "It's melting—like cotton candy," I said.

"Sure. That's the sugar in DNA," Smith said.

"Would it taste sweet?"

"No. DNA is an acid, and it's got salts in it. Actually, I've never tasted it."

Later, I got some dried calf DNA. I placed a bit of the fluff on my tongue. It melted into a gluey ooze that stuck to the roof of my mouth in a blob. The blob felt slippery on my tongue, and the taste of pure DNA appeared. It had a soft taste, unsweet, rather bland, with a touch of acid and a hint of salt. Perhaps like the earth's primordial sea. It faded away.

DNA FROM SIX DONORS who contributed their blood or semen was used for Celera's human-genome project. The donors included both men and women, and a variety of ethnic groups. Just one person, a man, supplied the DNA for First Assembly. Only Craig Venter and one other person at the company are said to know who the donors are. "I don't know who they are, but I wouldn't be surprised if one of them is Craig," Ham Smith remarked.

Craig Venter grew up in a working-class neighborhood on the east side of Millbrae, on the San Francisco peninsula. His family's house was near the railroad tracks. One of his favorite childhood activities, he says, was to play chicken on the tracks. In high school, he excelled in science and shop. He built two speed boats, and spent a lot of time surfing Half Moon Bay. He attended two junior colleges in a desultory way, but mostly he surfed, until he enlisted in the Navy. He had long blond hair and a crisp body then. He was a medical corpsman in Vietnam, and twice he was sentenced to the brig for disobeying orders.

Venter has a history of confrontation with government authorities. He told me that as an enlisted man in San Diego he was court-martialled for refusing a direct order given by an officer. "She happened to be a woman I was dating," Venter said. "We had a spat, and she ordered me to cut my hair. I refused." A friend of his, Ron Nadel, who was a doctor in Vietnam, recalls that one of Venter's blowups with authority involved "telling a superior officer to do something that was anatomically impossible." Venter worked for a year in the intensive-care ward at Da Nang hospital, where, he calculates, more than a thousand Vietnamese and American soldiers died during his shifts, many of them while the 1968 Tet offensive was going on. When he returned to the United States, Venter finished college and then earned a Ph.D. in physiology and pharmacology from the University of California at San Diego.

Venter is married to a molecular biologist, Claire Fraser, who is the president of The Institute for Genomic Research (TIGR, pronounced "Tiger"), in Rockville, a nonprofit institute that he and Fraser helped establish in 1992. In 1998, he endowed TIGR with half of his original stake in Celera—five per cent of the company. The gift is currently worth about a hundred and fifty million dollars, and it will be used to analyze the genomes of microbes that cause malaria and cholera and other diseases.

A few years ago, Venter developed a hole in his intestine, due to diverticulitis. He collapsed after giving a speech, and nearly died. He is fine now, but he blames stress caused by his enemies for his burst intestine. Venter has enemies of the first water. They are brilliant, famous, articulate, and regularly angry. At times, Venter seems to thrive on his enemies' indignation with an indifferent grace, like a surfer shooting a tubular wave, letting himself be propelled through their cresting wrath. At other times, he seems baffled, and says he can't understand why they don't like him.

One of Venter's most venerable enemies is James Watson, who, with Francis Crick and Maurice Wilkins, won the Nobel Prize in Medicine in 1962 for discovering the shape of the DNA molecule—what they called the double helix. Watson helped found the Human Genome Project, and he was the first head of the N.I.H. genome program. I visited him in his office at the Cold Spring Harbor Laboratory, on Long Island. The office is panelled in blond oak and has a magnificent eastward view across Cold Spring Harbor. Watson is now in his seventies. He has a narrow face, lopsided teeth, a frizz of white hair, sharp, restless eyes, a squint, and a dreamy way of speaking in sentences that trail off. He put his hands on his head and squinted at me. "In 1953, with our first paper on DNA, we never saw the possibility . . ." he said. He looked away, up at the walls. "No chemist ever thought we could read the molecule." But a number of biologists began to think that reading the human genome might just be possible, and by the mid-nineteen-eighties Watson had become convinced that the decryption of the genome was an important goal and should be pursued, even if it cost billions and took decades.

Watson appeared before Congress in May of 1987 and asked for an initial annual budget of thirty million dollars for the project. The original plan was to sequence the human genome by 2005, at a projected cost of about three billion dollars. The principal work of the project is now carried out by five major DNA-sequencing centers, as well as by a number of smaller centers around the world—all academic, nonprofit labs. The big centers include one at Baylor University in Texas, one at Washington University in St. Louis, the Whitehead Institute at M.I.T., the Joint Genome Institute of the Department of Energy, and the Sanger Centre, near Cambridge, England. The Wellcome Trust of Great Britain—the largest nonprofit medical-research foundation in the world—is funding the Sanger work, which is to sequence a third of the human genome. One of the founding principles of the Human Genome Project was the immediate release of all the human code that was found, making it available free of charge and without any restrictions on who could use it or what anyone could do with it.

In 1984, Craig Venter had begun working at the N.I.H., where he eventually developed an unorthodox strategy for decoding bits of genes. At the time, other scientists were painstakingly reading the complete sequence of each gene they studied. This process seemed frustratingly slow to Venter. He began isolating what are called expressed sequence tags, or E.S.T.s, which are fragments of DNA at the ends of genes. When the E.S.T.s were isolated, they could be used to identify genes in a rough way. With the help of a few sequencing machines, Venter identified bits of thousands of human genes. This was a source of unease at the N.I.H., because it was a kind of skimming rather than a complete reading of genes. Venter published his method in 1991 in an article in *Science*, along with partial sequences from about three hundred and fifty human genes. The method was not received well by many genomic scientists. It was fast, easy, and powerful, but it didn't look elegant, and some scientists seemed threatened by it. Venter claims that two of his colleagues, who are now heads of public genome centers, asked him not to publish his method or move forward with it for fear they would lose their funding for genome sequencing.

The N.I.H. decided to apply for patents on the gene fragments that Venter had identified. James Watson blew his stack over the idea of anyone trying to patent bits of genes, and he got into a hostile situation with the director of the N.I.H., Bernadine Healy, who defended the patenting effort. In July of 1991, during a meeting in Washington called by Senator Pete Domenici, of New Mexico, to review the genome program, Watson dissed Venter's methods. "It isn't science," he said, adding that the machines "could be run by monkeys."

It was a strange moment. The Senate hearing room was almost empty—few politicians were interested in genes then. But Craig Venter was sitting in the room. "Jim Watson was clearly referring to Craig as a monkey in front of a U.S. senator," another scientist who was there said to me. "He portrayed Craig as the village idiot of genomics." Venter seemed to almost thrash in his chair, stung by Watson's words. "Watson was the ideal father figure of genomics," Venter says. "And he was attacking me in the Senate, when I was relatively young and new in the field."

Today, James Watson insists that he wasn't comparing Craig Venter to a monkey. "It's the patenting of genes I was objecting to. That's why I used the word 'monkey'! I hate it!" he said to me. The patent office turned down the N.I.H.'s application, but a few years later, two genomics companies, Incyte and Human Genome Sciences, adopted the E.S.T. method for finding genes, and it became the foundation of their businesses—currently worth, combined, about seven billion dollars on the stock market. Incyte and Human Genome Sciences are Celera's main business competitors. Samuel Broder, the chief medical officer at Celera, who is a former director of the National Cancer Institute, said to me, heatedly, "None of the people who severely and acrimoniously criticized Craig for his E.S.T. method ever said they were personally sorry. They ostracized Craig and then went on to use his method with never an acknowledgment."

James Watson now says, "The E.S.T. method has proved immensely useful, and it should have been encouraged."

VENTER WAS INCREASINGLY UNHAPPY at the N.I.H. He had received a ten-million-dollar grant to sequence human DNA, and he asked for permission to use some of the money to do E.S.T. sequencing, but his request was denied by the genome project. Venter returned the grant money with what he says was a scathing letter to Watson. In addition, Claire Fraser had been denied tenure at the N.I.H. Her review committee (which was composed entirely of middle-aged men) explained to her that it could not evaluate her work independently of her husband's. At the time, Fraser and Venter had separate labs and separate research programs. Fraser considered suing the N.I.H. for sex discrimination.

Watson was forced to resign as head of the genome project in April of 1992, in part because of the dispute over patenting Craig Venter's work. That summer, Venter was approached by a venture capitalist named Wallace Steinberg, who wanted to set up a company that would use Venter's E.S.T. method to discover genes, create new drugs, and make money. "I didn't want to run a company, I wanted to keep doing basic research," Venter says. But Steinberg offered Venter a research budget of seventy million dollars over ten years—a huge amount of money, then, for biotech. Venter, along with Claire Fraser and a number of colleagues, left the N.I.H. and founded TIGR, which is a nonprofit organization. At the same time, Steinberg established a for-profit company, Human Genome Sciences, to exploit and commercialize the work of TIGR, which was required to license its discoveries exclusively to its sister company. Thus Venter got millions of dollars for research, but he had to hand his discoveries over to Human Genome Sciences for commercial development. Venter had one foot in the world of pure science and one foot in a bucket of money.

By 1994, the Human Genome Project was mapping the genomes of model organisms, which included the fruit fly, the roundworm, yeast, and *E. coli* (the organism that lives in the human gut), but no genome of any organism had been completed, except virus genomes, which are relatively small. Venter and Hamilton O. Smith (who was then at the Johns Hopkins School of Medicine) proposed speeding things up by using a technique known as whole-genome shotgun sequencing. In shotgunning, the genome is broken into small, random, overlapping pieces, and each piece is sequenced, or read. Then the jumble of pieces is reassembled in a computer that compares each piece to every other piece and matches the overlaps, thus assembling the whole genome.

Venter and Smith applied for a grant from the N.I.H. to shotgun-sequence the genome of a disease-causing bacterium called *H. influenzae,* or H. flu for short. It causes fatal meningitis in children. They proposed to do it in just a year. H. flu has 1.8 million letters of code, which seemed massive then (though the human code is two thousand times as long). The review panel at the N.I.H. gave Venter's proposal a low score, essentially rejecting it. According to Venter, the panel claimed that an attempt to shotgun-sequence a whole microbe was excessively risky and perhaps impossible. He appealed. The appeals process dragged on, and he went about shotgunning H. flu anyway. Venter and the TIGR team had nearly finished sequencing the H. flu genome when, in early 1995, a letter arrived at TIGR saying that the appeals committee had denied the grant on the ground that the experiment wasn't feasible. Venter published the H. flu genome a few months later in *Science*. Whole-genome shotgunning had worked. This was the first completed genome of a free-living organism.

It seems quite possible that Venter's grant was denied because of politics. The review panel seems to have hated the idea of giving N.I.H. money to TIGR to make discoveries that would be turned over to a corporation, Human Genome Sciences. It turned down the grant, in spite of the fact that "all the smart people knew the method was straightforward and would work," Eric Lander, the head of the genome center at M.I.T. and one of the leaders of the public project, said to me.

Around this time, Wallace Steinberg died of a heart attack, and his death provided a catalyst for a split between TIGR and Human Genome Sciences, which was run by a former AIDS researcher, William Haseltine. Venter and Haseltine were widely known to dislike each other. Venter sold his stock in Human Genome Sciences because of the rift between them, and after Steinberg died the relationship between the two organizations was formally ended.

LATE IN 1997, TIGR was doing some DNA sequencing for the Human Genome Project, and Venter began going to some of the project's meetings. That was when he started calling the heads of the public project's DNA-sequencing centers the Liars' Club, claiming that their predictions about when they would finish a task and how much it would cost were false. This did not win Venter many friends. But he seemed to have a point.

Francis Collins, a distinguished medical geneticist from the University of Michigan, had become the head of the N.I.H. genome program shortly after James Watson resigned in 1992. In early January, 1998, an internal budget projection from Collins's office somehow found its way to Watson (he seems to find out everything that's happening in molecular biology). This budget projection—it is not clear whether it was formal or was just an unofficial projection—was a document about eight pages long. It contained a graph marked "Confidential" indicating that Collins planned to spend only sixty million dollars per year on direct human-DNA sequencing through 2005. It also predicted that by that year—when the human genome was supposed to be completed—only 1.6 billion to 1.9 billion letters of human code would be sequenced; that is, slightly more than half of the human genome.

This upset Watson, and he decided to discuss it with Eric Lander. On January 17th, Watson travelled to Rockefeller University, on the East Side of Manhattan, where Lander was giving the prestigious Harvey Lecture. The two men met after the lecture at the faculty club at Rockefeller. They were dressed in black tie and were somewhat inebriated. Traditionally among medical people, the Harvey Lecture is given and listened to under the influence.

The Rockefeller faculty club overlooks a lawn and sycamore trees and the traffic of York Avenue. Watson and Lander sat down with cognacs at a small table in a dim corner of the room, on the far side of a pool table, where they could talk without being overheard. Also present and drinking cognac was a biologist named Norton Zinder, who is one of Watson's best friends. Zinder, like Watson, is a founder of the Human Genome Project. One of the older men brought up the confidential budget document with Lander, and both of them began to press him about it. They felt that it provided evidence that Collins did not intend to spend more than sixty million dollars a year on human-DNA sequencing—nowhere near enough to get the job done, they felt.

Watson evidently felt that Lander had influence with Francis Collins, and he urged him to try to persuade Collins to spend more on direct sequencing of human DNA, and to twist Congress's arm for more money.

Norton Zinder was somewhat impaired with cocktails. "This thing is potchkeeing along, going nowhere!" he said, hammering the little table and waving his arms as he spoke. For him, the issue was simple: he had had a quadruple coronary bypass, and he had been receiving treatments for cancer, and now he was afraid he would not live to see the deciphering of the human genome. This was intolerable. The human genome had begun to seem like a vision of Canaan to Norton Zinder, and he thought he wouldn't make it there.

Eric Lander did not view things the way the older biologists did. In his opinion, the problem was organizational. The Human Genome Project was "too bloody complicated, with too many groups." He felt the real problem was a lack of focus. He wanted the project to create a small, elite group that would do the major sequencing of human DNA—shock cavalry that would lead a charge into the human genome. Implicitly, he thought its leader should be Eric Lander.

The three men downed their cognacs with a sense of frustration. "I had essentially given up seeing the human genome in my lifetime," Zinder says.

AT ABOUT THE SAME MOMENT that Watson and his friends were lamenting the slowness of the public project, the Perkin-Elmer Corporation, which was a manufacturer of lab instruments, was secretly talking about a corporate reorganization. It controlled more than ninety per cent of the market for DNA-sequencing equipment, and it was developing the ABI Prism 3700. The Prism was then only a prototype sitting in pieces in a laboratory in Foster City, California, but already it looked as if it were going to be at least ten times faster than any other DNA-sequencing machine. Perkin-Elmer executives began to wonder just what it could do. One day Michael Hunkapillar, who was then the head of the company's instrument division, got out a pocket calculator and estimated that several hundred Prisms could whip through a molecule of human DNA in a few weeks, although only in a rough way. To fill in the gaps—places where the DNA code came out garbled or wasn't read properly by the machines—it would be necessary to sequence the molecule again and again. This is known as repeat sequencing, or manyfold coverage, and might take a few years. Hunkapillar persuaded the chief executive of Perkin-Elmer, Tony White, to restructure the business and create a genomics company.

In December, 1997, executives from Perkin-Elmer began telephoning Venter to see if he'd be interested in running the new company. He blew them off at first, but in early February, 1998, he went to California with a colleague, Mark Adams, to look at the prototype Prism. When they saw it, they immediately understood its significance. Before the end of that day, Venter, Adams, and Hunkapillar had laid out a plan for decoding the human genome. A month later, Norton Zinder, Watson's friend, flew to California to see the machine. "It was just a piece of equipment sitting on a table, but I said, 'That's it! We've got the genome!' " he recalled. Zinder joined Celera as a member of its board of advisers, and received stock in the company, which has considerably enriched him. ("The chemists have been cleaning up," he said to me. "Now biologists have their hands on the money, too.") Zinder and Watson have maintained their friendship but have agreed not to speak about Celera with each other. They evidently fear that one or both of them could have a stroke arguing about Craig Venter.

AT ELEVEN O'CLOCK in the morning on May 8, 1998, Craig Venter and Mike Hunkapillar walked into the office of Harold Varmus, who was then the director of the N.I.H., and announced the pending formation of a corporation, led by Venter, that was going to decode the human genome. (Celera did not yet have a name.) They proposed to Varmus that the company and the public project collaborate, sharing their data and—this point is enormously important to scientists—sharing the publication of the human genome, which meant sharing the credit and the glory for having done the work, including the unspoken possibility of a Nobel Prize. Varmus strongly suspected that this wasn't a sincere offer, and he told them that he needed time, particularly to check with Francis Collins. Later that same day, Venter and Hunkapillar drove to Dulles Airport, where they met Collins at the United Airlines Red Carpet Club, and again offered collaboration. Venter recalls that Collins seemed upset. Collins recalls that he merely asked Venter for time to consider the offer. Time was one thing Venter was not prepared to give.

Venter had alerted the *New York Times* to the story about the creation of the new company, and just an hour or so after the meeting with Collins he called the *Times* and told the paper it should run it. In the story, Venter announced that he would sequence the human genome by 2001—four years ahead of the public project—and he would do it, he claimed, for between a hundred and fifty and two hundred million dollars—less than a tenth of the projected cost of the public project. The *Times* reporter, Nicholas Wade, implied that the Human Genome Project might not meet its goals and might be superfluous.

Four days later, on May 12th, Venter and Hunkapillar went to the Cold Spring Harbor Laboratory, where a meeting of the heads of the Human Genome Project was taking place. Venter got up and told them, in effect, that they could stop working, since he was going to sequence the human genome *tout de suite.* Later that week, sitting beside Varmus and Collins at a press conference, Venter looked out at a room full of reporters and suggested that biology and society would be better off if the Human Genome Project shifted gears and moved forward to do the genome of the . . . mouse.

It was a fart in church of magnitude nine. "The mouse is essential for interpreting the human genome," Venter tried to explain, but that didn't help. In the words of one head of a sequencing center who was at the Cold Spring Harbor meeting, "Craig has a certain lack of social skills. He goes into that meeting thinking everyone is going to thank him for doing the human genome himself. The thing blew up into a huge explosion." The head of another center recalled, "Craig came up to me afterward, and he said, 'Ha, ha, I'm going to do the human genome. You should go do the mouse.' I said to him, 'You bastard. You *bastard,*' and I almost slugged him."

They felt that Venter was trying to stake out the human genome for himself as a financial asset while at the same time stealing the scientific credit. They felt that he was belittling their work. Venter said that he would make the genome available to the public but would charge customers who wanted to see and work with Celera's analyzed data.

James Watson was furious. He did not like the idea of having to pay money to Craig Venter for anything. Watson did not attend Venter's presentation, but he appeared in the lobby afterward, where he repeatedly said to people, "He's Hitler. This should not be Munich." To Francis Collins he said, "Are you going to be Churchill or Chamberlain?"

Venter left the meeting soon afterward, and he and Watson have exchanged only chilly greetings since.

The British leaders of the public project—John Sulston, the director of the Sanger Centre, and Michael Morgan, of the Wellcome Trust—reacted swiftly to Venter's announcement. They were in England, but they flew to the United States, and the next day arrived at Cold Spring Harbor, where they found things in disarray, if not in fibrillation, with scientists wondering if the Human Genome Project was about to die. To a standing ovation, Michael Morgan got up and read a Churchillian statement declaring that the Wellcome Trust would nearly double its funding for the public project, and would decode a full third of the human genome, and would challenge any "opportunistic" patents of the genome. "We were reacting, in part, to Craig's suggestion that we just close up shop and go home," Morgan says now.

Venter also announced that Celera would use the whole-genome shotgun method. The public project was using a more conventional method. John Sulston and Robert Waterston, the head of the sequencing center at Washington University, published a letter in *Science* asserting that Venter's method would be "woefully inadequate." Francis Collins was quoted in *USA Today* as saying that Celera was going to produce "the Cliffs Notes or the Mad Magazine version" of the human genome. Collins says now that his words were taken out of context, and he regrets the quote.

THE COMPANY FORGED from Perkin-Elmer amid the turmoil was the P.E. Corporation, which holds the P.E. Biosystems Group, the unit that makes the Prism machines, and Celera Genomics. Michael Hunkapillar, who is now the president of P.E. Biosystems, believed that he could sell a lot of machines to everyone, including the Human Genome Project. There was a fat profit margin in the chemicals the machines use. The chemicals cost far more than the machine over the machine's lifetime. This was the razor-blade principle: if you put razors in people's hands, you will make money selling blades.

In August, Incyte Pharmaceuticals announced that it was starting a human-genome project of its own. In September, James Watson quietly went to some key members of Congress and persuaded them to spend more money on the public project. At the same time, the leaders of the project announced a radical new game plan: they would produce a "working draft" of the human genome by 2001—a year ahead of when Venter said he'd be done—and a finished, complete version by 2003. An epic race had begun.

A couple of months ago, Michael Morgan, of the Wellcome Trust, was talking to me about Venter and what had happened with the creation of Celera. "From the first press release, Craig saw the public program as something he wanted to denigrate," Morgan said. "This was our first sign that Celera was setting out to undermine the international effort. What is it that motivates Craig? I think he's motivated by the same things that drive other scientists—personal ego, a degree of altruism, a desire to push human knowledge forward—but there must be something else that drives the guy. I think Craig has a huge chip on his shoulder that makes him want to be loved. I actually think Craig is desperate to win a Nobel Prize. He also wants to be very, very rich. There is a fundamental incompatibility there."

One day, I ran into a young player in the Human Genome Project. He believed in the worth and importance of the project, and said that he had turned down a job offer from Celera. He didn't have any illusions about human nature. He said, "Here's why everyone is so pissed at Craig. The whole project started when James Watson persuaded Congress to give him money for the human genome, and he turned around and gave it to his friends—they're the heads of centers today. It grew into a lot of money, and then the question was, Who was going to get the Nobel Prize? In the United States, there were seventeen centers in the project, and there was no quality control. It didn't matter how bad your data was, you just had to produce it, and people weren't being held accountable for the quality of their product. Then Celera appeared. Because of Celera, the N.I.H. was suddenly forced to consolidate its funding. The N.I.H. and Francis Collins began to dump more than eighty per cent of the money into just three centers—Baylor, Washington University, and M.I.T.—and they jacked everybody else. They had to do it, because they had to race

Celera, and they couldn't control too many players. So all but three centers were cut drastically, and some of the labs closed down. Celera was not just threatening their funding but threatening their very lives and everything they had spent years building. It's kind of sad. Now those people hang around meetings, and the leaders treat them like 'If you're really nice, we'll give you a little piece of the mouse.' That's the reason so many of them are so angry at Celera. It's easier for them to go after Craig than to go after Francis Collins and the N.I.H."

AT CELERA'S HEADQUARTERS in Rockville, I was shown how human DNA was shotgunned into small pieces when it was sprayed through a hospital nebulizer that cost a dollar-fifty. The DNA fragments were then introduced into *E. coli* bacteria, and grown in glass dishes. The bacteria formed brown spots—clones—on the dishes. Each spot had a different fragment of human DNA growing in it. The dishes were carried to a room where three robots sat in glass chambers the size of small bedrooms. Each robot had an arm that moved back and forth rapidly over a dish. Little needles on the arms kept stabbing down and taking up the brown spots.

Craig Venter stood watching the robots move. The room smelled faintly like the contents of a human intestine. "This used to be done by hand. We've been picking fifty-five thousand clones a day," he said. (Later, Celera got that rate up to a hundred and twenty thousand clones a day.) All the DNA fragments would eventually wind up in the Prism sequencing machines, and what would be left, at the end, was a collection of up to twenty-two million random fragments of sequenced human DNA. Then the river of shattered DNA would come to the computer, and to a computer scientist named Eugene Myers, who with his team devised the First Assembly.

Gene Myers has dark hair and a chiselled, handsome face. He wears glasses and a green half-carat emerald in his left ear and brown Doc Martens shoes. He also has a ruby and a sapphire that he will wear in his ear, instead of the emerald, depending on his mood. He is sensitive to cold. On the hottest days of summer, Myers wears a yellow Patagonia fibrepile jacket, and he keeps a scarf wrapped around his neck. "My blood's thin," he explained to me. He says the scarf is a reference to the DNA of whatever organism he happens to be working on. When I first met Myers, in the hot summer of 1999, he was keeping himself warm in his fruit-fly scarf. It had a black-and-white zigzag pattern. This spring, Myers started wearing his human scarf, which has a green chenille weave of changing stripes. He intended his scarf to make a statement about the warfare between Celera and the public project. "I picked green for my human scarf because I've heard that green is a positive, healing color," he said. "I really want all this bickering to go away." His office is a cubicle in a sea of cubicles, most of which are stocked with Nerf guns, Stomp Rockets, and plastic Viking helmets. Occasionally, Myers puts the "Ride of the Valkyries" on a boom box, and in a loud voice he declares war. Nerf battles sweep through Celera whenever the tension rises. Myers fields a compound double-action Nerf Lock-N-Load Blaster equipped with a Hyper*Sight. "Last week we slaughtered the chromosome team," he said to me.

Myers used to be a professor of computer science at the University of Arizona in Tucson. He specializes in combinatorial algorithms. This involves the arrangements and patterns of objects. One day in 1995, he got a telephone call from a geneticist named James Weber, at the Marshfield Medical Research Foundation in Wisconsin. Weber said he felt that whole-genome shotgunning would work for organisms that have very long DNA molecules, such as humans. He wondered if Myers could help him with the math.

Jim Weber submitted a proposal to the N.I.H. for a grant—twelve million dollars—to support a pilot study of the shotgun approach on the human genome. This might speed up the project dramatically, he suggested. Weber was invited to speak to the annual meeting of the heads of the project, held in Bermuda.

Weber was nervous about it, and wanted Gene Myers to go with him to help explain the math. "Jim asked them to invite me, but they didn't," Myers says. So on February 26, 1996, Jim Weber went alone to the meeting in Bermuda and tried to make a case for shotgunning the human genome. He found himself facing a U-shaped table with about forty people at it. "They trounced Jim," Myers said. "They said it wouldn't work. They said it would be full of holes. 'A Swiss-cheese genome'—that's the term we've often heard. The grant proposal was soundly rejected."

Jim Weber says that the Swiss-cheese analogy was not far off, but that "it would have been much better to get most of the human genome quickly, even with holes in it, so that people could start using the information to understand diseases and begin to find cures for them. It would have been better if the N.I.H. had funded a pilot study. Instead, Gene and his team went out and did it. That is a huge accomplishment."

Craig Venter was hanging around while I was talking with Myers. He came up to us and said, "They not only shot Gene down—they ridiculed him. They said he was a kook. We're going to prove that Gene was right, and we're going to prove that there's something fundamentally wrong with the system."

On September 9, 1999, Venter announced that Celera had completed the sequencing of the fruit fly's DNA, and had begun to run human DNA through its sequencing machines—there were now three hundred of them crammed into Building One in Rockville. The Command Center was up and running, and from then on Celera operated in high-speed mode. One day that fall, I talked with the company's information expert, a stocky man named Marshall Peterson. He took me to the computer room, in Building Two. To get into the room, Peterson punched in a security code and then placed his hand on a sensor, which read the unique pattern of his palm. There was a clack of bolts sliding back, and we pushed through the door.

A chill of cold air washed over us, and we entered a room filled with racks of computers that were wired together. "What you're looking at in this room is roughly the equivalent of America Online's network of servers," Peterson said. "We have fifty-five miles of fibre-optic cables running through this building." Workmen standing on ladders were installing many more cables in the ceiling. "The disk storage in this room is five times the size of the Library of Congress. We're getting more storage all the time. We need it."

He took me to the Command Center, where a couple of people were hanging around consoles. Some of the consoles had not had equipment installed in them yet. A big screen on the wall showed CNN Headline News. "I've got a full-time hacker working for me to prevent security breaches," Peterson said. "We're getting feelers over the Internet all the time—people trying to break into our system." Celera would be dealing with potentially valuable information about the genes of all kinds of organisms. Peterson thought that some of what he called feelers—subtle hacks and unfriendly probes—had been emanating from Celera's competitors. He said he could never prove it, though. Lately, the probes had been coming from computers in Japan. He thought it was American hackers co-opting the Japanese machines over the Internet.

By October 20th, forty days after Celera started running human DNA through its machines, the company announced that it had sequenced 1.2 billion letters of human code. The letters came in small chunks from all over the genome. Six days later, Venter announced that Celera had filed provisional patent applications for six thousand five hundred human genes. The applications were for placeholder patents. The company hoped to figure out later which of the genes would be worth patenting in earnest.

A gene patent gives its holder the right to make commercial products and drugs derived from the gene for a period of seventeen years. Pharmaceutical companies argue that patents are necessary, because without them businesses would never invest the hundreds of millions of dollars that are needed to develop a new drug and get it through the licensing process of the Food and Drug Administration. ("If you have a disease, you'd better hope someone patents the gene for it," Venter said to me.) On the other hand, parcelling out genes to various private companies could lead to what Francis Collins refers to as the "Balkanization of the human genome," a paralyzing situation that might limit researchers' access to genes.

Venter insists that Celera is an information company and that patenting genes is not its main goal. He has said that Celera will attempt to get patents on not more than about three hundred human genes. There is no question that Celera hopes to nail down some very valuable genes—billion-dollar genes, perhaps.

CELERA'S STOCK HAD DRIFTED since the summer, but around Halloween, as investors began to realize that the company was cranking out the human genome—and filing large numbers of placeholder patents—it jumped up to forty dollars a share. (The prices here are pre-split prices. Adjusted for today's prices, the stock moved up to twenty dollars.) On December 2nd, the Human Genome Project announced that it had deciphered most of the code on chromosome No. 22, the second-shortest chromosome in the human genome. This made the reading of the whole genome seem more imminent, and Celera's stock began a spectacular rise. It shot up that day by nine points, to close at over seventy dollars. Then, after the market's close on Thursday, December 16th, Jeff Fischer, a co-founder of the Web site called The Motley Fool, announced that he was buying shares of Celera for his own portfolio. It is called the Rule Breaker Portfolio, and it has famously delivered wealth—Fischer bought A.O.L. very early, for example. On that Friday morning, a great number of people tried to buy Celera, and they drove the stock up twenty points. It was on its way to the pre-split equivalent of more than five hundred dollars a share. That past summer, it had been trading at fourteen.

I went to visit Celera on Tuesday of the following week, and that morning the company's stock could not open for trading. Everyone wanted to buy it, and nobody wanted to sell it. While the stock was halted—at a hundred and one dollars a share—I wandered around. There was a feeling of paralysis in the air, and I sensed that not much work was getting done that day, except by the machines. Employees were checking the quote on the Internet and wondering what their net worth would be when the stock opened. The lobby now sported fish-eye security cameras. The walls smelled of fresh paint, and the floors had a new purple carpet with a pattern that resembled worms. They were meant to look like fragments of DNA.

I found Hamilton O. Smith in his lab, puttering around with human DNA in tiny test tubes, but his heart was not in the job. He was tired. He explained that he was renovating an old house that he and his wife had bought. He had stayed up all night ripping carpet out of the basement, because new carpets were due to arrive that morning. He had driven to work in his '83 Mercury Marquis. He owned thousands of shares of Celera.

Smith passed a computer, stopped, and brought up a quote. Celera had finally opened for trading. It had gapped up—jumped instantly upward—by thirteen points. It was at a hundred and fourteen. Smith's net worth had gapped up by something on toward a million dollars. "Is there no end to this?" he muttered.

Craig Venter came into Smith's lab and asked him to lunch. In the elevator, Smith said to him, "I can't stand it, Craig. The bubble will break." They sat down beside each other in the cafeteria and ate cassoulet from bowls on trays.

"This defies common sense," Smith said. "It's really impossible to put a value on this company."

"That's what we've been telling the analysts," Venter said.

LATER THAT DAY, I ended up in Claire Fraser's office at TIGR headquarters, a complex of semi-Mission-style buildings a couple of miles from Celera's offices and labs. Fraser is a tall, reserved woman with dark hair and brown eyes, and her voice has a faint New England accent. She grew up in Saugus, Massachusetts. In high school, she says, she was considered a science geek. "The only lower citizens were the nerdy guys in the audiovisual club. Of course, now they're probably in Hollywood." Her office has an Oriental rug on the floor and a table surrounded by Chippendale chairs. It was originally Venter's office. ("This is Craig's extravagant taste, not mine," she explained.) She wore an expensive-looking suit. Two poodles, Cricket and Marley, slept by a fireplace.

"Before genomics, every living organism was a black box," she said. "When you sequence a genome, it's like walking into a dark room and turning on a light. You see entirely new things everywhere."

Fraser placed a sheet of paper on the table. It contained an impossibly complicated diagram that looked like a design for an oil refinery. She explained that it was an analysis of the genome of cholera, a single-celled microbe that causes murderous diarrhea. TIGR scientists had finished sequencing the organism's DNA a few weeks earlier. Much of the picture, she said, was absolutely new to our knowledge of cholera. About a quarter of the genes of every microbe that had been decoded by TIGR were completely new to science, and were not obviously related to any other gene in any other microbe. To the intense surprise and wonder of the scientists, nature was turning out to be an uncharted sea of unknown genes. The code of life was far richer than anyone had imagined.

Fraser's eyes moved quickly over the diagram. "Yes . . . wow. . . . There may be important transporters here. . . . You see these transporters in other bacteria, and . . . I don't know . . . it looks like there could be potential for designing a new drug that could block them."

The phone rang. Fraser walked across her office, picked up the receiver, and said softly, "Craig? Hello. What? It closed at a hundred and twenty-five?" Pause. "I don't know how much it's worth—you're the one with the calculator."

Their net worth had jumped above a hundred and fifty million dollars that day.

Fraser drove home, and I followed her in my car. Their house is in the country outside Washington. It sits behind a security gate at the end of a long driveway. Venter arrived in a new Porsche. The car would do zero to sixty in five seconds, he said. In the vaulted front hall of the house there was a large stained-glass window showing branches of a willow tree, and there was a model of H.M.S. Victory in a glass case. A jumble of woodworking machines—a bandsaw, a table saw, a drill press—filled a shop attached to the garage. Venter has worked with wood since high school.

In the kitchen, Claire fixed dinner for the poodles, while Craig circled the room, talking. "We created close to two hundred millionaires in the company today. I think most of them had not a clue this would happen when they joined Celera. We have a secretary who became a millionaire today. She's married to a retired policeman. He went out looking to buy a farm." He popped a Bud Light and swigged it. "This could only happen in America. You've got to love this country." Claire fed the poodles.

There were no cooking tools in the kitchen that I could see. The counters were empty. The only food I noticed was a giant sack of dog food, sitting on top of an island counter, and two boxes of cold cereal—Quaker Oatmeal Squares and Total. In the guest bathroom, upstairs, there were no towels, and the walls were empty. The only decorative object in the bathroom was a cheap wicker basket piled with little soaps and shampoos they had picked up in hotels.

We went to a restaurant and ate steak. "We're in the Wild West of genomics," Venter said. "Celera is more than a scientific experiment; it's a business experiment. Our stock-market capitalization as of today is three and a half billion dollars. That's more than the projected cost of the Human Genome Project. I guess that's saying something. The combined market value of the Big Three genomics companies—Celera, Human Genome Sciences, and Incyte—was about twelve billion dollars at the end of today. This wasn't imaginable six months ago. The Old Guard doesn't have control of genomics anymore." He chewed steak, and looked at his wife. "What the hell are we going to do with all this money? I could play around with boats. . . ."

Claire started laughing. "My God, I couldn't live with you."

"The money's nice, but it's not the motivation," Venter said to me. "The motivation is sheer curiosity."

IN DECEMBER, 1999, Celera and the Human Genome Project discussed whether it would be possible to collaborate. There was one formal meeting, and there were many points of difference. Meanwhile, Celera's stock seemed to go into escape velocity from the earth. In January, it soared over two hundred dollars a share. Celera filed to offer more shares to the public, and declared a two-for-one stock split. Shortly after the split, on February 25th, the stock hit an all-time high of two hundred and seventy-six dollars a share (more than five hundred and fifty dollars, pre-split). Celera's stock-market value reached fourteen billion dollars, and Venter's worth surpassed seven hundred million dollars. It looked as if Venter could become a billionaire of biotechnology.

Then, on March 6th, newspapers carried reports that the discussions between Celera and the public project had collapsed. The main point of disagreement, according to officials at the public project, was that Celera wanted to keep control of intellectual property in the human genome. Celera intended to license its analyzed database to pharmaceutical companies and nonprofit research institutions, for payment. Celera said that it would let anyone use the data, but that any other company would be forbidden from reselling the data. The Human Genome people insisted that the period of restriction on the data could be for no more than a year, and after that the data should be totally public. Celera argued that it didn't want its competitors to resell the information and profit from Celera's work. Celera's stock began to drop.

On March 14th, President Clinton and Prime Minister Blair of Great Britain released a joint statement to the effect that all the genes in the human body "should be made freely available to scientists everywhere." The statement had been drafted with the help of Francis Collins and his staff, and had been in the works for a year. It was vague, but it looked like an Anglo-American smart bomb aimed at Celera, and it scared the daylights out of investors in biotechnology stocks, who feared that potentially lucrative patents on genes might be undermined by some new government policy.

On the day of the Clinton-Blair statement, Celera's stock went into a screaming nosedive. It dropped fifty-seven dollars in a matter of hours, amid trading halts and order imbalances. The other genomic stocks crashed in sympathy with Celera, and this, in turn, dragged down the Nasdaq, which that day suffered the second-largest point loss in its history. Short-sellers—people who profit from the decline of a stock—encrusted Celera like locusts. As of this writing, the Nasdaq has not recovered. Venter's mother telephoned him afterward, and said to him, "Craig, you've managed to do overnight what Alan Greenspan has been trying to do for years."

"It's not every day you get attacked by the President and the Prime Minister," Venter said to me late that night on the telephone. "I'm expecting a call from the Pope any day now, asking me to recant the human genome." He sounded wired and exhausted. "I feel a little like Galileo. They offered to have a barbecue with him, right? Look, I'm not likening myself to Galileo in terms of genius, but it is clear that the human genome is the science event of our time. I am going to publish the genome, and that's what the threat to the public order is. If Celera was keeping the genome a secret, the way Incyte and Human Genome Sciences are, you wouldn't hear a peep out of the government. Our publishing the genome makes a mockery of the fifteen years and billions of dollars the public project has spent on it."

Venter seemed particularly upset with the British part of the public project. "In my opinion," he said, "the Wellcome Trust is now trying to justify how, as a private charity, it gave what I think was well over a billion dollars to the Sanger Centre to do just a third of the human genome, largely at the expense of the rest of British medical science. Clinton and Blair took forty billion dollars out of the biotechnology industry today—that's how much was lost by investors. It was money that would pay for cures for cancer, and it was taken off the table, all because some bastards at the Wellcome Trust are trying to cover up their losses."

I called Michael Morgan, at the Wellcome Trust, to see what he had to say about this. "In hindsight, it is easy to ascribe to us Machiavellian powers that the Prince would have been proud of," he said dryly. "As for the allegation that I'm a bastard, I can easily disprove it using the technology of the Human Genome Project."

The day after the Clinton-Blair statement and the crash in biotech stocks, a White House spokesman made a point of telling reporters that the Administration supported the patenting of genes.

ON MARCH 24TH, Venter and his colleagues published a substantially complete genome of the fruit fly—*Drosophila*—in *Science.* It was also published by Celera on a CD, which Venter had placed on the chairs of thirteen hundred fly researchers at a conference in Pittsburgh. Venter emphasized the fact that the fly genome had been a collaboration with a publicly funded project. In other words, he was suggesting there was no real reason that the Human Genome Project couldn't collaborate with Celera, too. The fly project—known as the Berkeley Drosophila Genome Project—is headed by a fly geneticist named Gerald Rubin.

"One of the things I really like about Craig Venter is that he almost totally lacks tact," Rubin said to me. "If he thinks you are an idiot, he will say so. I find that way of dealing very enjoyable. Craig is like somebody who's using the wrong fork at a fancy dinner. He'll tell you what he thinks of the food, but he won't even think about what fork he's using. It was a great collaboration."

John Sulston, the head of the Sanger Centre, told the BBC that he felt Celera planned to "Hoover up all the public data, which we are producing, add some of their own, and sell it as a packaged product." He added, "The emerging truth is absolutely extraordinary. They really do intend to establish a complete monopoly position on the human genome for a period of at least five years," and he said, "It's something of a con job."

"Sulston essentially called us a fraud. It's like he's been bit by a rabid animal," Venter fumed.

"It's puzzling. To me, the whole fight defies rational analysis," Hamilton O. Smith said to me, shortly after his net worth had cratered in Celera's mudslide. "But the publicly funded labs are angry for reasons I can partly understand. We took it away from them. We took the big prize away from them, when they thought they would be the team that would do the whole human genome and go down in history. Pure and simple, they hate us."

———

On April 6th, Venter announced that Celera had finished the sequencing phase of the human genome, and was moving on to First Assembly. Celera had produced some eighteen million fragments of the first genome, perhaps Craig Venter's. Soon afterward, on an unseasonably warm day, while the cherry trees were in full blossom, I visited Celera to see how the assembly was going. I found Gene Myers in his cubicle, looking chilled. He was bundled up in his yellow fibrepile jacket and his green human scarf. He and his team had started running chunks of human DNA code through the computers over the weekend. The first run had resulted in a mess—something was wrong with the software. They had done some tweaks, and they were running a few more chunks of code. It would take months to assemble the whole thing.

"Assembly is pretty boring," he said, somewhat apologetically.

Myers said that Celera would be using all of the Human Genome Project's human DNA code—which was published on the GenBank Web site—and would tear it into fragments and compare them with Celera's DNA code, and then the software that his team had written would try to assemble all the fragments into a whole human genome. At the same time, Celera was coolly telling the public project that its scientists could see Celera's data but only if they came to look at the data on Celera's computer. The collaboration had never come to pass.

Minutes later, one of Myers's people, a computer scientists named Knut Reinert, hurried in, and told him that the first assembled human-genome sequence had just come out of the computers. Myers put the "Ride of the Valkyries" on the boom box, and fifteen people tried to crowd into Reinert's cubicle.

Myers bent over Reinert's shoulder and said, "We got it! We got the first one! This is the first assembled human sequence we've gotten out of nature!"

What appeared on the screen was a mathematical diagram of a stretch of human DNA. It showed arrows going in various directions, connecting dots together. "The picture looks like a Super Bowl debriefing," one Celera programmer remarked.

They talked about it for a few minutes, and then everyone drifted back to work. That day, Celera's stock dropped another twenty per cent.

In early May, another company got into the business of the human genome. DoubleTwist, Inc., announced that it had teamed up with Sun Microsystems to compete with Celera. DoubleTwist and Sun were offering an analyzed database of the human genome to anyone for a fee, using the data from the Human Genome Project, not from Celera. The price was six hundred and fifty thousand dollars for a database that would be updated regularly.

At the same time, Celera's stock had gone down below a hundred dollars a share. Many investors had recently bought the secondary offering, paying two hundred and twenty-five dollars a share for it. This brought on a slew of class-action lawsuits against Celera, filed by law firms specializing in shareholder suits. There were various claims, sparked by the fact that the secondary investors had lost sixty per cent of their money. These lawsuits will probably be consolidated, and Celera will either settle or fight them in court.

As for the science, knowledgeable observers believed that, in the end, Celera had actually spent about half a billion dollars to sequence the human genome—three hundred million more than Venter had originally predicted. Was it worth it? I asked many biologists about this, and most of them spoke the way scientists do when they believe that a great door has been opened, and light is shining deep into nature, suggesting the presence of rooms upon rooms that have never been seen before. There was also a clear sense that the door would not have been opened so soon if Craig Venter and Celera had not given it a swift kick.

"We can thank Venter in retrospect," James Watson said, leaning back and smiling and squinting at the ceiling. "I was worried he could do it, and that would stop public funding of the Human Genome Project. But if an earthquake suddenly rattled through Rockville and destroyed Celera's computers, it wouldn't make much difference." He stood up, and offered me the door.

Eric Lander, who professes to like Venter, said, "Having the human genome is like having a Landsat map of the earth, compared to a world where the map tapers off into the unknown, and says, 'There be dragons.' It's as different a view of human biology as a map of the earth in the fourteen-hundreds was compared to a view from space today." As for the war between Celera and the public project, he said, "At a certain level, it is just boys behaving badly. It happens to be the most important project in science of our time, and it has all the character of a schoolyard brawl."

Norton Zinder, Watson's friend, who had feared that he would die before he saw the human genome, said that he felt marvellous. "I made it. Now I've gotta stay alive for four more years, or I won't get all my options in Celera." Zinder, who is a vigorous-seeming older man, was sprawled in a chair in his office overlooking the East River, gesturing with both hands raised. He shifted gears and began to look into the future. "This is the beginning of the beginning," he said. "The human genome alone doesn't tell you crap. This is like Vesalius. Vesalius did the first human anatomy." Vesalius published his work in 1543, an anatomy based on his dissections of cadavers. "Before Vesalius," Zinder went on, "people didn't even know they had hearts and lungs. With the human genome, we finally know what's there, but we still have to figure out how it all works. Having the human genome is like having a copy of the Talmud but not knowing how to read Aramaic."

PETER J. BOYER

DNA on Trial

FROM *THE NEW YORKER*

Each year, volunteers from the second- and third-year classes at the Benjamin N. Cardozo School of Law, in lower Manhattan, are selected to participate in a program called the Innocence Project, whose mission is to exonerate, through DNA testing, people who have been wrongly convicted of crimes. For the students, there is much effort and no pay, but there is invaluable real-life instruction on the workings—and the failings—of the criminal-justice system. The project, which was inaugurated in the early nineteen-nineties, has played a role in thirty-nine exonerations, and after every triumph the students deconstruct the case to show where the system failed. In the summer of 1999, their efforts produced what the project considers a prize study in miscarried justice: the case of Calvin C. Johnson, Jr.

Johnson, who spent nearly sixteen years in Georgia's state prisons, where he was serving a life sentence for rape, aggravated sodomy, and burglary, gained his freedom in June. Although DNA exonerations are hardly uncommon anymore, Johnson's story became national news, because it seemed the perfect allegory for the native Southern iniquity regarding race: Johnson is black, his accuser is white, and he was convicted by an all-white jury that reached its verdict in less than an hour. (The *Times,* for instance, saw in the case "the racial underpinnings of so many convictions in Georgia and the rest of the Deep South.")

On the day of his release, Johnson stood in the courthouse where he had been sentenced to prison and accepted a handshake from the man who had sent him there, Clayton County District Attorney Robert Keller. It was a satisfying scene, duly noted by the press, but Keller continued to believe that Johnson's prosecution and conviction had been entirely warranted. This view was shared by other law-enforcement officials, including Paul Howard, who had prosecuted Johnson in an earlier case, and Carole Wall, another of Johnson's prosecutors, who says of him, "I really do think that he's a serial rapist."

Keller had considered fighting Johnson's exoneration bid, but the prospect of taking on the Innocence Project was daunting. The project is run by the noted defense lawyers Barry Scheck and Peter Neufeld, who became national figures after their work in behalf of O. J. Simpson's defense team, and the Johnson case, with its racial dynamics, had the potential for political calamity. Also, there was the DNA. The science that exonerated Calvin Johnson is the marvel of forensics, and the Innocence Project's DNA expert, a forensic scientist named Edward Blake, is recognized as one of the world's best—so highly regarded that Keller's own forensics experts didn't even think it necessary to verify his findings. In the end, Keller assented and Johnson was freed.

At the Innocence Project, Calvin Johnson's case was supervised by Peter Neufeld, who viewed it as a grave injustice wrought by a flawed system—"a fabulous model for the lingering legacy of slavery," as he puts it. Neufeld, who is forty-nine, became a lawyer because he saw the law as a means to effect social change, and he is proud of his professional fidelity to his social conscience. After graduating from New York University School of Law, Neufeld went to work in the Legal Aid office in the South Bronx, where he met another earnest young lawyer on a social mission, Barry Scheck. Neufeld and Scheck became accomplished defense lawyers and colleagues as well, developing a particular interest in what Neufeld calls the "intersection between science and the law"—that is, forensics. Neufeld did pioneering work in forensic psychology; he was one of the first lawyers to successfully use "battered woman's syndrome" as a defense against murder. He chose cases according to a strict political code, refusing, for example, to represent accused rapists who claimed in their defense that the victims had consented to sex, even if he believed them to be innocent, because "I didn't want to be part of the process of trying to lobby jurors to develop a different approach toward the politics of sex."

Inevitably, Neufeld and Scheck were drawn to the new technology of DNA "fingerprinting," recognizing in it a powerful tool for exonerating people who had been wrongly convicted and imprisoned. In its probative value, it was such a monumental leap beyond conventional serology, or blood typing, that the two men imagined nothing less than a transformation of the criminal-justice system. To Neufeld, DNA identification is "the gold standard of innocence." Scheck has called it a "magical black box that suddenly produces the truth."

The lawyers found that applying DNA testing to long-settled criminal cases was a tricky business, though, because the evidence was fragile and old; in many instances, it no longer existed. But they went to work, with the help of students at Cardozo, where Scheck is a professor and Neufeld teaches part time, and began to win exonerations, based on the mesmerizing power of DNA proof. Word of the Innocence Project's work spread, creating a flash fire of hope, and soon the project had a pile of requests several hundred cases high. The lawyers were selective—biological material had to be available, and the defense had to have been that the accused had been wrongly identified by the victim. (Seventy per cent of the project's cases are dropped, because the evidence has been lost or destroyed.)

By the time Neufeld agreed to take on Calvin Johnson's case, some sixty men in America had been exonerated and freed from prison through DNA testing. "We think there'll be hundreds and hundreds exonerated in the next few years," Neufeld says.

The Innocence Project, however, is not what made Peter Neufeld a nationally famous lawyer. What accomplished that, to his chagrin, was his role on the "dream team" of lawyers who successfully defended O. J. Simpson against the charge that he murdered his ex-wife, Nicole, and her friend Ron Goldman. Neufeld, an affable New Yorker, grows visibly taut at the mention of the Simpson case, and he will ask a reporter not to mention this "matter that was litigated in the Los Angeles Superior Court." In his office, the only obvious reminder of the case is a mock Oscar on a bookshelf bearing the inscription "To the Simpson Defense Team"—a gift from an admirer in Los Angeles.

And yet the Simpson case is surely relevant to the matter of DNA as the "gold standard" of innocence or guilt. In the criminal trial, the prosecution presented a trove of credible DNA evidence—blood in Simpson's Bronco, on his socks, on the fence outside Nicole Brown Simpson's house, among other things—that connected Simpson to the crime. Neufeld and Scheck's job in the Simpson defense was to attack the DNA presentation, and, by employing their own fluency in DNA forensics, they reduced the prosecution's DNA case to one of practical irrelevance. Because the actual science of the DNA identification was unassailable, Scheck and Neufeld attacked the procedures used to collect and store the evidence, challenged the competency of the forensic specialists who handled it, and nourished the idea (however dubious) that the case against Simpson was the result of a frameup by the Los Angeles Police Department.

Their efforts were central to, and perhaps even decisive in, Simpson's acquittal, but there was a larger result, too: if the DNA evidence in the Simpson case was not enough to compel a corresponding verdict, then DNA could not be a "gold standard of innocence." Scheck and Neufeld demonstrated that in the hands of skillful lawyers operating in the adversarial context of a courtroom DNA becomes just another piece of evidence.

As it happened, the case of Calvin Johnson was not a simple matter of remedying old prejudices with new science. Georgia law-enforcement officials felt that they had good reasons for arresting Johnson, and the jury that convicted him likewise felt that it had good reasons for doing so. Although it was DNA evidence that ultimately freed Johnson, Robert Keller, the prosecutor, could in good conscience have fought his bid for freedom with the same sort of legal attack that prevailed against the prosecution's DNA evidence in Los Angeles Superior Court.

CALVIN JOHNSON, JR., hardly seemed a likely candidate for trouble with the law. He was a college graduate, with a degree in communications, who worked at Delta Air Lines. He was handsome and articulate, the product of a solid Atlanta family of proud achievers who hoped that he might become a successful broadcaster. But after college he fell in with a wild crowd, and he began finding mischief that soon became real trouble. Around midnight on April 13, 1981, Johnson drove to College Park, a mostly white, working-class town at the edge of the sprawling Atlanta airport, and slipped through a window into a darkened apartment. A neighbor happened to spot him, and she notified police, who arrived a few minutes later to discover Johnson rummaging through the kitchen. He was wearing cutoff jeans, gloves, and—the patrolman who arrested him noted—no underwear. Johnson was carrying a buck knife that he used at his job, at an airport warehouse.

The patrolman who caught Johnson in the act handcuffed him and drove him to the College Park police station, where he was booked on burglary and concealed-weapon charges and locked in a cell. A few hours later, the day shift arrived at the station. Reading the night log and noting the burglary arrest, Detective Lewis Harper and Sergeant G. J. Bencale brought Johnson to the detectives' room and advised him of his rights. Johnson signed the form, but declined to make any statement until he had seen a lawyer. The two officers took him to his cell, and sat down to review his arrest record. Then they did what policemen do: they theorized.

There had been a series of sexual assaults in College Park, the most recent four nights before. A College Park woman named Katherine Lowe had come home from dinner with a friend around midnight, and when she entered her apartment she was grabbed from behind by a man with a knife, who had pried open a sliding glass door. He led her to the bedroom and raped her. The assailant then asked her for money and took ten dollars. Lowe told police that he was a black man of medium build, that he had a mustache, and that he spoke softly. She said that he wore no underwear.

It struck the officers that the burglary suspect they had just talked to—the guy with no underwear—matched the physical description that Lowe had given of her assailant. She had guessed that he weighed a hundred and seventy pounds (Johnson weighed a hundred and sixty); she said he had a mustache (Johnson had a mustache); she said he was soft-spoken (so was Johnson). There was a discrepancy: Katherine Lowe believed that her assailant had been uncircumcised. Nevertheless, the rape had occurred in the same neighborhood as Johnson's burglary, and at the same time of night, and the assailant had also entered through a window. The night before Johnson was arrested, the woman who reported his break-in and a friend had seen a black man who looked like Johnson standing naked behind her apartment, with his clothes in one hand. A few weeks earlier, another woman, Lisa Givens, reported that a black man matching Johnson's description had entered her home, robbed her, and tried to rape her. (The names of all rape victims and witnesses in this article have been changed.) There had been no leads in these cases, and Johnson was beginning to seem very much like a lead.

Detective Harper got a warrant to obtain samples of Calvin Johnson's blood, hair, and saliva, to be tested against the samples taken from Katherine Lowe at the hospital on the night she was raped. There was no DNA technology in 1981—the best science was conventional serology. The tests showed that Johnson was of the same blood type, Type O, as Lowe's rapist, and hair taken from Johnson was of the same type as hairs retrieved from the crime scene. But because the forensic science was relatively imprecise—more than forty per cent of the population has Type O blood—any case would stand or fail on the victim's identification of her attacker.

A Polaroid photograph was taken of Calvin Johnson, and that night the detective who was working on the Katherine Lowe case, E. S. Meares, prepared a photographic lineup and asked Lowe to come to the station. Meares showed her Johnson's picture, along with those of five other black males of roughly the same age and aspect, and asked Lowe if she recognized her assailant. She picked Johnson's photograph.

The case fell to Paul Howard, a rising young prosecutor in the Fulton County district attorney's office. Howard, who is black, believed that he had a strong case, and a Fulton grand jury indicted Johnson on charges of rape, aggravated sodomy, attempted rape, burglary (two counts), armed robbery, and carrying a concealed weapon.

The distraught Johnson family hired a lawyer, Mark Kadish, to defend Johnson, but before the case moved to trial Howard came across an irregularity in the identification process which caused him to doubt whether he could win a conviction on the sex charges, so he dropped them. That left only the burglary and weapon charges, and since Johnson had been caught in the act Kadish agreed to a plea bargain. Johnson pleaded guilty to burglary and possession of a concealed weapon, and in September, 1981, he began serving an eight-year sentence.

In retrospect, it seems possible that Johnson might have been better off if he had been tried for the rape of Katherine Lowe and the other sex crimes. At a trial, he at least stood a chance of acquittal; as it was, he remained convicted of rape in the minds of the law-enforcement officials who one day soon would again decide his fate.

PAUL HOWARD, who is now the Fulton County District Attorney (the first black in Georgia's history to hold that office), says he believes that the evidence strongly implicated Johnson in the Katherine Lowe rape. "If you look at him getting caught in that apartment, where it was located, the time period, the identification, the hairs, the sperm, there's a good chance he would have been convicted," Howard says. Yet he has never publicly explained why he dropped the sex charges, except to vaguely cite a "problem" with the evidence. Neither the police nor Johnson's attorneys nor Johnson himself knew what that problem was.

But a close reading of the police records of the Katherine Lowe investigation reveals a critical police error—an error that would have made a conviction at trial virtually impossible. Once Detective Meares had arranged for Katherine Lowe to view a photographic lineup and she had chosen Johnson's picture, that would have been enough for Howard to bring charges, and possibly for a jury to convict. But Meares, perhaps in the enthusiasm of breaking a difficult case, then did something that is outside accepted practice. He had spoken to Johnson himself, and he knew that Johnson's manner of speaking was distinct from that of the usual suspects who ran through the police station. It occurred to Meares that Katherine Lowe's identification would be even stronger if she could hear Johnson speak. A supplementary police report shows that Meares escorted the victim downstairs to the jail, where she confronted Johnson. "Victim carried on a conversation with perpetrator," the report reads. "Victim identified voice. Also stated that she remembered his big eyes and that she was positive that was the subject that raped her."

In criminal-defense jargon, what Meares did was called a "one-on-one showup," a procedure that is greatly discouraged, because it is so suggestive. If a victim is asked by a police officer to identify a suspect who is presented alone, rather than in a group, she may infer that the police have already determined the suspect's guilt. Katherine Lowe's jailhouse meeting with Johnson would have legally tainted her entire identification, and without that the case against Johnson would likely have collapsed. Johnson's attorney never found out about the "showup," because the sex charges were dropped. Had the case gone forward, he says, he would certainly have moved the court to suppress Lowe's identification of Johnson.

Did Calvin Johnson rape Katherine Lowe? The legal answer is no, because a police blunder kept the question from the test of a jury, and what can't be known is what Calvin Johnson's defense might have been. That unresolved question, however, became the wellspring of everything that came afterward.

IN GEORGIA'S overcrowded prisons, burglars were a low priority, and in December, 1982, Johnson was paroled, after serving fifteen months. He went home to live with his parents and, everyone hoped, to start a new life. He took a job as a courier, and rode to work each morning with his father. In the afternoons, he worked out at a gym, continuing a routine he had begun in prison. His college girlfriend had waited for him, and they began to talk about marriage.

In the early morning of March 9, 1983, a woman named Eve Cooke told police that she had been raped and sodomized in her apartment by an intruder, a black man wearing gloves, who had slipped through an unlocked door and looped a belt around her neck and used it like a choke collar, at one point tightening it so fiercely that she passed out. The intruder had placed a towel, and then a pillowcase, over her head, allowing her only momentary glimpses of his face. But the attacker talked a great deal, Cooke said, and she remembered his voice clearly: "He didn't talk like a Southern black and he didn't talk street lingo or jive. It was smooth, smart, soft-spoken, and more educated with his words."

The College Park police knew instantly that there was a serial rapist loose. Two nights before, another woman, Martha Hudson, had been awakened by a rapist whose method was nearly identical to that described by Cooke. Hudson, who lived about a mile from Cooke, on the Fulton County side of College Park, described her assailant as a black man who wore a glove and put a belt around her neck, using it like a choke collar. During this three-day period, two other women in the neighborhood reported being accosted by a black man, but they escaped unharmed.

Among those working on the rape cases were Detective Meares and Detective Harper, who had figured Johnson as the assailant in the rape of Katherine Lowe, two years earlier. By the time of the Eve Cooke rape, Harper and Meares had already begun to speculate that Johnson, who had returned from prison ten weeks earlier, might be the rapist.

After the Cooke rape, Meares signed a report that noted, "Calvin was developed as a suspect. Calvin was arrested in April of 1981 for rape. His m.o. used in 4–81 is similar in nature" to that of the rape of Martha Hudson. A few days later, Detective Harper, who had led a search of Johnson's apartment in the 1981 case, requested a warrant to search the Johnsons' home for evidence in the Cooke and Hudson rapes. In his affidavit seeking the search warrant, Harper wrote, "This officer is personally familiar with Calvin Johnson, having arrested Johnson in April 1981 for similar type crimes and . . . Johnson did enter a plea of guilty to some of the charges as a result of this arrest." The police arrived at the Johnsons' home, and Johnson was cuffed and taken away. Calvin Johnson, Sr., who was home at the time, says he remembers one of the policemen saying to his son as they took him away, "We've got you this time, and you're going for a long, long time."

Jury selection for the trial in the rape of Eve Cooke began in November, 1983, in Clayton County Superior Court, in Jonesboro. The county's population was overwhelmingly white (ninety-two per cent) and mostly working class. Of the forty-two citizens in the jury pool, only a handful were black, and none of them made the final cut.

Prosecuting the case was the Clayton County district attorney, Robert Keller, who, at thirty-six, was already a six-year veteran in the job. Defending

Johnson was Akil Secret, a young attorney who had been recommended to the Johnsons by a local judge. As Secret would point out more than once during the trial, he and his client were the only black men before the bench.

Because of the nature of the crime—there were no witnesses to the attack, no fingerprints—the trial turned on the issue of identification, and there was cause for doubt. Eve Cooke had picked Johnson out of a photographic lineup, but she had picked someone else at a live lineup conducted at the Fulton County police station. She testified that she had picked the wrong person at the live lineup because looking at Johnson was too much for her to bear: "I just pushed my eyes away from him and picked someone else."

Martha Hudson, the Fulton County victim, was allowed to testify in the Clayton case because of a Georgia rule allowing testimony from other cases if it demonstrates a pattern of "similar transactions." Hudson had identified Johnson in a live lineup, but she had failed to choose his picture in a photo lineup. The two other women who said that they had been accosted by a black man around the time of the rapes also testified as identifying witnesses. Marie Webb said that on March 7th—several hours before the Hudson rape—she had stepped out of her shower and was confronted by a "clean and educated-sounding" black man, and that she locked herself in the bathroom and then ran out through the front door. Webb, who lived two doors away from Hudson, identified Johnson in a photograph. The last identifying eyewitness was Jennifer Townes, who testified that on the night Eve Cooke was raped—March 9th—a black man had knocked on her door and asked to come in, saying that he wanted to call a cab. She said that she came face to face with the man when she peered through her window and saw him staring back at her. She identified Johnson in the photographic lineup.

All four women, who were white, identified Johnson in court as the man who had accosted or raped them. Each of them said that she was particularly struck by Johnson's eyes. Hudson was a forceful witness. When she was asked to identify the man who had raped her, she looked directly at Johnson and said, "It's the man with the beard sitting right there. You'll get your time." When she had finished her testimony and was returning to her seat, she walked past Johnson at the defendant's table. Lurching toward him, she said, "You stupid bastard!" Secret asked for a recess, which the judge granted.

Secret worked to undermine the identifications, pointing out that none of the women had reported that the man they encountered had had a beard, yet Johnson had worn a full beard two months before the attacks, as was indicated by his photo identification from work. Although the Georgia Crime Lab had identified Johnson's blood type as being the same as that of Cooke's rapist, Secret noted that forty per cent of black males shared that type. And there was some forensic evidence that Secret considered exculpatory: three pubic hairs among Eve Cooke's bedclothes were tested, and were found to belong to a black man, but not to Johnson. Since Cooke testified that she had never had a black person, man or woman, inside her home as a guest, Secret suggested that the hairs must have come from her rapist. Keller's explanation was that the hairs must have come from a laundromat or a public restroom.

The trial lasted for three days, and the jury deliberated for forty-five minutes before announcing that it had found Johnson guilty of rape, aggravated sodomy, and burglary. One of the jurors later said that the victims' testimony had been the most compelling evidence in the case, and that Martha Hudson's outburst clearly demonstrated that there was "no doubt in her mind" that Johnson was guilty. The judge sentenced Johnson to life in prison for the rape conviction, and to fifteen-year terms for the burglary and aggravated-sodomy charges.

Before the second trial, for the rape of Martha Hudson, the Fulton County district attorney's office offered Johnson a plea bargain, but he refused to admit any guilt, and so that trial began in Fulton County Superior Court, in Atlanta, in September, 1984.

There is no transcript of the trial, but Michael Hauptman, who replaced Secret as Johnson's lawyer, says that it was a virtual replay of the Clayton County trial, with the exception of the racial makeup of the jury—seven blacks and five whites.

Carole Wall, the Fulton County prosecutor, believes that if racial bias was a factor in the Clayton County conviction, race was played effectively in the other direction in Fulton County. "Mike Hauptman, who is my dear friend, uses the racial issue very well," she says. " 'Cops are out to get the only black man in the area,' and so on. He does the racial issue. He'll beat you to death on the racial issue. And he's so smart. When he questioned those College Park cops, it was like clubbing a bunny."

The Fulton County jury deliberated overnight, and returned the next day with its verdict—not guilty. But Johnson, of course, was already in prison, serving a life term.

IN 1994, JOHNSON, who had become something of a jailhouse preacher, helping to lead his fellow-convicts in regular prayer services and Bible study, got a chance at parole, but in order to get it he would be required to acknowledge the crimes for which he was convicted, and he refused to do so.

Two years earlier, Johnson had read a news article about new DNA technology that was being used in forensics to match criminals with their crimes, and that could also be used to clear the wrongly accused, even in long-closed cases, if the evidence had not been lost or destroyed. Johnson got in touch with James Bonner, a public-aid lawyer who worked for the Prisoner Legal Counseling Project, at the University of Georgia School of Law, and Bonner agreed to try to help, although he thought that Johnson had little hope. His only legal recourse was to file what is called an extraordinary motion for a new trial based on new evidence. The problem was, there was no new evidence. In effect, his motion, if certified by the court, was a means by which he could try to hunt for new evidence from within the old evidence through DNA testing. "It was predicated upon the results of a DNA test that we didn't yet have, and hoped we could have, but weren't sure we ever could get," Bonner says.

The evidence, if it still existed, was controlled by the office of the Clayton County district attorney, Robert Keller. To Bonner's surprise, Keller agreed to the test. "They were basically just asking us if we'd do it," Keller says, "and our position was, if the sample was available, that would be the appropriate thing to do."

The evidence was available, but only as a result of happenstance. In Georgia, physical evidence introduced at trial is maintained by the court reporter, who serves at the pleasure of the judge. Stephen E. Boswell, the judge in Johnson's case, had long since retired, and his replacement did not rehire Boswell's reporter, J. Dan Black. In clearing out his things from the courthouse, Black was throwing away all the unclaimed evidence from cases that were no longer active—including the evidence from Johnson's case. Someone from Keller's office noticed the evidence bags from Johnson's case and asked Keller if they should be saved. There was no legal reason to retain the evidence—Johnson's request for DNA testing hadn't yet been made—but Keller had ordered that the bags be placed in a box in his evidence vault.

The box included Eve Cooke's torn underwear, her stained bedsheets, a blue towel that her assailant put over her head, cotton-swab sticks from the "rape kit" that was used to examine her on the night of the assault, glass slides bearing specimens swabbed from her vagina and cervix, and small packages containing hairs found at the scene and those taken from Johnson after he was arrested. Johnson's sister Judy, who worked at the United States Department of Transportation in Washington, found a laboratory, called Genetic Design, in Greensboro, North Carolina, that agreed to perform the DNA testing. In October, 1995, the box and all its contents were removed from the evidence vault in Georgia and, accompanied by a Clayton County investigator, transported to Greensboro.

Michael DeGuglielmo, a Genetic Design DNA expert, decided to test two of the items in the box—the blue towel and portions of the vaginal swabs. He extracted DNA from those items, hoping to compare it to Johnson's DNA, but he didn't have a fresh specimen from Johnson. Testing was suspended until a blood sample from Johnson could be provided. Because of bureaucratic inertia, or perhaps because prison authorities were unenthusiastic about Johnson's quest, it took nearly a year to get the blood sample.

Finally, in late 1996, Johnson's blood was sent to Genetic Design. By this time, though, DeGuglielmo had left the company for another firm. Three months later, the division of Genetic Design which concentrated on identity testing was sold to a national chain, LabCorp, headquartered in Burlington, North Carolina.

Johnson and his family knew none of this. "I called one day, and Genetic Design was shut down," Judy Johnson recalls. "I thought, My God, where's DeGuglielmo? I couldn't find him." Meanwhile, as Johnson's family searched for another attorney, a childhood friend of his told him about the Innocence Project. Johnson wrote to Barry Scheck, had the transcript of his trial sent, filled out an extensive questionnaire required by the project, and was accepted as a client.

BY ANOTHER BIT of luck, the box of evidence from Johnson's case was rescued after the Genetic Design division was sold to LabCorp. In March, 1997, seventeen months after the evidence was released by Keller's office, the box was transported from Greensboro to the new corporate owners' headquarters. There the material sat in storage for seven more months, and was then transported to the LabCorp forensic-science unit, in Research Triangle Park, an hour away. A senior director of that facility, Dr. Marcia Eisenberg, who had worked on several cases with Neufeld and Scheck before, says, "This was a case where we get this box of evidence, and we're trying to get Barry and Peter to dig up enough history on what is in this box. We're faxing them Xerox copies of bags in a box, and trying to get them to match them up to the original court documents. I mean, this is everybody's case from hell." Not only was the material now fifteen years old but it had been sliced and diced to pieces in two earlier rounds of testing—the original serology tests for the trial, and the tests at Genetic Design, which were never completed. The vaginal and cervical swabs, which once looked like Q-Tips, had been reduced to tattered sticks without cotton on the ends. LabCorp decided to test material on the vaginal slide—a glass plate on which body fluids from the victim are smeared for microscopic examination—along with portions of the towel, bedsheet, and Eve Cooke's underwear.

On November 11, 1997, LabCorp reported its findings to the Innocence Project. It had indeed been able to find genetic material on the items it tested, but only from the victim; it could not find sufficient material from the assailant even to compare with Johnson's blood sample. The long-awaited test, in other words, had failed to exonerate Johnson.

That might well have ended it for Johnson. Keller could have gone to court, declared Johnson's innocence quest a failure, and tried to put a stop to any further testing. Whether or not Keller would have done this can't be known, because Keller was not told that LabCorp's tests had been conducted, or that the results had been inconclusive. Nor did Neufeld inform Keller what he was about to do next.

Neufeld believed that if anyone in the world could find the DNA answer he was looking for in the old and infinitesimally small sample that was left it was a longtime friend and colleague, Dr. Edward Blake, a partner at Forensic Science Associates, in Richmond, California, near Berkeley.

"If we type this slide and there's no result," Neufeld told Blake, "then Mr. Johnson spends the rest of his life in prison."

In the world of forensic science, Neufeld says, "there are a bunch of .300 hitters, and then there's Ted Williams; Ed Blake is Ted Williams." Blake, whose rare gifts in the laboratory are broadly acknowledged, would be the first to agree with Neufeld's Ted Williams analogy, and he offers one of his own to describe what is distinct about his work. "Jack Benny will pick up a violin, and he'll make the music that Jack Benny makes," he says, "and Jascha Heifetz will pick up the violin and make the music that he makes."

While the science is the same for all forensic specialists, LabCorp failed to find male DNA through its testing, Blake says, because "of the skill with which the analysis was done, to be quite candid with you." Actually, the difference lies at least partly in the laboratories themselves. LabCorp is a publicly held company with labs nationwide, and, by company policy, its scientists will perform only tasks that they have specifically been asked (and paid) to perform. Blake, whose lab Neufeld describes as a "boutique operation," will do whatever it takes to obtain an accurate, definitive result.

In rape cases, the crucial step is a procedure called "differential extraction," which is the process of separating the male portion of a sample from the female portion. This is necessary because a stain on the victim's undergarments, for example, will have material that comes from both the male attacker and the female victim. Unless the male and female cells are separated, the DNA contained inside them will mix together, and confuse the identification test that comes later.

What makes separation possible is the nature of a sperm cell: its walls are tougher than the walls of other cells. In a test tube, the tiny speck of evidence is bathed in a sort of chemical detergent, which is strong enough to break open the cells of the non-sperm material, releasing the non-male DNA inside. The tube containing the whole mixture is placed on a centrifuge, which spins until the material in the tube has separated into two layers, with the sperm cells settling on the bottom, like sediment in a bottle of wine. The non-male material is decanted into a second tube, and what remains in the first tube—presumably sperm cells—is bathed again, in a stronger detergent, breaking down the tough sperm-cell walls and releasing the DNA inside for testing.

It is in this differential-extraction process that LabCorp failed. It is a delicate procedure, and Dr. Blake is known for his ability to coax results from it, in part because he takes extra steps that maximize his chances for success. Blake says that the key to this part of the process is caution, using just the right amount of chemical detergent in the first phase to wash away the non-sperm matter without disturbing the sperm-cell walls. To separate the solution into male and female layers, Blake uses a centrifuge that is angled horizontally, rather than at the customary forty-five degrees. He does this to increase the likelihood that the sperm cells will actually settle on the bottom of the tube rather than splashing up along the sides, where they might get decanted away with the non-sperm material. After he has completed the separation, Blake takes another extra step: he puts the material at the bottom of the first tube back under the microscope to make certain that the sperm is there and that non-sperm material has been removed.

Such a painstaking routine pays dividends. In Johnson's case, for example, Blake examined the remnants of the swab sticks (which LabCorp didn't examine) and found a tiny fibre of DNA-laden cotton embedded in one stick.

BLAKE'S SUCCESS in the lab has made him not only a mythic figure in forensics but a coveted asset in a legal fight, on either side of the aisle. His confidence in the lab sometimes extends to strong feelings about the legal case itself, a tendency that some find discordant with the scientist's mandate for dispassionate inquiry. Blake worked with Neufeld and Scheck on O. J. Simpson's defense, even though he would be the first to concede that the scientific test-results offered by the prosecution were unassailable. The dream-team defense's mission, after all, was not to illuminate scientific truth (such as the 6.8-billion-to-one certainty that blood found on Simpson's socks belonged to his murdered ex-wife); its purpose was to attack the prosecution's methods until the science itself, as Blake puts it, "became irrelevant."

When Blake gets involved in a legal fight, he can seem as if he's "on a mission," says Jim Catterson, the district attorney for Suffolk County, New York. Catterson and Blake were adversaries in a rape case that could be a textbook study in the paradox of DNA in the courtroom: for all DNA's intrinsic immutability, advocates in a legal case routinely persuade themselves (and sometimes a jury) to disbelieve the science. In the case of an accused rapist named Kerry Kotler, the reliability of DNA evidence was literally played both ways by both sides, and alternately proved to be both Kotler's salvation and his undoing.

The Kotler case was one of the first taken up by the Innocence Project, and it has some striking parallels to the case of Calvin Johnson. But it is more like the O. J. Simpson case in one key regard: Peter Neufeld would rather not talk about it. When I mentioned Kotler to Neufeld, he began lawyering, saying that "it was a weird case" and shouldn't be included in any story about the Innocence Project.

Kotler, who is white, was a middle-class kid prone to trouble, a runaway and a car thief. In 1981, when he was in his early twenties, he was convicted on a statutory-rape charge. While free on probation, he was arrested for the knife-point rape of a thirty-three-year-old East Farmingdale, Long Island, housewife named Patricia Gould. The authorities were especially eager to find and punish the rapist, because Mrs. Gould had been raped three years earlier by a masked intruder whom they believed to be the same assailant. In the second rape, the man greeted Mrs. Gould as she returned from the grocery store, and, before running a carpenter's knife over her skin, taunted her by saying, "I came back for another visit. We're going to do it all over again."

Like Calvin Johnson, Kotler protested his innocence, and there were some problems with the police investigation here, too, including a police report that seemed to the defense to have been altered in order to make it easier to place Kotler at the scene of the crime. But, as with Johnson, the police were wedded to a theory that featured Kotler as the prime suspect, and there was compelling evidence that seemed to incriminate him. The best forensic science of the time, conventional serology, narrowed the universe of possible suspects to three per cent of the population, and that three per cent included Kotler. (The jury was not allowed to hear this.) There was also the testimony of the victim, who identified Kotler in two police lineups, in a photograph, and by voice, when he and several other men were instructed to say the menacing words "We're going to do it all over again." The jury believed her, and in 1982 Kotler was sentenced to a maximum of fifty years in prison.

Like Calvin Johnson, Kotler filed an appeal from prison, and it was unsuccessful. Then, in the late nineteen-eighties, he heard about DNA-evidence testing and found out about the Innocence Project. The evidence in his case was dispatched to Blake's lab.

This time, though, Blake could report no findings that warranted instant exoneration. In the early nineteen-nineties, DNA testing allowed scientists a very limited spectrum of alleles, or genetic markers, for comparison tests. As it happened, Kotler's tested alleles were the same as those of the victim, Patricia Gould. (About ten per cent of the population shared this type.) The evidence specimen, taken from the rape kit and Gould's garments eleven years earlier, revealed the presence of the genetic trait shared by Gould and Kotler, but it also contained another marker, which neither Gould nor Kotler possessed. In Blake's view, that could mean only one of two things: that Kotler was innocent, and the marker he found came from Gould's assailant; or that Kotler might have been the assailant, but only if Gould had had sexual relations with another partner near the time of the assault, thus explaining the mystery marker found in the specimen.

To Neufeld and Scheck, Blake's results seemed conclusive enough to exonerate Kotler: Blake had found in the sample a genetic marker that Kotler didn't possess, which must mean that he was innocent. (Patricia Gould had sworn under oath that she had not had any consensual sexual relations in the time period which would make a mixed sample possible.) Blake leaned toward the same conclusion, but he could not be unequivocal, and his report stated that Kotler "appears" to have been eliminated as the "primary source" of the semen.

Catterson, in the district attorney's office, refused to agree to a new hearing for Kotler. There was an obvious answer to the stalemate, which was to test a sample from Gould's husband. If he possessed the unidentified marker, Kotler could not be excluded, because there was no way of knowing whether the other marker, the one he shared with the victim, was his or Gould's. But if Gould's consensual partner did not possess the marker, that had to mean, according to Kotler's lawyers, that it came from the assailant, and it could not have been Kotler, because he didn't possess that particular trait.

On this point the case veered from the province of scientific inquiry into the realm of human relations. Gould told prosecutors that she had had sex with no one but her husband, so the matter could be easily resolved by taking a blood sample from him and testing it for the unidentified marker. Catterson had no reason to doubt Gould, and since he was a tough-on-crime politician, his sympathies were all with the victim. He saw no reason to disturb her further. For a year, Catterson opposed the testing of the husband. Finally, a judge ordered that it be done, and the sample was sent to Blake, who found that it did not contain the marker. His conclusion: Kotler could not have been Gould's assailant.

Catterson still firmly believed in Kotler's guilt, just as the police and prosecutors in Calvin Johnson's case continued to believe in his. Catterson was particularly affected by a meeting he had with Gould, who said she was so terrified of Kotler that she and her family had moved out of state. She told him that someone once left a menacing note on her car windshield making a reference to her daughter; she was sure that it was Kotler, who was out on bail at the time. On another occasion, she said that even after she moved away she received a telephone call from someone she was certain was Kotler, letting her know that she had been found. "I thought Kotler was the Devil incarnate," Catterson says. "He tried to put the hit on one of the reporters in the courtroom. He enjoys making women squirm. He enjoys torturing people mentally. He likes to get the edge on women."

When presented with Blake's new test results, the district attorney's office resisted. The prosecutors requested that their own experts review the findings and, when those experts agreed with Blake, raised the question of contamination: Couldn't the evidence sample have been contaminated somehow during the eleven years it spent in an evidence vault, or while it was being conveyed from one laboratory to the next? The experts said that that was highly unlikely. Catterson then questioned their reliability; he believed (and believes) that they were "in awe of Blake." The D.A.'s office consulted a DNA specialist in Boston, who confirmed Blake's findings. Meanwhile, Catterson was getting pounded by Scheck and Neufeld in court and excoriated in the press, most fiercely by the newspaper columnist Jim Dwyer, who recently wrote a book about DNA exonerations with Neufeld and Scheck entitled *Actual Innocence*, which will be published in February 2000. The news coverage became so intense that a flustered Catterson resorted to defending himself by blurting to a reporter, "I am no reactionary cretin."

Finally, in December, 1992, Catterson gave up the fight and declared in court that he would not contest Kotler's motion to set aside his guilty verdict. "Catterson went kicking and screaming," Neufeld says now, "and he was backed into a corner where the press was going to strangle him."

After nearly eleven years in prison, Kerry Kotler went free, but he was not inclined to forgive and forget. He accused the police and Catterson of misconduct, and Scheck, speaking to reporters, said of the Suffolk justice system, "This place was what Mississippi must have been like in the sixties." The press noticed heroic qualities in Kotler, dutifully reporting his assertions of law-enforcement malfeasance and passing along such tidbits as his lawyers' suggestion that Kotler had the stuff of a good lawyer, and that he might even attend law school. He appeared on ABC's "Primetime Live" program, and galloped on horseback on the beach while reporters took notes.

And, unlike Calvin Johnson, Kotler had the prospect of financial remuneration for the injustice he had suffered. Scheck and Neufeld pressed Kotler's case under New York's Unjust Conviction and Imprisonment Act. Kotler did not enter law school, as it turned out, but went to work as a commercial fisherman and diver.

For his part, Jim Catterson refused to apologize for Kotler's conviction, and he did not surrender his belief that a guilty man had been set free. In fact, Catterson believes that future developments in DNA testing will eventually validate the prosecution's position. Ed Blake says that he has seen this response from prosecutors before; he calls it the "white-hat syndrome." He explains, "I don't know a prosecutor that I've ever worked with who doesn't believe, as far deep in his heart as you can ferret, that he advocates on behalf of justice, truth, and the American way. He firmly believes that he rides a white horse, shoots silver bullets, and wears a white hat. And for somebody like that to come to terms with having put an innocent person behind bars for a significant portion of his life is a very, very difficult thing to grapple with."

Then, in the early hours of an August morning in 1995, something happened that caused a dramatic reversal of roles in the Kotler case. A twenty-year-old female college student left a Hampton Bays bar and was driving toward William Floyd Parkway when she was pulled over by a man who flashed his headlights and showed her a badge. After telling her to follow him in her car to a site about a mile away, the man told the young woman that she was driving erratically, and that she would have to take a breath test. When she got out of her car, the man pulled a knife and forced her into his car, then drove her to the woods and raped her. After the attack, the woman later told police, the assailant did something strange: he got a plastic water bottle he'd brought from his car and douched his victim—an attempt, police concluded, to wash away evidence.

The young woman was able to give only a partial description of her attacker, but she did know what model car he was driving—a white Pontiac Grand Am—and she was able to remember several numbers on its license plate. The police ran a computer check on that information and discovered that there was only one such car registered on Long Island. They were questioning the woman who owned the car when Kerry Kotler drove up in it—she was his girlfriend. "They seized the car and started working the case," Catterson says. "And, remember, we had his blood. We couldn't believe that this guy would be so arrogant. You could have knocked me over with a feather when I heard this guy pulled this stupid stuff again. We had him by the balls."

Nevertheless, as the new rape investigation proceeded, Neufeld and Scheck continued to push Kotler's suit for financial reparations. (In July of 1997, they won a $1.5 million award.)

In the new investigation, two pre-indictment lab tests linked Kotler's DNA to the crime scene, but Kotler's new attorney, Jack Litman, wasn't satisfied. In early 1997, the judge agreed to Litman's request for an independent lab test of Kotler's choosing. The evidence was gathered and brought to California to be tested by Ed Blake. This time, Blake's results were unambiguous. "I would say without any hesitation whatsoever," Blake says now, "that there is absolutely no doubt that the source of the sperm in the second case is Kerry Kotler."

Faced with such odds, Litman, an old friend of Neufeld's, decided to mount a Simpson defense. That is, given the fact that Kotler's DNA was all over the crime scene (his semen was found on the victim's clothes and on the swabs from her rape kit), Litman had to prove that someone other than Kerry Kotler had placed it there, and the Simpson model suggested itself: the cops did it. Litman developed the theory that the police, driven by their animus toward Kotler, had looted his trash for used condoms and then spread his semen over the victim's clothes and the specimens from the rape kit.

To say the least, it was a long-shot defense. In his favor, though, Kotler had one surprising witness—Ed Blake. While working on Kotler's exoneration in the first case, Blake had developed a pointed opinion about the prosecution's reluctance to set Kotler free: they were in the grip of the white-hat syndrome, and it seemed entirely possible to him that the authorities would frame Kotler.

The fact that the assailant in the second case had douched the victim puzzled Blake. The police and prosecutors thought the explanation was obvious: the attacker, perhaps someone who knew about the probative reach of forensic science, wanted to wash away the evidence. "But remember what a douching device primarily is," Blake says. "It's primarily a device for introducing something, not for removing something." He came to believe in the possibility that police, or some agent acting for them, could have used the plastic water bottle to plant Kotler's semen at the scene.

Blake based this view on what he found in Kotler's semen, testifying that it contained massive amounts of staphylococcus bacteria, which, he said, could have grown so rampantly only in a liquid state.

The jury didn't believe it. Kotler was convicted in July, 1997, and sent back to prison to serve a sentence of seven to twenty-one years.

Looking back, Neufeld still insists that Kotler's exoneration was right—"We've never been wrong in any of our exonerations. . . . We haven't been"—but he acknowledges that the ironies of the case are inescapable. "What is true," Neufeld says a bit grudgingly, "is that the same science that exonerated him on the other rape brought about his undoing in a subsequent matter four years after he was released from prison."

Catterson, however, has no difficulty at all explaining what happened in the case of Kerry Kotler: a rapist got his due. He insists that he is interested only in justice, not vindication, but he cannot resist a dig at the exoneration crusaders. "This is an interesting case, because it chips away at the clay feet of the Neufelds of this world," Catterson says. "Isn't that ironic? Hoist by their own petard. It's really—it's justice. Finally, somebody's got justice in this world."

IN THE FALL of 1998, three years after Calvin Johnson's evidence box was sent to North Carolina to be examined, Blake tested a cervical slide from Eve Cooke's rape kit and the tiny cotton fibre of DNA that he had managed to find on the remnants of the swab stick. On November 20th, he reported to Neufeld and Scheck that Johnson "is eliminated as a potential source of the spermatozoa." A few days later, Neufeld wrote to Robert Keller, Johnson's prosecutor, giving him the results and asking that Keller respond when he had had a chance to consult with his own forensic specialists about the findings.

For Keller, Neufeld's letter was a bolt from the sky. The last he knew, the Calvin Johnson evidence was in North Carolina, being tested by LabCorp. Now he suddenly found out that the box was in Richmond, California, having been sent there by Johnson's attorney without consulting Keller's office, to be examined by a forensic scientist of Peter Neufeld's choosing. And this scientist said that Calvin Johnson was innocent.

Keller didn't know what to make of Blake's report, or of the assertion that

LabCorp's testing had been inconclusive. He went to his expert, Dr. George Herrin, of the Georgia Crime Lab, and asked about Blake. The technicians at the crime lab knew all about Blake, and they believed in his results. "Blake's reputation with our Georgia Crime Lab was such that our guys basically told us, 'If Blake said it was O.K., it was O.K.,' " Keller says.

This presented Keller with a dilemma. His conviction of Calvin Johnson in 1983 had been a big victory, both in the courtroom and with his constituency, and he still believed it was a righteous conviction. But things had changed, not only legally, with the new DNA evidence, but politically, too. Clayton County had been mostly white and mostly working class in 1983, when Keller obtained Johnson's conviction. Since then, Atlanta's boom had spilled into Clayton County, recasting its demography, to become sharply more middle class and multiracial. To fight the exoneration of a black man whose innocence now seemed to have been validated by DNA, and whose champion was the famous Peter Neufeld, of New York City, posed an unappealing political prospect.

Keller says that he and his assistants considered their options. They could have attacked the DNA evidence, particularly regarding the sinuous journey that it had taken from the time it left Keller's vault until Blake reported his results. The evidence was in the defense's control after it left North Carolina, and certainly could have been subject to tampering or accidental contamination or mislabelling. Even Blake suggested that a fresh blood sample be obtained from the victim, Eve Cooke, to insure that the evidence he examined was, in fact, from her case. It is difficult to imagine that, had their roles been reversed, Neufeld would not have harassed Keller about the evidence chain.

Also, Keller could have attacked the conclusions of the DNA findings themselves. Eve Cooke had had consensual sex the day before her attack, raising the possibility that the semen tested by Blake came not from her assailant but from her consensual sexual partner. It is a fairly common scenario suggested by prosecutors, and Blake and Neufeld derisively call it "the unindicted co-ejaculator theory." But tests showed that the DNA in the tested specimen was probably from a black man, and Eve Cooke had testified that she'd never had a black person in her home. In any case, Keller says, his office was not able to find Eve Cooke, which made everything else moot. "We didn't have a victim," he says.

Keller did ask for one more test. Although he had argued at the trial that the pubic hairs found in the bedclothes were irrelevant to the case—they had been picked up at a laundromat or a public restroom, he's said—he now wanted them tested. One was sent to LabCorp, which determined that it belonged to the same person whose DNA Blake had found among the other evidence. That decided it for Keller, who put aside all the other questions he had.

CALVIN JOHNSON was twenty-five when he was sent to prison for the rape conviction, and he was released a few weeks before his forty-second birthday. After coming home, Johnson was showered with job offers and interview requests, and for a while he maintained a busy speaking schedule. He was flown by private plane to speak at a church in Alabama, and he even ministered in Uganda.

Unlike New York, Georgia does not have a law providing compensation for people who have been wrongfully imprisoned, but Neufeld is preparing a "private bill" that would pay Johnson from the state treasury for his time and anguish. In the meantime, Johnson has a job (as a Transit Authority station manager) and a new girlfriend, and is living in his parents' home, a cramped apartment in a faded little town called Hapeville, near the airport. The College Park police station is two miles away. I asked Johnson if he is worried about new troubles with the police. "I look at it like this," he said. "They got DNA now, you know? So I'll be O.K."

That is the allure of DNA evidence—its suggestion of absolute immutability. But the Kotler case and, indeed, Johnson's own suggest that whatever truth DNA holds for criminal justice is revealed only through the lens of human advocates. DNA exonerated Kerry Kotler, and then it convicted him again, over the objections of a scientist who became his advocate. DNA freed Johnson, but Georgia police and prosecutors still have not exonerated him completely in their minds.

Lewis Harper, who is now the deputy chief of police in College Park, is among those who continue to believe in the rightness of Johnson's arrest. He and his officers know that Johnson has returned to the neighborhood. "It's something that we can always keep in the back of our head," Harper says. "If we start having these problems again, we can certainly rely on our memory and look at anything that might be related."

John Terborgh

In the Company of Humans

FROM *NATURAL HISTORY*

In *The Biophilia Hypothesis,* biologist Edward O. Wilson addresses the psychological and evolutionary reasons humans are attracted to animals. My own experience as a field biologist has exposed me time and time again to convincing evidence that many humans are indeed powerfully drawn to animals. For more than three decades, I have spent part of every year in the Peruvian Amazon, where I have been privileged to visit villages belonging to half a dozen premodern tribes. Nearly every household has included pets, and even though many of these pets ultimately wind up in the supper pot, the villagers treat them with obvious affection. Among the animals selected are birds of assorted sizes and habits, tortoises, iguanas, and mammals (especially primates but also peccaries, agoutis, and coatimundis). Many have been captured as juveniles, usually by hunters who shot the mother; the young are then raised by humans, who sometimes even suckle them until they can be weaned.

This story has a flip side, however. Under certain circumstances, wild animals are drawn to people. Not always do they flee or recoil from humans; instead, it has often seemed to me, the animals quietly observe them as if attempting to judge their intentions. Then, if the people appear to be nonthreatening, various kinds of interactions become possible.

But before I begin to elaborate on why animals may choose to associate with humans, perhaps I should review the circumstances in which animals of different species are drawn to one another. In the forests around the Cocha Cashu Biological Station in Peru's Manu National Park, such associations are common. In one type, called the beater syndrome, one species unintentionally makes food available to another by creating a disturbance as it moves through the habitat. The beater syndrome is a form of commensalism—the unilateral transfer of benefits from one species to another at little or no cost to the benefactor.

One of the beneficiaries of the beater syndrome in Manu is the rare *huanganapescco,* known in English as the rufous-vented ground cuckoo. In Quechua, the language of the Incas, *pescco* means "bird," and *huangana* is the local name for the white-lipped peccary, a New World mammal similar to a pig. These peccaries are half again as large as the more familiar collared peccary, and they travel in imposing herds that can number in the hundreds, blackening the forest floor with their massed bodies and filling the air with a cacophony of bleating, rumbling, and clacking. Using a tactic similar to that of the cattle egret, the *huanganapescco* positions itself amid a herd of peccaries and keeps a keen eye out for the lizards, frogs, and arthropods routed by the animals' hooves. Meanwhile, overhead, several woodcreepers cling to tree trunks, ready to snatch insects that take wing to avoid being trampled.

A more intimate form of commensalism, termed the cleaner syndrome, involves direct body contact between the associates, implying both trust and recognition. Viewers of nature programs on television are familiar with the cleaner wrasse, a small coral reef fish that makes a living by nipping parasites off larger fish. The most famous terrestrial cleaner syndrome involves the colorful tick birds of the African savanna. These birds forage exclusively on the backs and legs of large mammals, where they dine on parasites, principally ticks. The Amazonian counterpart of Africa's tick bird is the giant cowbird, which forages independently most of the time but deticks capybaras and tapirs when opportunity knocks. Obviously comfortable with the relationship, capybaras (the largest living rodents) are unfazed when cowbirds alight on their heads and begin to peck around their eyes and ears.

Sometimes birds of a feather flock together for less transparent reasons, as one unplanned "natural" experiment showed. Years ago, as a graduate student at the University of California, Berkeley, ornithologist Pete Myers was studying sanderlings—the pale little sandpipers that frenetically chase waves up and down beaches along both coasts of North America. During the first winter of his study, based at California's Point Reyes National Seashore, Myers observed that sanderlings, when not foraging, roosted amicably in large flocks on sandbars. But when the tide was propitious, they spread out along the beaches and set up individual territories, chasing away any rival sanderlings that ventured too close. This behavior provided gratifying confirmation of the then-new theory of optimal foraging, which held that the highest feeding rate could be attained by individuals that maintained exclusive rights to a foraging area. In this case, the area was a strip of beach about a hundred yards long.

The following year, Myers encountered an entirely unanticipated situation. Instead of spreading out and confronting their neighbors in hostile face-offs, foraging sanderlings bunched together in tight little flocks. Many birds feeding in a small area quickly deplete the prey, however, lowering individual foraging success and compelling a flock to keep moving in quest of fresh sites. Clearly the birds were paying a price for their newfound togetherness. The question was, What had inspired them to change their behavior so profoundly?

Myers soon discovered that a merlin (a small falcon) had taken up residence that winter at Point Reyes. Although the merlin was usually out of sight, the sanderlings never forgot that a predator was in the vicinity. Membership in a flock meant that each individual gained the advantage of more eyes and ears to detect the approaching predator.

The costs of joining a flock, school, or herd can be lower when an animal joins a group composed predominantly of other species. Consorting with aliens, as it were, offers all the advantages of foraging in a group, while it minimizes competition with other individuals of the same species. This probably explains why flocks of birds consisting of many species, but no more than a few individuals of any one, are so commonplace around the world.

By now, the reader is surely wondering what all this has to do with what I call homophilia (literally, a friendly feeling toward humans) in animals. In fact, it has a great deal to do with the rest of my story, which begins with trumpeters.

Distant relatives of cranes, trumpeters are long-legged, chicken-sized birds that glean fallen fruit from the ground. Unlike other birds that live on the forest floor, trumpeters are not particularly shy and readily habituate to the presence of humans. One day, while observing monkeys feeding in a giant fig tree, I understood why.

Shortly after the monkeys began to eat—sloppily dropping nearly as much fruit as they consumed—a group of trumpeters showed up on the forest floor beneath them. Soon an agouti (a large tropical rodent) appeared and began to feed among the trumpeters, which were unperturbed by its presence. Before long, a group of collared peccaries joined the crowd. Again the trumpeters showed no reaction at all.

I later learned that various terrestrial mammals routinely join feeding trumpeters, presumably to benefit from various loud alarm calls that the birds make against such animals as jaguars, bush dogs, eagles, and snakes, as well as from their habit of posting sentinels whenever other group members are feeding. To the trumpeters, I realized, a person is just one more large but nonthreatening mammal come to join the group.

All these animals appear able to recognize a good thing when they see it, and my many years at Manu have convinced me that our little research station—by providing opportunities for safety that some animals decide to take advantage of—is a bit like the group under that fig tree. The station's unobtrusive buildings bring the scientists into unusually close contact with the inhabitants of the surrounding forest. Every year, certain individual birds and mammals linger near the station, often strolling in open view through the clearing or perching right in front of a building, hardly more than an arm's reach away.

The species that have shown such boldness are extraordinarily diverse in their habits and diets and thus seem to have no common denominator. Among those that have been drawn into our midst are tinamous—plump, partridgelike birds notorious for their shyness. A small path that leads from one group of our buildings to another apparently cuts through areas frequented by these birds. Observing that humans passed by at frequent intervals without adverse consequences, several tinamous grew so comfortable with our presence that they would sometimes stand in the middle of the path and fail to budge when someone approached. Occasionally I found it necessary to make a verbal request before a tinamou would step aside so that I could pass.

Over the years, I have sometimes sought refuge from the hubbub of the station in a screened tent in the forest. One morning, after I had sat down at my desk, a movement in the tent caught my eye. It was one of the "tame" tinamous. The bird did not seem to be the least bothered about sharing the cramped space of the tent with me. It calmly strolled around inside for a few minutes and then let itself out through a crack at the bottom of the door.

Tinamous have demonstrated the flexibility of animal behavior in other ways as well. Several years ago, we kept chickens at the research station. (We were studying ocelots at the time and needed chickens to lure the cats into our traps.) I happened to glance idly at a group of foraging chickens one day and was thunderstruck to see two tinamous scratching and pecking among them.

They were doing two things tinamous never do (or so I had thought): participating in a social group of foraging birds and exposing themselves to an open sky that might have contained raptors. This scene of interspecific amity was repeated day after day as the tinamous took advantage of the safety in numbers provided by our chickens.

One of the habituated birds I remember best was a piping guan—a chicken-sized bird normally found only in the highest treetops—that chose to nest only two yards from a building under construction. As the guan calmly sat on her nest, a team of carpenters erected beams and nailed them into place almost eyeball to eyeball with the unflappable bird. Even the chainsaw didn't disturb her. Eventually the guan's three eggs hatched into downy chicks, and for many days afterward she remained within a few yards of our buildings while she tended her growing brood.

Birds have not been the only creatures at our site to seek the company—or at least the nearness—of humans. Perhaps the most remarkable were Howeird and Moreweird, two subadult male red howler monkeys. Howlers are among the most distinctive and characteristic primates of the New World tropical forest. The adult males' roars are often so loud and startling that first-time visitors are convinced they are in the immediate presence of a jaguar. In reality, however, few animals could be less threatening than these languid vegetarians that spend much of every day lounging in the canopy digesting leaves.

One extraordinary day a number of years ago, primatologist Patricia Wright was doing her laundry when a furry red limb suddenly intruded into her field of vision. Looking up with a start, she confronted a howler monkey backing down the very tree to which the washboard was attached. Transfixed in surprise, Pat stood motionless as the monkey proceeded down to the ground under the washboard, where it set about eating soil that had been soaked in wash water. So unconcerned was the howler by Pat's presence that at one point it rested its hand on her shoe.

Howler monkeys are well known to engage in geophagy, or earth eating, though the reason they do it remains unclear. One idea is that the soil provides certain mineral elements lacking in their diet of fruit and leaves. Another hypothesis is that clay minerals in the soil help alleviate the effects of some of the toxins that must inevitably be ingested by an animal that consumes leaves.

After Howeird had broken the ice with Pat, he began to hang around the station buildings, often resting in the rafters under the open roof when Pat was inside. Although she never fed him, Howeird persisted in following her around. When she went to her tent for the night, Howeird was right behind. Not wishing to spend the night with a monkey in her tent, Pat would quickly slip in and zip the door behind her. Undiscouraged, Howeird would climb up a small tree that overhung her tent and spend the night there.

After several days of this behavior, another subadult male howler, duly named Moreweird, joined Pat and Howeird. The three of them were nearly inseparable until the day the two monkeys just vanished into the forest, never to be seen again. Their departure was as unexpected as their arrival. Pat, who enjoyed imagining that true love had brought the howlers out of the forest to her, was soon forced to accept that the attraction was something more mundane: the two monkeys departed right after her bottle of lemon-scented detergent ran out and was replaced by one of a different brand. A dejected Pat had to admit that they didn't love her after all; they only loved her detergent.

LAST YEAR, something nearly as remarkable happened—less amusing but deeply touching. At lunch one day, a student announced a very unusual sighting: a lone *huangana.* Finding one of these big peccaries all by itself and away from its herd was unprecedented in our experience. The student had encountered the animal half a mile to the north of the station and noted that it appeared sick and lame. Late that afternoon, another researcher met the same animal only 300 feet from the station. It had been standing in the middle of Trail 1, the main thoroughfare between the station and our port on the Manu River. When confronted by the approaching human, the *huangana* hobbled a few feet off the trail and stood there while the researcher passed by.

The next day several people saw it, always standing at the edge or in the middle of Trail 1. The animal could not have selected a busier place to reside; many people, often in noisy groups, go back and forth to the port every day. Yet the *huangana* chose to settle precisely here. For the first few days it seemed to be in decline, limping badly and responding listlessly to the blandishments of nervous researchers who didn't want to get too close to a potentially dangerous animal.

After perhaps a week, the *huangana* appeared more alert and was steadier on its feet, although I don't know what it could have been finding to eat during all that time. Had the animal wanted to distance itself from further contact with humans, it could easily have done so. But it remained in the middle of Trail 1 by day, and by night, we discovered, it quietly bedded down just a few feet from an investigator's tent.

I can think of no other way to interpret the *huangana*'s behavior except to imagine that it "wanted" to be close to humans. Why else would it have walked half a mile in an enfeebled condition to be near us?

The *huangana* is a prime example of a species that seeks safety in numbers. Its archenemy, the jaguar, never launches a frontal attack on a herd, because adult peccaries defend themselves with long, saberlike tusks that could easily disembowel a big cat. Instead, the jaguar stalks the herd in the hope of being able to assault a juvenile or a peripheral individual and subdue it before the others react. A lone *huangana* is thus in a very vulnerable situation. From a jaguar's point of view, such an animal is a freebie.

Our peccary must have decided that the risk of consorting with humans was less than the one it faced by remaining alone in the forest. Perhaps it had noticed that the jaguar was seldom in the vicinity of the station. Whatever its reasoning, the *huangana* was right. Its vigor and agility steadily improved until, one day, a herd of its species crossed Trail 1 and our peccary was gone.

As a scientist, I am admonished to be unrelenting in my skepticism and to demand the highest standards of evidence before drawing conclusions. Above all, I should resist any temptation to construct anthropomorphic interpretations. What I have recounted here are anecdotes—isolated occurrences of an essentially unrepeatable, and thus scientifically untestable, nature.

Nevertheless, having spent a lifetime observing animals in the wild, I have come to the conclusion that many birds and mammals are highly observant, that they are able to weigh very abstract risks, and that they can reach conclusions based on the assessed balance of those risks and then take appropriate action.

Night and day, the Amazonian rainforest teems with predators. No animal, except perhaps a top carnivore, can afford to be unmindful of the omnipresent threat of predation. If animals can be said to think about anything, heading the list must be how to conduct their lives in a way that minimizes exposure to predators—since, of course, only living animals can pass along their genes.

Whether a particular bird or mammal is territorial or social is commonly regarded as characteristic of the species. Pete Myers's sanderlings, however, demonstrated a capacity for radically altering their behavior in direct response to an increased threat of predation. So did the habituated tinamous at our research station when they perceived that by consorting with chickens, they could forage in the open at reduced risk. Monkeys are similarly opportunistic in their choice of companions. Never at ease when alone, bachelor males routinely seek the company of other species of monkey. Howeird and Moreweird apparently decided that Pat could provide some sort of protection against predators. To be alone is to be vulnerable, because no animal is able to maintain vigilance 100 percent of the time.

I am not suggesting that animals have the same intrinsic affinity for people that E. O. Wilson claims people have for animals. But when under the threat of predation, many animals do have an affinity for other animals, whether of their own or of different species. Having been taught as a child that nearly all animals instinctively avoid people, I was pleasantly surprised to learn that animals can occasionally overcome their inhibitions and see us as benign. My colleagues and I at Manu are gratified when the birds and mammals with which we share the forest choose to draw near, even if it is only to use us as foils against their enemies.

JAMES SCHWARTZ

Death of an Altruist

FROM *LINGUA FRANCA*

In a ruthless, Darwinian world, human heartlessness is easy to explain. After all, natural selection eliminates the weak and rewards the strong. Unselfish behavior, on the other hand, is baffling. Compassion, kindness, and loyalty ought to be weeded out almost as soon as they arise.

To explain the evolution of altruism, Charles Darwin suggested that natural selection could act on groups as well as individuals—an idea known as group selection. Within a tribe, it could hardly be doubted that "selfish and treacherous parents" would have the most children, he wrote in his 1871 classic, *The Descent of Man.* On the other hand, he explained, tribes including many members "ready to give aid to each other and sacrifice themselves for the common good would be victorious over other tribes."

But Darwin never fully developed his ideas about group selection, and his heirs continue to argue over it. The late William D. Hamilton, widely regarded as the most important evolutionary thinker since Darwin, proposed an enormously influential alternative to group selection called the theory of nepotistic altruism. This is the idea that all apparently altruistic behavior is directed exclusively toward genetic relatives. True selflessness, in Hamilton's opinion, almost never happened. Among human beings, the two examples that came most readily to Hamilton's mind were Mother Teresa and his friend the brilliant, iconoclastic American scientist George Price.

Despite his remarkable scientific achievements and the intense drama of his personal life, George Price has remained a relatively obscure figure in the history of science. And yet he played a key role in shaping the conceptual basis of sociobiology and its offshoot, evolutionary psychology. He was the first to apply the principles of game theory to the analysis of animal conflicts, and he discovered an elegant formula to describe evolutionary change that both simplified and improved Hamilton's theory of nepotistic altruism. Furthermore, Price's formula provides a rigorous mathematical framework in which to understand group selection.

In his last years, even Price himself lost interest in securing recognition for his scientific achievement. Altruism, which had begun as an intellectual problem, became an all-consuming personal concern. In the midst of an extraordinary burst of scientific creativity in the summer of 1970, Price abruptly converted from militant atheist to fundamentalist Christian. After his conversion, he combined his work in genetics with a passionate interest in biblical exegesis. Later, he decided that the mission of a true Christian was to help his fellow man, and he spent increasing amounts of time aiding homeless alcoholics and the elderly in his adopted city of London. As he ministered to others, his own life disintegrated into chaos. He died in 1975.

In the years before his own untimely death in the spring of 2000, William Hamilton often wrote and spoke about Price in an effort to draw attention to his old friend's ideas. Several months before he died, Hamilton shared with me his lengthy correspondence with Price. With the help of these unpublished letters and Hamilton's recollections, as well as additional aid from Price's two daughters and many friends and colleagues, I have tried to piece together the tumultuous story of this extraordinary man.

A PHYSICAL CHEMIST and a journalist before he studied genetics, George Price took a circuitous path to evolutionary biology. Born in 1922, he was only four years old when his father died. His mother, a former opera singer and actress, struggled to keep the family's lighting company afloat through the Depression, and it was difficult for her to support George and his older brother, Edison. After attending public school in New York City, George went to the University of Chicago, where he earned a Ph.D. in chemistry, and later worked on the Manhattan Project. In 1947, he married Julia Madigan, and they had two daughters. Three years later, the family moved to Minneapolis, where Price worked as a medical researcher at the University of Minnesota. Because George was a fire-spitting atheist and Julia a devout Christian, their relationship was contentious from the beginning. After eight years, the marriage ended in acrimony.

In 1955, the year of his divorce, Price published his first magazine piece, a long article in *Science* in which he questioned the quality of the evidence used to demonstrate ESP. It was the first of several highly visible, and often far-sighted, forays into journalism. The following year Price published "How to Speed Up Invention" in *Fortune* magazine. In the age of punched cards and Teletypes, the article described in detail a hypothetical "design machine," which would feature a graphic display, a cursorlike light pen, and a mouselike device to rotate, shrink, and enlarge shapes.

In 1957, Price sent Senator Hubert Humphrey an early draft of an essay he was writing for *Life* titled "Arguing the Case for Being Panicky." The article warned that a decline in U.S. military strength could lead to catastrophe, and it so impressed Humphrey that the two men went on to exchange dozens of letters. Price told Humphrey about several novel foreign policy ideas he had derived from game theory. He suggested, for example, that the United States offer to buy every Russian citizen two pairs of high-quality shoes (at a cost of $2 billion) in exchange for the liberation of Hungary. The same year, Price began to write a book on the Soviet threat for Doubleday, which he abandoned in 1959 when he concluded that the Soviet position on disarmament was more commendable than the U.S. position. He returned to scientific research by accepting a post at IBM, where he worked on the development of mainframe computers and the mathematical modeling of free markets.

But Price's career soon took a dramatic turn. In 1966, he was treated for thyroid cancer. Price believed that the doctor, an old friend of his from Chicago, badly mishandled the operation. The surgery, added to earlier nerve damage caused by polio, left Price's shoulder partially paralyzed. The misfortune earned Price a generous insurance settlement, and he decided to use the money to finance one final career shift.

IN NOVEMBER 1967, Price sailed on the *Queen Elizabeth,* pleased to be the sole occupant of a stateroom meant for two. In England, bankrolled by his insurance settlement, he rented a large flat in an affluent section of London near Oxford Circus. In his first letter to his grown daughters, he reports humorously about his exploration of the city: "Dear Babies, I have seen quite a lot of London so far, including the British Museum library, the Museum of Natural History library, the University of London library, the University College library, the Wellcome Historical Medical library, and Science and Technology library. Soon I hope to visit the Royal College of Surgeons library and Royal Zoological Society library."

Among the articles Price read in his library visits was William D. Hamilton's now-classic study "The Genetical Evolution of Social Behavior." A watershed in the history of evolutionary biology, it would be Price's point of entry into the field.

According to traditional Darwinism, natural selection is the survival of the fittest—with "the fittest" defined as those organisms who leave the most descendants. It was easy for Darwin to account for the evolution of traits that directly benefit individuals, such as good eyesight. However, it was harder to explain the evolution of behavior that benefits fellow organisms while lowering an individual's own fitness. In the early 1960s, the most common explanation was the theory of group selection. It was believed that natural selection acted on two levels, favoring better adapted individuals (who left more progeny), on the one hand, and better adapted groups, on the other. A selfish individual would tend to leave more offspring than an altruist, but a group that contained many altruists would tend to grow faster than a group that didn't.

Hamilton rejected this notion altogether. He insisted that the natural selection of individuals was responsible for all significant evolutionary adaptations, including self-sacrificing behavior. The problem, as Hamilton saw it, was that evolutionary biologists had been defining fitness too narrowly. What mattered was not merely the number of offspring an individual had but his "inclusive fitness." In addition to an individual's own progeny, you had to consider the progeny of his relatives, fractionally weighted according to how closely they were related. Hamilton's argument depended on the "gene's eye" point of view—the idea that it was the survival of particular genes, rather than the survival of the individuals who carried them, that was crucial. This was the argument that Richard Dawkins would develop with dazzling lucidity in his 1976 book, *The Selfish Gene.*

In March 1968, Price wrote Hamilton requesting a reprint of the paper. "The mathematics in it is a bit formidable to absorb in library reading," he explained. Price's initial observations were nonetheless so perceptive that Hamilton was inspired to write two dense, handwritten pages in return. Hamilton described his recent work and disappointed Price with the news that he would be unavailable for further discussion: He was about to depart for a nine-month research trip in Brazil.

Left to his own devices, Price set out to find a simpler, more direct way of achieving Hamilton's results. That summer, he wrote Hamilton in Brazil to report that he had found "a more transparent (though less rigorous) derivation and formulation of the main result of your paper." This was an understatement. He hadn't merely simplified Hamilton's math. He had reconceptualized the mathematics of how genes flourish or fade under natural selection, and he had done it with a startlingly simple equation. The equation was so pared down that it could apply to any form of selection, "from selecting a book or a radio station to genetic selection," as Price later put it in a letter to a friend.

At the heart of Price's equation is the mathematical concept of covariance—a statistical measure of the relationship between any two sets of data. Take, for example, the rainfall and temperature in a given place over thirty days. If rainfall is independent of temperature, then the covariance of rainfall and temperature will be close to zero. If days with more than average rain also tend to have higher-than-average temperatures, then the covariance will be positive. But if days with more rain than usual tend to have lower temperatures than usual, then the covariance will be negative.

How can a covariance measure natural selection? Price's equation describes the change in a gene's frequency from one generation to the next. Specifically, it relates that change to the covariance between an individual's possession of the gene and the number of children he or she has. If having the gene leads to having more offspring, the frequency of the gene will increase in the next generation.

It sounds so straightforward as to be almost tautological. But Price's mathematics was strikingly original. Furthermore, in the process of inventing a new algebra of natural selection, Price also devised a more sophisticated and accurate way of understanding relatedness. In his own work, Hamilton had assumed a population that was infinitely large. But Price saw that calculations of relatedness turn out differently in a small population of closely related individuals. In a finite population, individual A can be *negatively* related to individual B, in the sense of sharing fewer genes with B than with the members of the group on average. Negative relatedness changed the predictions of Hamilton's theory in subtle but significant ways. First, it made clear that if you have a gene that benefits your group, that gene will survive only if it is of more benefit to your close relatives than to your distant ones. More disturbingly, it implied that it could be adaptive for animals to harm themselves in order to harm others more. Price called this "spiteful behavior": It might pay for an animal to take the life of a "negatively related" neighbor even at some cost to itself.

Price's approach bore no relation to previous work. At first, even Price was convinced that his equation was too simple to be new, so he decided to check it with an expert at the Galton Laboratory at University College, London, the famous department of human genetics named in honor of Darwin's cousin Sir Francis Galton. When he visited the Galton Labs in June 1968, Price asked to speak with a mathematical geneticist and was taken to Cedric Smith, a respected biostatistician. "Smith said it was very interesting, very pretty, and he had never seen anything like it before," Price later wrote to his mother. After quizzing him about his work, Smith brought Price to the department chairman. Eighty minutes later, Price had an honorary appointment, an office, and keys. He left walking on air.

WITH A TITLE, an office, and the support of certified experts, Price set off with new confidence to write up his results. He believed that university backing would help him get funding and make it easier to publish his work. But outside of work, it was a difficult period for Price. His shoulder was troubling him again, and he was anxious over money, since he had spent most of his insurance settlement. To top it off, his mother had fallen seriously ill.

In March 1969, his mother's health took a precipitous downturn, and he flew to New York to see her once more before she died. During his visit, he met Richard Lewontin, a young and highly regarded population geneticist. If anyone was in a position to understand Price's new selection mathematics, it was Lewontin, the most mathematical of the young geneticists. But the meeting did not go well; and Price blamed himself for failing to convey his new approach. After he returned to England in early May, however, things started to look up. He learned that he had become a grandfather, and in July he received a grant from Great Britain's *Science* Research Council.

Hamilton meanwhile had returned from Brazil, and in late July 1969, Price contacted him again. Nearly a year had elapsed since their last communication. In his letter, Price gently explained that Hamilton's formulation of nepotistic altruism did not work quite the way Hamilton thought it did. He generously offered to spare Hamilton the awkwardness of being corrected publicly. "I did want—in view of your friendly correspondence, because I respected your work, and because everyone makes a mistake now and then—to publish in a way that would not embarrass you," Price wrote. He left a telephone number and the times he could be reached.

In his 1996 book, *Narrow Roads of Gene Land,* Hamilton recalled his first telephone conversation with Price. "His voice was squeaky and condescending, rather guarded, on the phone," Hamilton wrote. In the conversation, Price described his covariance equations as "surprising for me too—quite a miracle." Price also asked if Hamilton had noticed that the equation could describe group selection as well as individual selection. Hamilton was skeptical. After all, it was through his rejection of group selection that he had arrived at the theory of nepotistic altruism in the first place.

But Price would eventually cause Hamilton to rethink his position. To understand why, we have to back up for a moment. Price's equation was in fact slightly more complex than the rendition of it above. In order to assess the fortunes of a particular gene, it's not enough to know whether having that gene correlates with having more children. It's also important to know how likely it is that your children will actually inherit that gene. And so there was a second term in Price's equation, assessing the likelihood of a particular gene's transmission. In most cases, children inherit a parent's genes at random—that is, the gene does not affect the viability of the sperm or egg—and the second term can be safely ignored.

But it is thanks to this second term that Price's equation sheds such powerful light on group selection. As Price hinted to Hamilton on the phone, it is possible to redefine the two terms in his equation by shifting them one notch up the population structure. Instead of applying to *individuals and their sperm or eggs,* the equation would apply to *groups and the individuals they contain.* The forces of selection acting at both the group and the individual level appear side by side in the revamped formula, making it possible to compare their relative strengths. A century after Darwin first wrote about group selection, Price's equation spelled out the precise conditions under which the interests of the group could trump the interests of the individual.

As it happened, one of the first well-documented examples of group selection had recently been reported. In Australia, scientists had introduced a disease virus in an attempt to control the native rabbit population. Although the virus was at first an extremely effective rabbit killer, over time it became considerably less virulent. Within any given rabbit, viruses that multiplied faster had the advantage over viruses that multiplied more slowly. In other words, on the individual level—in this case, the virus level—natural selection favored faster-multiplying (therefore more lethal) viruses. But when a rabbit died, the viruses inside died with it. (The virus was spread by mosquito, and mosquitoes do not bite dead rabbits.) Thus the faster a group of viruses multiplied, the more likely it was that their rabbit host would die before the virus could be transmitted to another rabbit. At the group level—at the level of the rabbit host—natural selection favored less lethal, slower-multiplying viruses. Price's equation could explain this evolution of an "altruistic" virus—a virus that multiplied less greedily so that its rabbit host could live longer.

Despite the strain in their first conversation, Hamilton quickly grasped the significance of Price's covariance equation and soon found himself won over. "I am enchanted with the formula derived in your manuscript," he wrote to Price in December 1969. A month later he'd begun to rethink the issue of group selection. "In its general form, I can see how one might use your formula to investigate group selection," he wrote.

Although Price's equation was strikingly original, its publication, which would be Price's first in his new field, was by no means assured. Hamilton, who had felt isolated and unappreciated while working out his theory of nepotistic altruism, was anxious to help his friend avoid a similar fate. Together they devised a clever strategy to break into *Nature,* one of the premier science journals. Price would submit his paper on the mathematics of natural selection first. One week later, Hamilton would submit a paper that depended on Price's formula to re-derive his theory of inclusive fitness.

It came as no surprise when Price's paper was returned immediately. The editors had not seen fit to send it out for review. No less surprising, the paper by Hamilton, a well-established name, was accepted without delay. According to plan, Hamilton wrote *Nature* to withdraw his paper. He explained that he had made use of a "powerful new method," and he could not in good conscience publish his results until the method he used was published. The plan went off without a hitch; *Nature* promptly reconsidered. Price's "Selection and Covariance" was received on November 12, 1969, and published on August 1, 1970. Befitting its entirely original approach, the paper appeared without citations.

EARLY IN THE SUMMER of 1970, at the age of forty-seven, Price underwent a sudden religious conversion. "On June 7th I gave in and admitted that God existed," he explained to friends. He viewed his conversion as a logical necessity, the result of a series of coincidences that had befallen him. After calculating the odds of their occurrence and finding them to be "astronomically low," he was convinced that there had been supernatural manipulation. One week later, he attended his first service at All Souls at Langham, a particularly evangelical branch of the Church of England, located around the corner from his apartment.

Over the course of the next year, Price's scientific work was accompanied by a new passion—biblical exegesis. Adopting a highly literal approach to the Bible, Price set out to reconcile discrepancies among the four Gospels. Nearly a year after his conversion, he completed a fifty-page article, "The Twelve Days of Easter," which proposed to replace the traditional eight-day Holy Week with a new chronology. He believed he had resolved several of the long-standing puzzles of biblical scholarship.

He sent the article to Hamilton, who was impressed by Price's reconstruction of the events of Easter week and encouraged him to publish it. But he did not accept Price's arguments for the existence of God, nor did he convert, as Price had hoped he would. In a letter, Hamilton likened his resistance to Christianity to "the Irishman who was asked whether he liked oysters and he replied, no, he didn't like oysters and he was glad he didn't like them because if he did he'd be eating them all the time when he hated the damned things."

Hamilton also attempted a more serious response to Price's new thinking about Christianity, prophecy, and free will. "Why should we respect Moses if he was just a puppet carrying out maneuvers to foreshadow the crucifixion, and why respect Jesus for following a canon that he was bound to follow anyway? A plot so elaborate would make life meaningless if one believed in it—and ugly too."

Price waited four months before replying. "The question is not whether you like it but whether it is true," he wrote. "What difference does it make whether you approve of it or not? Do you think that is something that I wanted to believe in?"

Meanwhile, Price's second major breakthrough in evolutionary biology was at last about to see the light of day. Back in 1968, the same summer he happened upon his covariance equation, Price had become intrigued by the fact that male animals of the same species rarely fight to the death. In July of that year, anxious that his windfall insurance payment was running out, Price had worked night and day to complete an article setting forth a new idea, namely, that game theory might help evolutionary biologists understand animal conflict. At the time, Price hoped that a quick and dazzling academic paper would pave the way for profitable magazine sales.

Price's key insight had been to see that the genetically optimal behavior for an animal could depend on the behavior of other animals. In a population made up of animals genetically programmed to make war, for example, an animal programmed to retreat from a threat might actually be at an advantage. On the other hand, in a population of less aggressive males, a confrontational male would have the advantage. Price had titled his paper "Antlers, Intraspecific Combat, and Altruism" and sent it off to *Nature* on the last day of July 1968. The following February, he learned that *Nature* had accepted the article, provided it was shortened. But several years passed, and Price never bothered to undertake the revision.

As it happened, the reviewer of Price's paper had been John Maynard Smith, the head of the biology department at the newly created University of Sussex. Maynard Smith saw the potential of Price's unpublished idea, and he wanted to use it in a paper he was writing. In 1971, he wrote a letter asking for permission to thank Price for showing him an "unpublished manuscript." Price wrote back that he would rather Maynard Smith referred to a discussion between them and not a manuscript. "If one mentions an unpublished manuscript," Price explained, somewhat mysteriously, "then someone might wonder whether it was used with permission."

Price's concern was a delicate reference to a long-running, acrimonious dispute between Maynard Smith and Hamilton, whose paper on nepotistic altruism Maynard Smith had reviewed for the *Journal of Theoretical Biology* a few years earlier. It was Hamilton's belief that Maynard Smith had in effect stolen his idea. "His account of the matter," Price explained to Maynard Smith in the fall of 1972, "is that you refereed his 1964 paper for the *Journal of Theoretical Biology,* and required a major revision . . . that caused a nine-month delay in publication and meanwhile you sent *Nature* a letter with the term 'kin selection' that has received much of the credit for the idea."

"I seem to have this fate of getting ideas from other people's manuscripts when I referee them," the genial, white-haired Maynard Smith explains today. Asked about Hamilton in particular, he is somewhat more defensive: "I wasn't trying to steal his idea, or I don't think I was, so it wasn't conscious." At first, Price doubted Maynard Smith's integrity and suspected him of delaying the "Antlers" paper as he had Hamilton's. But he changed his mind after meeting Maynard Smith. In fact, Maynard Smith was scrupulous about crediting Price—he offered to make Price a co-author of the paper he was writing.

In the fall of 1972, "The Logic of Animal Conflict," co-written by John Maynard Smith and George Price, was accepted by *Nature.* On receiving this news, Price wrote to Maynard Smith that "I think this is the happiest and best outcome of refereeing I've ever had: to become co-author with the referee of a much better paper than I could have written by myself." The paper was one of the first to set forth the ideas of evolutionary game theory, ideas that in the years since have been used to analyze everything from reciprocal grooming in African antelopes to egg swapping among hermaphroditic fish.

IN JUNE 1972, the three-year grant that Price had obtained from the *Science* Research Council came to an end, and he chose not to seek to renew it. He preferred to give more time to his Christian work and less to mathematical genetics. By the fall, Price had decided to live according to his literal interpretation of the teaching of Jesus. Inspired by Jesus' advice in the Sermon on the Mount to take no heed of the morrow, Price was pushing himself to the brink of disaster. He was almost joyous in anticipation of the extreme deprivation that his faith had brought upon him. "I am now down to exactly fifteen pence," he wrote to Maynard Smith that October. "I look forward eagerly to when that fifteen pence will be gone."

A month later, Price's diet consisted of one pint of milk a day, and he was weak from malnutrition. He stopped taking the thyroxine pills that his thyroid cancer had rendered necessary to his survival. Price believed that if God wanted him to continue living, He would provide the missing hormone. In early December, Price seems to have attempted suicide; in any case, he ended up in a hospital. An alert doctor noticed that he was suffering from myxedema, a result of his thyroxine deficiency, and provided the missing hormone without informing the patient. Taking this as a sign that he was meant to go on living, Price followed the doctor's orders and started taking his pills again.

Price made his final revisions to "The Logic of Animal Conflict" the following February. In a cover letter, he explained to Maynard Smith that he had made a few changes to accommodate his newfound belief in creationism. "I think I found wordings that you won't object to, and that won't shock *Nature*'s readers by making them suspect what I believe," he wrote.

Later that month, Price's religious crisis deepened. In what he described as "an encounter with Jesus," he saw that he had misunderstood the real nature of Christianity and that his true duty was the care and love of people rather than biblical study. He began to devote himself to the needy. He helped out at various old people's homes and gave away all his money to homeless alcoholics on the street, often inviting them to stay in his flat. He also set out to make amends for the failures in his private life, apologizing to his elder daughter, Annamarie, for deserting her and being a poor father. For a short period, he hoped to remarry his former wife, Julia, and reunite the family in London.

Toward the end of June 1973, Price gave up his comfortable flat and lived as an itinerant. Once the model IBM employee, short-haired and suited, Price now let his hair grow out and dressed in sneakers and colorful shirts with an aluminum cross around his neck. He gave away the last of his possessions, including his watch and coat, and lived hand to mouth. In a letter to Hamilton, he explained that he was living according to Luke 6:30: "Give to every man that asketh of thee; and of him that taketh away thy goods ask not again."

Elizabeth Mansell, who worked as a manager of an old people's home where Price volunteered, recalls his arriving on Christmas Eve 1973 "like an angel coming in." She and Price stayed up past midnight wrapping Christmas presents for the old people, and she overslept the next morning. Rushing down in a panic, she found Price had fed and dressed all twenty-one residents.

In the little time that remained after doing his Christian work, Price continued his genetics research at the Galton Labs, where Cedric Smith, his first sponsor, had procured him a one-year Medical Research Council stipend. But by the end of the year, he was forced to avoid the lab because his charity cases were causing disturbances there. In one incident, a belligerent alcoholic (whose abused wife Price had been protecting) pissed publicly on the front steps of the genetics building, smashed a bike lamp, and scattered the contents of a student's satchel while shouting obscenities. Fortunately, Price's recent paper with Maynard Smith in *Nature* had earned him credit with the lab authorities. As he wrote to his daughter Annamarie, "I expect that one cover-illustrated lead article in *Nature* compensates for one urination at the front entrance to the building."

In March 1974, Price took up temporary residence in the home of an elderly woman whom he'd helped. By keeping his whereabouts secret from his charity cases, he hoped to get some work done on a joint project on sexual selection with Hamilton. In June, he took a job as a night office cleaner. In a letter to Hamilton, he wrote, "I thought it was about the first honest work I'd done in my life—working for others rather than for my own amusement or advance." In August, Price gave up his night job and moved into a commune located in six deserted buildings a mile north of his first flat.

This move would mark the last major transition in his life. He wrote to his daughter Kathleen that after the summer's low in money and social prestige, he was "heading back up" and had started acquiring possessions again, which he was now slower to give away. He wrote to Hamilton that he believed "Jesus wants me to do less about helping others and give more attention to sorting out my own problems." To another friend he wrote that honesty perhaps meant confessing to "one's deepest selfish desires." He had fallen in love with a woman in his commune, and he hoped to move back to the States. In November, he confided to his brother, Edison, that he had given up his social work, returned to a more conventional Christianity, and was considering marriage.

Meanwhile, Price was beginning to receive some long-overdue recognition. Five years after their New York meeting, Richard Lewontin wrote Price: "It has taken me a long time to come around to understanding the work you have been doing, which I was too stupid to appreciate when you first showed it to me." In October, Price also received a long and detailed letter from the eminent American population geneticist James Crow. Like Lewontin, Crow expressed chagrin at having been so slow to appreciate the significance of Price's work.

Just before Christmas, Price visited the Hamilton family for a little more than a week at their home outside London. When he left on December 19, he seemed to have recovered much of his good spirits, and Hamilton had almost persuaded him to resume full-time work in genetics. The Hamiltons were going to spend Christmas in Ireland, but it was agreed that Price would stay with them again after the New Year. However, on his return to London, Price's spirits dropped precipitously. On January 2, one of his Christian acquaintances wrote to urge him to contact the Samaritans. But Price's life had spun out of control. He was found dead in his squatter's tenement on January 6, 1975. The nail scissors he had used to cut his throat were on the floor beside him.

THE SCIENTIFIC COMMUNITY has been slow to appreciate Price's contributions, perhaps because he never pursued academic recognition strenuously. The application of Price's covariance formula to group selection, for example, has far-reaching implications, but Price never drew attention to them. Shortly before his own death in March 2000, Hamilton said, "It's as if you've discovered the calculus and put it into one of your obscure papers but never explained to people how useful it was." In a 1975 paper, Hamilton had tried to popularize Price's cryptic treatment of the subject. But Hamilton's paper got little attention, and Price's genuinely novel approach remained largely overlooked.

That may be changing. In 1995, twenty years after Price's death, the theoretical biologist Steven Frank of the University of California at Irvine wrote a paper reviewing Price's remarkable contributions to evolutionary theory in the *Journal of Theoretical Biology.* The journal also presented a previously unpublished manuscript by Price giving a complete account of his unified theory of selection mathematics. (Two other complete Price manuscripts remain to be published.) Price's equation also played a prominent role in *Unto Others,* an important new book on altruism by David Sober and David Sloan Wilson.

For his groundbreaking insight into the evolution of altruism, Price merits a special place in the history of evolutionary biology. The painful irony is that his struggle to extinguish all selfish motives in his own life nearly prevented him from achieving it. In *Narrow Roads of Gene Land,* Hamilton recalled a conversation between them that took place when Hamilton was first beginning to appreciate how Price's discovery might change the understanding of group selection. "I thought you would see that," Price said.

"Then why aren't you working on it yourself, George?" Hamilton asked.

"Oh, yes. Cedric wants me to also. . . . But I have so many other things to do," Price replied. "Population genetics is not my main work, as you know. But perhaps I should pray, see if I am mistaken."

ERNST MAYR

Darwin's Influence on Modern Thought

FROM *SCIENTIFIC AMERICAN*

Clearly, our conception of the world and our place in it is, at the beginning of the 21st century, drastically different from the zeitgeist at the beginning of the 19th century. But no consensus exists as to the source of this revolutionary change. Karl Marx is often mentioned; Sigmund Freud has been in and out of favor; Albert Einstein's biographer Abraham Pais made the exuberant claim that Einstein's theories "have profoundly changed the way modern men and women think about the phenomena of inanimate nature." No sooner had Pais said this, though, than he recognized the exaggeration. "It would actually be better to say 'modern scientists' than 'modern men and women,' " he wrote, because one needs schooling in the physicist's style of thought and mathematical techniques to appreciate Einstein's contributions in their fullness. Indeed, this limitation is true for all the extraordinary theories of modern physics, which have had little impact on the way the average person apprehends the world.

The situation differs dramatically with regard to concepts in biology. Many biological ideas proposed during the past 150 years stood in stark conflict with what everybody assumed to be true. The acceptance of these ideas required an ideological revolution. And no biologist has been responsible for more—and for more drastic—modifications of the average person's worldview than Charles Darwin.

Darwin's accomplishments were so many and so diverse that it is useful to distinguish three fields to which he made major contributions: evolutionary biology; the philosophy of science; and the modern zeitgeist. Although I will be focusing on this last domain, for the sake of completeness I will put forth a short overview of his contributions—particularly as they inform his later ideas—to the first two areas.

A Secular View of Life

DARWIN FOUNDED a new branch of life science, evolutionary biology. Four of his contributions to evolutionary biology are especially important, as they held considerable sway beyond that discipline. The first is the non-constancy of species, or the modern conception of evolution itself. The second is the notion of branching evolution, implying the common descent of all species of living things on earth from a single unique origin. Up until 1859, all evolutionary proposals, such as that of naturalist Jean-Baptiste Lamarck, instead endorsed linear evolution, a teleological march toward greater perfection that had been in vogue since Aristotle's concept of *Scala Naturae,* the chain of being. Darwin further noted that evolution must be gradual, with no major breaks or discontinuities. Finally, he reasoned that the mechanism of evolution was natural selection.

These four insights served as the foundation of Darwin's founding of a new branch of the philosophy of science, a philosophy of biology. Despite the passing of a century before this new branch of philosophy fully developed, its eventual form is based on Darwinian concepts. For example, Darwin introduced historicity into science. Evolutionary biology, in contrast with physics and chemistry, is a historical science—the evolutionist attempts to explain events and processes that have already taken place. Laws and experiments are inappropriate techniques for the explication of such events and processes. Instead one constructs a historical narrative, consisting of a tentative reconstruction of the particular scenario that led to the events one is trying to explain.

For example, three different scenarios have been proposed for the sudden extinction of the dinosaurs at the end of the Cretaceous: a devastating epidemic; a catastrophic change of climate; and the impact of an asteroid, known as the Alvarez theory. The first two narratives were ultimately refuted by evidence incompatible with them. All the known facts, however, fit the Alvarez theory, which is now widely accepted. The testing of historical narratives implies that the wide gap between science and the humanities that so troubled physicist C. P. Snow is actually nonexistent—by virtue of its methodology and its acceptance of the time factor that makes change possible, evolutionary biology serves as a bridge.

The discovery of natural selection, by Darwin and Alfred Russel Wallace, must itself be counted as an extraordinary philosophical advance. The principle remained unknown throughout the more than 2,000-year history of philosophy ranging from the Greeks to Hume, Kant and the Victorian era. The concept of natural selection had remarkable power for explaining directional and adaptive changes. Its nature is simplicity itself. It is not a force like the forces described in the laws of physics; its mechanism is simply the elimination of inferior individuals. This process of nonrandom elimination impelled Darwin's contemporary, philosopher Herbert Spencer, to describe evolution with the now familiar term "survival of the fittest." (This description was long ridiculed as circular reasoning: "Who are the fittest? Those who survive." In reality, a careful analysis can usually determine why certain individuals fail to thrive in a given set of conditions.)

The truly outstanding achievement of the principle of natural selection is that it makes unnecessary the invocation of "final causes"—that is, any teleological forces leading to a particular end. In fact, nothing is predetermined. Furthermore, the objective of selection even may change from one generation to the next, as environmental circumstances vary.

A diverse population is a necessity for the proper working of natural selection. (Darwin's success meant that typologists, for whom all members of a class are essentially identical, were left with an untenable viewpoint.) Because of the importance of variation, natural selection should be considered a two-step process: the production of abundant variation is followed by the elimination of inferior individuals. This latter step is directional. By adopting natural selection, Darwin settled the several-thousand-year-old argument among philosophers over chance or necessity. Change on the earth is the result of both, the first step being dominated by randomness, the second by necessity.

Darwin was a holist: for him the object, or target, of selection was primarily the individual as a whole. The geneticists, almost from 1900 on, in a rather reductionist spirit preferred to consider the gene the target of evolution. In the past 25 years, however, they have largely returned to the Darwinian view that the individual is the principal target.

For 80 years after 1859, bitter controversy raged as to which of four competing evolutionary theories was valid. "Transmutation" was the establishment of a new species or new type through a single mutation, or saltation.

"Orthogenesis" held that intrinsic teleological tendencies led to transformation. Lamarckian evolution relied on the inheritance of acquired characteristics. And now there was Darwin's variational evolution, through natural selection. Darwin's theory clearly emerged as the victor during the evolutionary synthesis of the 1940s, when the new discoveries in genetics were married with taxonomic observations concerning systematics, the classification of organisms by their relationships. Darwinism is now almost unanimously accepted by knowledgeable evolutionists. In addition, it has become the basic component of the new philosophy of biology.

A most important principle of the new biological philosophy, undiscovered for almost a century after the publication of *On the Origin of Species,* is the dual nature of biological processes. These activities are governed both by the universal laws of physics and chemistry and by a genetic program, itself the result of natural selection, which has molded the genotype for millions of generations. The causal factor of the possession of a genetic program is unique to living organisms, and it is totally absent in the inanimate world. Because of the backward state of molecular and genetic knowledge in his time, Darwin was unaware of this vital factor.

Another aspect of the new philosophy of biology concerns the role of laws. Laws give way to concepts in Darwinism. In the physical sciences, as a rule, theories are based on laws; for example, the laws of motion led to the theory of gravitation. In evolutionary biology, however, theories are largely based on concepts such as competition, female choice, selection, succession and dominance. These biological concepts, and the theories based on them, cannot be reduced to the laws and theories of the physical sciences. Darwin himself never stated this idea plainly. My assertion of Darwin's importance to modern thought is the result of an analysis of Darwinian theory over the past century. During this period, a pronounced change in the methodology of biology took place. This transformation was not caused exclusively by Darwin, but it was greatly strengthened by developments in evolutionary biology. Observation, comparison and classification, as well as the testing of competing historical narratives, became the methods of evolutionary biology, outweighing experimentation.

I do not claim that Darwin was single-handedly responsible for all the intellectual developments in this period. Much of it, like the refutation of French mathematician and physicist Pierre-Simon Laplace's determinism, was "in the air." But Darwin in most cases either had priority or promoted the new views most vigorously.

The Darwinian Zeitgeist

A 21st-century person looks at the world quite differently than a citizen of the Victorian era did. This shift had multiple sources, particularly the incredible advances in technology. But what is not at all appreciated is the great extent to which this shift in thinking indeed resulted from Darwin's ideas.

Remember that in 1850 virtually all leading scientists and philosophers were Christian men. The world they inhabited had been created by God, and as the natural theologians claimed, He had instituted wise laws that brought about the perfect adaptation of all organisms to one another and to their environment. At the same time, the architects of the scientific revolution had constructed a world-view based on physicalism (a reduction to spatiotemporal things or events or their properties), teleology, determinism and other basic principles. Such was the thinking of Western man prior to the 1859 publication of *On the Origin of Species.* The basic principles proposed by Darwin would stand in total conflict with these prevailing ideas.

First, Darwinism rejects all supernatural phenomena and causations. The theory of evolution by natural selection explains the adaptedness and diversity of the world solely materialistically. It no longer requires God as creator or designer (although one is certainly still free to believe in God even if one accepts evolution). Darwin pointed out that creation, as described in the Bible and the origin accounts of other cultures, was contradicted by almost any aspect of the natural world. Every aspect of the "wonderful design" so admired by the natural theologians could be explained by natural selection. (A closer look also reveals that design is often not so wonderful—see "Evolution and the Origins of Disease," by Randolph M. Nesse and George C. Williams; *Scientific American,* November 1998.) Eliminating God from science made room for strictly scientific explanations of all natural phenomena; it gave rise to positivism; it produced a powerful intellectual and spiritual revolution, the effects of which have lasted to this day.

Second, Darwinism refutes typology. From the time of the Pythagoreans and Plato, the general concept of the diversity of the world emphasized its invariance and stability. This viewpoint is called typology, or essentialism. The seeming variety, it was said, consisted of a limited number of natural kinds (essences or types), each one forming a class. The members of each class were thought to be identical, constant, and sharply separated from the members of other essences.

Variation, in contrast, is nonessential and accidental. A triangle illustrates essentialism: all triangles have the same fundamental characteristics and are sharply delimited against quadrangles or any other geometric figures. An intermediate between a triangle and a quadrangle is inconceivable. Typological thinking, therefore, is unable to accommodate variation and gives rise to a misleading conception of human races. For the typologist, Caucasians, Africans, Asians or Inuits are types that conspicuously differ from other human ethnic groups. This mode of thinking leads to racism. (Although the ignorant misapplication of evolutionary theory known as "social Darwinism" often gets blamed for justifications of racism, adherence to the disproved essentialism preceding Darwin in fact can lead to a racist viewpoint.)

Darwin completely rejected typological thinking and introduced instead the entirely different concept now called population thinking. All groupings of living organisms, including humanity, are populations that consist of uniquely different individuals. No two of the six billion humans are the same. Populations vary not by their essences but only by mean statistical differences. By rejecting the constancy of populations, Darwin helped to introduce history into scientific thinking and to promote a distinctly new approach to explanatory interpretation in science.

Third, Darwin's theory of natural selection made any invocation of teleology unnecessary. From the Greeks onward, there existed a universal belief in the existence of a teleological force in the world that led to ever greater perfection. This "final cause" was one of the causes specified by Aristotle. After Kant, in the *Critique of Judgment,* had unsuccessfully attempted to describe biological phenomena with the help of a physicalist Newtonian explanation, he then invoked teleological forces. Even after 1859, teleological explanations (orthogenesis) continued to be quite popular in evolutionary biology. The acceptance of the *Scala Naturae* and the explanations of natural theology were other manifestations of the popularity of teleology. Darwinism swept such considerations away.

(The designation "teleological" actually applied to various different phenomena. Many seemingly end-directed processes in inorganic nature are the simple consequence of natural laws—a stone falls or a heated piece of metal cools because of laws of physics, not some end-directed process. Processes in living organisms owe their apparent goal-directedness to the operation of an inborn genetic or acquired program. Adapted systems, such as the heart or kidneys, may engage in activities that can be considered goal seeking, but the systems themselves were acquired during evolution and are continuously fine-tuned by natural selection. Finally, there was a belief in cosmic teleology, with a purpose and predetermined goal ascribed to everything in nature. Modern science, however, is unable to substantiate the existence of any such cosmic teleology.)

Fourth, Darwin does away with determinism. Laplace notoriously boasted that a complete knowledge of the current world and all its processes would enable him to predict the future to infinity. Darwin, by comparison, accepted the universality of randomness and chance throughout the process of natural selection. (Astronomer and philosopher John Herschel referred to natural selection contemptuously as "the law of the higgledy-piggledy.") That chance should play an important role in natural processes has been an unpalatable thought for many physicists. Einstein expressed this distaste in his statement, "God does not play dice." Of course, as previously mentioned, only the first step in natural selection, the production of variation, is a matter of chance. The character of the second step, the actual selection, is to be directional.

Despite the initial resistance by physicists and philosophers, the role of contingency and chance in natural processes is now almost universally acknowledged. Many biologists and philosophers deny the existence of universal laws in biology and suggest that all regularities be stated in probabilistic terms, as nearly all so-called biological laws have exceptions. Philosopher of science Karl Popper's famous test of falsification therefore cannot be applied in these cases.

Fifth, Darwin developed a new view of humanity and, in turn, a new anthropocentrism. Of all of Darwin's proposals, the one his contemporaries found most difficult to accept was that the theory of common descent applied to Man. For theologians and philosophers alike, Man was a creature above and apart from other living beings. Aristotle, Descartes and Kant agreed on this sentiment, no matter how else their thinking diverged. But biologists Thomas Huxley and Ernst Haeckel revealed through rigorous comparative anatomical study that humans and living apes clearly had common ancestry, an assessment that has never again been seriously questioned in science. The application of the theory of common descent to Man deprived man of his former unique position.

Ironically, though, these events did not lead to an end to anthropocentrism. The study of man showed that, in spite of his descent, he is indeed unique among all organisms. Human intelligence is unmatched by that of any other creature. Humans are the only animals with true language, including grammar and syntax. Only humanity, as Darwin emphasized, has developed genuine ethical systems. In addition, through high intelligence, language and long parental care, humans are the only creatures to have created a rich culture. And by these means, humanity has attained, for better or worse, an unprecedented dominance over the entire globe.

Sixth, Darwin provided a scientific foundation for ethics. The question is frequently raised—and usually rebuffed—as to whether evolution adequately explains healthy human ethics. Many wonder how, if selection rewards the individual only for behavior that enhances his own survival and reproductive success, such pure selfishness can lead to any sound ethics. The widespread thesis of social Darwinism, promoted at the end of the 19th century by Spencer, was that evolutionary explanations were at odds with the development of ethics.

We now know, however, that in a social species not only the individual must be considered—an entire social group can be the target of selection. Darwin applied this reasoning to the human species in 1871 in *The Descent of Man.* The survival and prosperity of a social group depends to a large extent on the harmonious cooperation of the members of the group, and this behavior must be based on altruism. Such altruism, by furthering the survival and prosperity of the group, also indirectly benefits the fitness of the group's individuals. The result amounts to selection favoring altruistic behavior.

Kin selection and reciprocal helpfulness in particular will be greatly favored in a social group. Such selection for altruism has been demonstrated in recent years to be widespread among many other social animals. One can then perhaps encapsulate the relation between ethics and evolution by saying that a propensity for altruism and harmonious cooperation in social groups *is* favored by natural selection. The old thesis of social Darwinism—strict selfishness—was based on an incomplete understanding of animals, particularly social species.

The Influence of New Concepts

LET ME NOW TRY to summarize my major findings. No educated person any longer questions the validity of the so-called theory of evolution, which we now know to be a simple fact. Likewise, most of Darwin's particular theses have been fully confirmed, such as that of common descent, the gradualism of evolution, and his explanatory theory of natural selection.

I hope I have successfully illustrated the wide reach of Darwin's ideas. Yes, he established a philosophy of biology by introducing the time factor, by demonstrating the importance of chance and contingency, and by showing that theories in evolutionary biology are based on concepts rather than laws. But furthermore—and this is perhaps Darwin's greatest contribution—he developed a set of new principles that influence the thinking of every person: the living world, through evolution, can be explained without recourse to supernaturalism; essentialism or typology is invalid, and we must adopt population thinking, in which all individuals are unique (vital for education and the refutation of racism); natural selection, applied to social groups, is indeed sufficient to account for the origin and maintenance of altruistic ethical systems; cosmic teleology, an intrinsic process leading life automatically to ever greater perfection, is fallacious, with all seemingly teleological phenomena explicable by purely material processes; and determinism is thus repudiated, which places our fate squarely in our own evolved hands.

To borrow Darwin's phrase, there is grandeur in this view of life. New modes of thinking have been, and are being, evolved. Almost every component in modern man's belief system is somehow affected by Darwinian principles.

GREG CRITSER

Let Them Eat Fat

FROM *HARPER'S MAGAZINE*

Not long ago, a group of doctors, nurses, and medical technicians wheeled a young man into the intensive care unit of Los Angeles County-USC Medical Center, hooked him to a ganglia of life-support systems—pulse and respiration monitors, a breathing apparatus, and an IV line—then stood back and collectively stared. I was there visiting an ailing relative, and I stared, too.

Here, in the ghastly white light of modern American medicine, writhed a real-life epidemiological specter: a 500-pound twenty-two-year-old. The man, whom I'll call Carl, was propped up at a 45-degree angle, the better to be fed air through a tube, and lay there nude, save for a small patch of blood-spotted gauze stuck to his lower abdomen, where surgeons had just labored to save his life. His eyes darted about in abject fear. "Second time in three months," his mother blurted out to me as she stood watching in horror. "He had two stomach staplings, and they both came apart. Oh my God, my boy . . ." Her boy was suffocating in his own fat.

I was struck not just by the spectacle but by the truth of the mother's comment. This *was* a boy—one buried in years of bad health, relative poverty, a sedentary lifestyle, and a high-fat diet, to be sure, but a boy nonetheless. Yet how surprised should I have been? That obesity, particularly among the young and the poor, is spinning out of control is hardly a secret. It is, in fact, something that most Americans can agree upon. Along with depression, heart disease, and cancer, obesity is yet another chew in our daily rumination about health and fitness, morbidity and mortality. Still, even in dot-com America, where statistics fly like arrows, the numbers are astonishing. Consider:

- Today, one fifth of all Americans are obese, meaning that they have a body mass index, or BMI, of more than 30. (BMI is a universally recognized cross-measure of weight for height and stature.) The epidemiological figures on chronic corpulence are so unequivocal that even the normally reticent dean of American obesity studies, the University of Colorado's James O. Hill, says that if obesity is left unchecked almost all Americans will be overweight within a few generations. "Becoming obese," he told the *Arizona Republic*, "is a normal response to the American environment."
- Children are most at risk. At least 25 percent of all Americans now under age nineteen are overweight or obese. In 1998, Dr. David Satcher, the new U.S. surgeon general, was moved to declare childhood obesity to be epidemic. "Today," he told a group of federal bureaucrats and policymakers, "we see a nation of young people seriously at risk of starting out obese and dooming themselves to the difficult task of overcoming a tough illness."
- Even among the most careful researchers these days, "epidemic" is the term of choice when it comes to talk of fat, particularly fat children. As William Dietz, the director of nutrition at the Centers for Disease Control, said last year, "This is an epidemic in the U.S. the likes of which we have not had before in chronic disease." The cost to the general public health budget by 2020 will run into the hundreds of billions, making HIV look, economically, like a bad case of the flu.

Yet standing that day in the intensive care unit, among the beepers and buzzers and pumps, epidemic was the last thing on my mind. Instead I felt heartbreak, revulsion, fear, sadness—and then curiosity: Where did this boy come from? Who and what had made him? How is it that we Americans, perhaps the most health-conscious of any people in the history of the world, and certainly the richest, have come to preside over the deadly fattening of our youth?

THE BEGINNING of an answer came one day in the fall of 1999, in the same week that the Spanish language newspaper *La Opinión* ran a story headlined "Diabetes epidemia en latinos," when I attended the opening of the newest Krispy Kreme doughnut store in Los Angeles. It was, as they say in marketing circles, a "resonant" event, replete with around-the-block lines, celebrity news anchors, and stern cops directing traffic. The store, located in the heart of the San Fernando Valley's burgeoning Latino population, pulsed with excitement. In one corner stood the new store's manager, a young Anglo fellow, accompanied by a Krispy Kreme publicity director. Why had Krispy Kreme decided to locate here? I asked.

"See," the manager said, brushing a crumb of choco-glaze from his fingers, "the idea is simple—accessible but not convenient. The idea is to make the store accessible—easy to get into and out of from the street—but just a tad away from the—eh, mainstream so as to make sure that the customers are presold and very intent before they get here," he said, betraying no doubts about the company's marketing formula. "We want them intent to get at least a dozen before they even think of coming in."

But why this slightly non-mainstream place?

"Because it's obvious . . ." He gestured to the stout Mayan doñas queuing around the building. "We're looking for all the bigger families."

Bigger in size?

"Yeah." His eyes rolled, like little glazed crullers. "*Bigger in size.*"

Of course, fast food and national restaurant chains like Krispy Kreme that serve it have long been the object of criticism by nutritionists and dietitians. Despite the attention, however, fast-food companies, most of them publicly owned and sprinkled into the stock portfolios of many striving Americans (including mine and perhaps yours), have grown more aggressive in their targeting of poor inner-city communities. One of every four hamburgers sold by the good folks at McDonald's, for example, is now purchased by inner-city consumers who, disproportionately, are young black men.

In fact, it was the poor, and their increasing need for cheap meals consumed outside the home, that fueled the development of what may well be the most important fast-food innovation of the past twenty years, the sales gimmick known as "supersizing." At my local McDonald's, located in a lower-middle-income area of Pasadena, California, the supersize bacchanal goes into high gear at about five P.M., when the various urban caballeros, drywalleros, and jardineros get off work and head for a quick bite. Mixed in is a sizable element of young black kids traveling between school and home, their economic status apparent by the fact that they've walked instead of driven. Customers are cheerfully encouraged to "supersize your meal!" by signs saying, "If we don't recommend a supersize, the supersize is free!" For an extra seventy-nine cents, a kid ordering a cheeseburger, small fries, and a small Coke will get said cheeseburger plus a supersize Coke (42 fluid ounces versus 16, with free refills) and a supersize order of french fries (more than double the weight of a regular order). Suffice it to say that consumption of said meals is fast and, in almost every instance I observed, very complete.

But what, metabolically speaking, has taken place? The total caloric content of the meal has been jacked up from 680 calories to more than 1,340 calories. According to the very generous U.S. dietary guidelines, 1,340 calories represent more than half of a teenager's recommended daily caloric consumption, and the added calories themselves are protein-poor but fat- and carbohydrate-rich. Completing this jumbo dietetic horror is the fact that the easy availability of such huge meals arrives in the same years in which physical activity among teenage boys and girls drops by about half.

Now consider the endocrine warfare that follows. The constant bombing of the pancreas by such a huge hit of sugars and fats can eventually wear out the organ's insulin-producing "islets," leading to diabetes and its inevitable dirge of woes: kidney, eye, and nerve damage; increased risk of heart disease; even stroke. The resulting sugar-induced hyperglycemia in many of the obese wreaks its own havoc in the form of glucose toxicity, further debilitating nerve endings and arterial walls. For the obese and soon to be obese, it is no overstatement to say that after supersized teen years the pancreas may never be the same. Some 16 million Americans suffer from Type 2 diabetes, a third of them unaware of their condition. Today's giggly teen burp may well be tomorrow's aching neuropathic limb.

Diabetes, by the way, is just the beginning of what's possible. If childhood obesity truly is "an epidemic in the U.S. the likes of which we have not had before in chronic disease," then places like McDonald's and Winchell's Donut stores, with their endless racks of glazed and creamy goodies, are the San Francisco bathhouses of said epidemic, the places where the high-risk population indulges in high-risk behavior. Although open around the clock, the Winchell's near my house doesn't get rolling until seven in the morning, the Spanish-language talk shows frothing in the background while an ambulance light whirls atop the Coke dispenser. Inside, Mami placates Miguelito with a giant apple fritter. Papi tells a joke and pours ounce upon ounce of sugar and cream into his 20-ounce coffee. Viewed through the lens of obesity, as I am inclined to do, the scene is not so *feliz.* The obesity rate for Mexican-American children is shocking. Between the ages of five and eleven, the rate for girls is 27 percent; for boys, 23 percent. By fourth grade the rate for girls peaks at 32 percent, while boys top out at 43 percent. Not surprisingly, obesity-related disorders are everywhere on display at Winchell's, right before my eyes—including fat kids who limp, which can be a symptom of Blount's disease (a deformity of the tibia) or a sign of slipped capital femoral epiphysis (an orthopedic abnormality brought about by weight-induced dislocation of the femur bone). Both conditions are progressive, often requiring surgery.

The chubby boy nodding in the corner, waiting for his Papi to finish his *café,* is likely suffering from some form of sleep apnea; a recent study of forty-one children with severe obesity revealed that a third had the condition and that another third presented with clinically abnormal sleep patterns. Another recent study indicated that "obese children with obstructive sleep apnea demonstrate clinically significant decrements in learning and memory function." And the lovely but very chubby little girl tending to her schoolbooks? Chances are she will begin puberty before the age of ten, launching her into a lifetime of endocrine bizarreness that not only will be costly to treat but will be emotionally devastating as well. Research also suggests that weight gain can lead to the development of pseudotumor cerebri, a brain tumor most common in females. A recent review of 57 patients with the tumor-like condition revealed that 90 percent were obese. This little girl's chances of developing other neurological illnesses are profound as well. And she may already have gallstones: obesity accounts for up to 33 percent of all gallstones observed in children. She is ten times more likely than her non-obese peers to develop high blood pressure, and she is increasingly likely to contract Type 2 diabetes, obesity being that disease's number-one risk factor.

Of course, if she is really lucky, that little girl could just be having a choco-sprinkles doughnut on her way to school.

What about poor rural whites? Studying children in an elementary school in a low-income town in eastern Kentucky, the anthropologist Deborah Crooks was astonished to find stunting and obesity not just present but prevalent. Among her subjects, 13 percent of girls exhibited notable stunting; 33 percent of all kids were significantly overweight; and 13 percent of the children were obese—21 percent of boys and 9 percent of girls. A sensitive, elegant writer, Crooks drew from her work three important conclusions: One, that poor kids in the United States often face the same evolutionary nutritional pressures as those in newly industrializing nations, where traditional diets are replaced by high-fat diets and where labor-saving technology reduces physical activity. Second, Crooks found that "height and weight are cumulative measures of growth . . . reflecting a sum total of environmental experience over time." Last, and perhaps most important, Crooks concluded that while stunting can be partially explained by individual household conditions—income, illness, education, and marital status—obesity "may be more of a community-related phenomenon." Here the economic infrastructure—safe playgrounds, access to high-quality, low-cost food, and transportation to play areas—was the key determinant of physical-activity levels.

Awareness of these national patterns of destruction, of course, is a key reason why Eli Lilly & Co., the $75 billion pharmaceutical company, is now building the largest factory dedicated to the production of a single drug in industry history. That drug is insulin. Lilly's sales of insulin products totaled $357 million in the third quarter of 1999, a 24 percent increase over the previous third quarter. Almost every leading pharmaceutical conglomerate has like-minded ventures under way, with special emphasis on pill-form treatments for non-insulin-dependent forms of the disease. Pharmaceutical companies that are not seeking to capture some portion of the burgeoning market are bordering on fiduciary mismanagement. Said James Kappel of Eli Lilly, "You've got to be in diabetes."

Wandering home from my outing, the wondrous smells of frying foods wafting in the air, I wondered why, given affluent America's outright fetishism about diet and health, those whose business it is to care—the media, the academy, public-health workers, and the government—do almost nothing. The answer, I suggest, is that in almost every public-health arena, the need to address obesity as a class issue—one that transcends the inevitable divisiveness of race and gender—has been blunted by bad logic, vested interests, academic cant, and ideological chauvinism.

Consider a 1999 story in the *New York Times* detailing the rise in delivery-room mortality among young African-American mothers. The increases were attributed to a number of factors—diabetes, hypertension, drug and alcohol abuse—but the primary factor of obesity, which can foster both diabetes and hypertension, was mentioned only in passing. Moreover, efforts to understand and publicize the socioeconomic factors of the deaths have been thwarted. When Dr. Janet Mitchell, a New York obstetrician charged with reviewing several recent maternal mortality studies, insisted that socioeconomics were the issue in understanding the "racial gap" in maternal mortality, she was unable to get government funding for the work. "We need to back away from the medical causes," she told the *Times,* clearly exasperated, "and begin to take a much more ethnographic, anthropological approach to this tragic outcome."

In another example, a 1995 University of Arizona study reported that young black girls, who are more inclined toward obesity than white girls, were also far less likely to hold "bad body images" about themselves. The slew of news articles and TV reports that followed were nothing short of jubilant, proclaiming the "good news." As one commentator I watched late one evening announced, "Here is one group of girls who couldn't care less about looking like Kate Moss!" Yet no one mentioned the long-term effects of unchecked weight gain. Apparently, when it comes to poor black girls the media would rather that they risk diabetes than try to look like models.

"That's the big conundrum, as they always say," Richard MacKenzie, a physician who treats overweight and obese girls in downtown L.A., told me recently. "No one wants to overemphasize the problems of being fat to these girls, for fear of creating body-image problems that might lead to anorexia and bulimia." Speaking anecdotally, he said that "the problem is that for every one affluent white anorexic you create by 'overemphasizing' obesity, you foster ten obese poor girls by downplaying the severity of the issue." Judith Stern, a professor of nutrition and internal medicine at UC Davis, is more blunt. "The number of kids with eating disorders is positively dwarfed by the number with obesity. It sidesteps the whole class issue. We've got to stop that and get on with the real problem."

Moreover, such sidestepping denies poor minority girls a principal, if sometimes unpleasant, psychological incentive to lose weight: that of social stigma. Only recently has the academy come to grapple with this. Writing in a recent issue of the *International Journal of Obesity,* the scholar Susan Averett looked at the hard numbers: 44 percent of African-American women weigh more than 120 percent of their recommended body weight yet are less likely than whites to perceive themselves as overweight.[1] Anglo women, poor and otherwise, registered higher anxiety about fatness and experienced far fewer cases of chronic obesity. "Social stigma may serve to control obesity among white women," Averett reluctantly concluded. "If so, physical and emotional effects of greater pressure to be thin must be weighed against reduced health risks associated with overweight and obesity." In other words, maybe a few more black Kate Mosses might not be such a bad thing.

While the so-called fat acceptance movement, a very vocal minority of super-obese female activists, has certainly played a role in the tendency to deny the need to promote healthy thinness, the real culprits have been those with true cultural power, those in the academy and the publishing industry who have the ability to shape public opinion. Behind much of their reluctance to face facts is the lingering influence of the 1978 bestseller, *Fat Is a Feminist Issue,* in which Susie Orbach presented a nuanced, passionate look at female compulsive eating and its roots in patriarchal culture. But although Orbach's observations were keen, her conclusions were often wishful, narcissistic, and sometimes just wrong. "Fat is a social disease, and fat is a feminist issue," Orbach wrote. "Fat is not about self-control or lack of will power. . . . It is a response to the inequality of the sexes."[2]

Perhaps so, if one is a feminist, and if one is struggling with an eating disorder, and if one is, for the most part, affluent, well-educated, and politically aware. But obesity itself is preeminently an issue of class, not of ethnicity, and certainly not of gender. True, the disease may be refracted though its concentrations in various demographic subgroupings—in Native Americans, in Latinos, in African Americans, and even in some Pacific Island Americans—but in study after study, the key adjective is *poor:* poor African Americans, poor Latinos, poor whites, poor women, poor children, poor Latino children, etc. From the definitive *Handbook of Obesity:* "In heterogeneous and affluent societies like the United States, there is a strong inverse correlation of social class and obesity, particularly for females." From *Annals of Epidemiology:* "In white girls . . . both TV viewing and obesity were strongly inversely associated with household income as well as with parental education."

Yet class seems to be the last thing on the minds of some of our better social thinkers. Instead, the tendency of many in the academy is to fetishize or "postmodernize" the problem. Cornell University professor Richard Klein, for example, proposed in his 1996 book, *Eat Fat,* "Try this for six weeks: Eat fat." (Klein's mother did and almost died from sleep apnea, causing Klein to reverse himself in his epilogue, advising readers: "Eat rice.") The identity politics of fat, incidentally, can cut the other way. To the French, the childhood diet has long been understood as a serious medical issue directly affecting the future of the nation. The concern grew directly from late-nineteenth-century health issues in French cities and the countryside, where tuberculosis had winnowed the nation's birth rate below that of the other European powers. To deal with the problem, a new science known as *puériculture* emerged to educate young mothers about basic health and nutrition practices. Long before Americans and the British roused themselves from the torpor of Victorian chub, the French undertook research into proper dietary and weight controls for the entire birth-to-adolescence growth period. By the early 1900s, with birth rates (and birth weights) picking up, the *puériculture* movement turned its attention to childhood obesity. Feeding times were to be strictly maintained; random snacks were unhealthy for the child, regardless of how "natural" it felt for a mother to indulge her young. Kids were weighed once a week. All meals were to be supervised by an adult. As a result, portion control—perhaps the one thing that modern obesity experts can agree upon as a reasonable way to prevent the condition—very early became institutionalized in modern France. The message that too much food is bad still resounds in French child rearing, and as a result France has a largely lean populace.

WHAT ABOUT THE SO-CALLED Obesity Establishment, that web of researchers, clinicians, academics, and government health officials charged with finding ways to prevent the disease? Although there are many committed individuals in this group, one wonders just how independently minded they are. Among the sponsors for the 1997 annual conference of the North American Association for the Study of Obesity, the premier medical think tank on the subject, were the following: the Coca-Cola Company, Hershey Foods, Kraft Foods, and, never to be left out, Slim Fast Foods. Another sponsor was Knoll Pharmaceuticals, maker of the new diet drug Meridia. Of course, in a society where until recently tobacco companies sponsored fitness pageants and Olympic games, sponsorship hardly denotes corruption in the most traditional sense. One would be hard-pressed to prove any kind of censorship, but such underwriting effectively defines the parameters of public discussion. Everybody winks or blinks at the proper moment, then goes on his or her way.

Once upon a time, however, the United States possessed visionary leadership in the realm of childhood fitness. Founded in 1956, the President's Council on Youth Fitness successfully laid down broad-based fitness goals for all youth and established a series of awards for those who excelled in the effort. The council spoke about obesity with a forthrightness that would be political suicide today, with such pointed slogans as "There's no such thing as stylishly stout" and "Hey kid, if you see yourself in this picture, you need help."

By the late 1980s and early 1990s, however, new trends converged to undercut the council's powers of moral and cultural suasion. The ascendancy of cultural relativism led to a growing reluctance to be blunt about fatness, and, aided and abetted by the fashion industry's focus on baggy, hip-hop-style clothes, it became possible to be "stylishly stout." Fatness, as celebrated on rap videos, was now equated with wealth and power, with identity and agency, not with clogging the heart or being unable to reach one's toes. But fat inner-city black kids and the suburban kids copying them are even more disabled by their obesity. The only people who benefit from kids being "fat" are the ones running and owning the clothing, media, food, and drug companies. In upscale corporate America, meanwhile, being fat is taboo, a surefire career-killer. If you can't control your own contours, goes the logic, how can you control a budget or a staff? Look at the glossy business and money magazines with their cooing profiles of the latest genius entrepreneurs: to the man, and the occasional woman, no one, I mean *no one,* is fat.

Related to the coolification of homeboyish fat—perhaps forcing its new status—is the simple fact that it's hard for poor children to find opportunities to exercise. Despite our obsession with professional sports, many of today's disadvantaged youth have fewer opportunities than ever to simply shoot baskets or kick a soccer ball. Various measures to limit state spending and taxing, among them California's debilitating Proposition 13, have gutted school-based physical-education classes. Currently, only one state, Illinois, requires daily physical education for all grades K–12, and only 19 percent of high school students nationwide are active for twenty minutes a day, five days a week, in physical education. Add to this the fact that, among the poor, television, the workingman's baby sitter, is now viewed at least thirty-two hours a week. Participation in sports has always required an investment, but with the children of the affluent tucked away either in private schools or green suburbias, buying basketballs for the poor is not on the public agenda.

HUMAN NATURE and its lazy inclinations aside, what do America's affluent *get* out of keeping the poor so fat? The reasons, I'd suggest, are many. An unreconstructed Marxist might invoke simple class warfare, exploitation fought through stock ownership in giant fast-food firms. The affluent know that the stuff will kill them but need someone (else) to eat it so as to keep growing that retirement portfolio. A practitioner of vulgar social psychology might argue for "our" need for the "identifiable outsider." An economist would say that in a society as overly competitive as our own, the affluent have found a way to slow down the striving poor from inevitable nipping at their heels. A French semiotician might even say that with the poor the affluent have erected their own walking and talking "empire of signs." This last notion is perhaps not so far-fetched. For what do the fat, darker, exploited poor, with their unbridled primal appetites, have to offer us but a chance for we diet- and shape-conscious folk to live vicariously? Call it boundary envy. Or, rather, boundary-free envy. And yet, by living outside their boundaries, the poor live within ours; fat people do not threaten our way of life; their angers entombed in flesh, they are slowed, they are softened, they are *fed.*

Meanwhile, in the City of Fat Angels, we lounge through a slow-motion epidemic. Mami buys another apple fritter. Papi slams his second sugar and cream. Another young Carl supersizes and double supersizes, then supersizes again. Waistlines surge. Any minute now, the belt will run out of holes.

Notes

1. Certainly culture plays a role in the behavior of any subpopulation. Among black women, for example, obesity rates persist despite increases in income. A recent study by the National Heart, Lung, and Blood Institute concludes that obesity in black girls may be "a reflection of a differential social development in our society, wherein a certain lag period may need to elapse between an era when food availability is a concern to an era of affluence with no such concern." Other observers might assert that black women find affirmation for being heavy from black men, or believe themselves to be "naturally" heavier. Such assertions do not change mortality statistics.

2. At the edges of the culture, the inheritors of Susie Orbach's politics have created Web sites called FaT GIRL and Largesse: the Network for Size Esteem, which claim that "dieting kills" and instruct how to induce vomiting in diet centers as protest.

ANDREW SULLIVAN

The He Hormone

FROM *THE NEW YORK TIMES MAGAZINE*

It has a slightly golden hue, suspended in an oily substance and injected in a needle about half as thick as a telephone wire. I have never been able to jab it suddenly in my hip muscle, as the doctor told me to. Instead, after swabbing a small patch of my rump down with rubbing alcohol, I push the needle in slowly until all three inches of it are submerged. Then I squeeze the liquid in carefully, as the muscle often spasms to absorb it. My skin sticks a little to the syringe as I pull it out, and then an odd mix of oil and blackish blood usually trickles down my hip.

I am so used to it now that the novelty has worn off. But every now and again the weirdness returns. The chemical I am putting in myself is synthetic testosterone: a substance that has become such a metaphor for manhood that it is almost possible to forget that it has a physical reality. Twenty years ago, as it surged through my pubescent body, it deepened my voice, grew hair on my face and chest, strengthened my limbs, made me a man. So what, I wonder, is it doing to me now?

There are few things more challenging to the question of what the difference between men and women really is than to see the difference injected into your hip. Men and women differ biologically mainly because men produce 10 to 20 times as much testosterone as most women do, and this chemical, no one seriously disputes, profoundly affects physique, behavior, mood and self-understanding. To be sure, because human beings are also deeply socialized, the impact of this difference is refracted through the prism of our own history and culture. But biology, it is all too easy to forget, is at the root of this process. As more people use testosterone medically, as more use testosterone-based steroids in sports and recreation and as more research explores the behavioral effects of this chemical, the clearer the power of that biology is. It affects every aspect of our society, from high divorce rates and adolescent male violence to the exploding cults of bodybuilding and professional wrestling. It helps explain, perhaps better than any other single factor, why inequalities between men and women remain so frustratingly resilient in public and private life. This summer, when an easy-to-apply testosterone gel hits the market, and when more people experience the power of this chemical in their own bodies, its social importance, once merely implicit, may get even harder to ignore.

My own encounter with testosterone came about for a simple medical reason. I am H.I.V.-positive, and two years ago, after a period of extreme fatigue and weight loss, I had my testosterone levels checked. It turned out that my body was producing far less testosterone than it should have been at my age. No one quite knows why, but this is common among men with long-term H.I.V. The usual treatment is regular injection of artificial testosterone, which is when I experienced my first manhood supplement.

At that point I weighed around 165 pounds. I now weigh 185 pounds. My collar size went from a 15 to a 17½ in a few months; my chest went from 40 to 44. My appetite in every sense of that word expanded beyond measure. Going from napping two hours a day, I now rarely sleep in the daytime and have enough energy for daily workouts and a hefty work schedule. I can squat more than 400 pounds. Depression, once a regular feature of my life, is now a distant memory. I feel better able to recover from life's curveballs, more persistent, more alive. These are the long-term effects. They are almost as striking as the short-term ones.

Because the testosterone is injected every two weeks, and it quickly leaves the bloodstream, I can actually feel its power on almost a daily basis. Within hours, and at most a day, I feel a deep surge of energy. It is less edgy than a double espresso, but just as powerful. My attention span shortens. In the two or three days after my shot, I find it harder to concentrate on writing and feel the need to exercise more. My wit is quicker, my mind faster, but my judgment is more impulsive. It is not unlike the kind of rush I get before talking in front of a large audience, or going on a first date, or getting on an airplane, but it suffuses me in a less abrupt and more consistent way. In a word, I feel braced. For what? It scarcely seems to matter.

And then after a few days, as the testosterone peaks and starts to decline, the feeling alters a little. I find myself less reserved than usual, and more garrulous. The same energy is there, but it seems less directed toward action than toward interaction, less toward pride than toward lust. The odd thing is that, however much experience I have with it, this lust peak still takes me unawares. It is not like feeling hungry, a feeling you recognize and satiate. It creeps up on you. It is only a few days later that I look back and realize that I spent hours of the recent past socializing in a bar or checking out every potential date who came vaguely over my horizon. You realize more acutely than before that lust is a chemical. It comes; it goes. It waxes; it wanes. You are not helpless in front of it, but you are certainly not fully in control.

Then there's anger. I have always tended to bury or redirect my rage. I once thought this an inescapable part of my personality. It turns out I was wrong. Late last year, mere hours after a T shot, my dog ran off the leash to forage for a chicken bone left in my local park. The more I chased her, the more she ran. By the time I retrieved her, the bone had been consumed, and I gave her a sharp tap on her rear end. "Don't smack your dog!" yelled a burly guy a few yards away. What I found myself yelling back at him is not printable in this magazine, but I have never used that language in public before, let alone bellow it at the top of my voice. He shouted back, and within seconds I was actually close to hitting him. He backed down and slunk off. I strutted home, chest puffed up, contrite beagle dragged sheepishly behind me. It wasn't until half an hour later that I realized I had been a complete jerk and had nearly gotten into the first public brawl of my life. I vowed to inject my testosterone at night in the future.

That was an extreme example, but other, milder ones come to mind: losing my temper in a petty argument; innumerable traffic confrontations; even the occasional slightly too prickly column or e-mail flame-out. No doubt my previous awareness of the mythology of testosterone had subtly primed me for these feelings of irritation and impatience. But when I place them in the larger context of my new testosterone-associated energy, and of what we know about what testosterone tends to do to people, then it seems plausible enough to ascribe some of this increased edginess and self-confidence to that biweekly encounter with a syringe full of manhood.

TESTOSTERONE, ODDLY ENOUGH, is a chemical closely related to cholesterol. It was first isolated by a Dutch scientist in 1935 from mice testicles and successfully synthesized by the German biologist Adolf Butenandt. Although testosterone is often thought of as the definition of maleness, both men and women produce it. Men produce it in their testicles; women produce it in their ovaries and adrenal glands. The male body converts some testosterone to estradiol, a female hormone, and the female body has receptors for testosterone, just as the male body does. That's why women who want to change their sex are injected with testosterone and develop male characteristics, like deeper voices, facial hair and even baldness. The central biological difference between adult men and women, then, is not that men have testosterone and women don't. It's that men produce much, much more of it than women do. An average woman has 40 to 60 nanograms of testosterone in a deciliter of blood plasma. An average man has 300 to 1,000 nanograms per deciliter.

Testosterone's effects start early—really early. At conception, every embryo is female and unless hormonally altered will remain so. You need testosterone to turn a fetus with a Y chromosome into a real boy, to masculinize his brain and body. Men experience a flood of testosterone twice in their lives: in the womb about six weeks after conception and at puberty. The first fetal burst primes the brain and the body, endowing male fetuses with the instinctual knowledge of how to respond to later testosterone surges. The second, more familiar adolescent rush—squeaky voices, facial hair and all—completes the process. Without testosterone, humans would always revert to the default sex, which is female. The Book of Genesis is therefore exactly wrong. It isn't women who are made out of men. It is men who are made out of women. Testosterone, to stretch the metaphor, is Eve's rib.

The effect of testosterone is systemic. It engenders both the brain and the body. Apart from the obvious genital distinction, other differences between men's and women's bodies reflect this: body hair, the ratio of muscle to fat, upper-body strength and so on. But testosterone leads to behavioral differences as well. Since it is unethical to experiment with human embryos by altering hormonal balances, much of the evidence for this idea is based on research conducted on animals. A Stanford research group, for example, as reported in Deborah Blum's book *Sex on the Brain*, injected newborn female rats with testosterone. Not only did the female rats develop penises from their clitorises, but they also appeared fully aware of how to use them, trying to have sex with other females with merry abandon. Male rats who had their testosterone blocked after birth, on the other hand, saw their penises wither or disappear entirely and presented themselves to the female rats in a passive, receptive way. Other scientists, theorizing that it was testosterone that enabled male zebra finches to sing, injected mute female finches with testosterone. Sure enough, the females sang. Species in which the female is typically more aggressive, like hyenas in female-run clans, show higher levels of testosterone among the females than among the males. Female sea snipes, which impregnate the males, and leave them to stay home and rear the young, have higher testosterone levels than their mates. Typical "male" behavior, in other words, corresponds to testosterone levels, whether exhibited by chromosomal males or females.

Does this apply to humans? The evidence certainly suggests that it does, though much of the "proof" is inferred from accidents. Pregnant women who were injected with progesterone (chemically similar to testosterone) in the 1950's to avoid miscarriage had daughters who later reported markedly tomboyish childhoods. Ditto girls born with a disorder that causes their adrenal glands to produce a hormone like testosterone rather than the more common cortisol. The moving story, chronicled in John Colapinto's book *As Nature Made Him,* of David Reimer, who as an infant was surgically altered after a botched circumcision to become a girl, suggests how long-lasting the effect of fetal testosterone can be. Despite a ruthless attempt to socialize David as a girl, and to give him the correct hormonal treatment to develop as one, his behavioral and psychological makeup was still ineradicably male. Eventually, with the help of more testosterone, he became a full man again. Female-to-male transsexuals report a similar transformation when injected with testosterone. One, Susan/Drew Seidman, described her experience in *The Village Voice* last November. "My sex-drive went through the roof," Seidman recalled. "I felt like I had to have sex once a day or I would die. . . . I was into porn as a girl, but now I'm *really* into porn." For Seidman, becoming a man was not merely physical. Thanks to testosterone, it was also psychological. "I'm not sure I can tell you what makes a man a man," Seidman averred. "But I know it's not a penis."

The behavioral traits associated with testosterone are largely the cliché-ridden ones you might expect. The Big T correlates with energy, self-confidence, competitiveness, tenacity, strength and sexual drive. When you talk to men in testosterone therapy, several themes recur. "People talk about extremes," one man in his late 30's told me. "But that's not what testosterone does for me. It makes me think more clearly. It makes me think more positively. It's my Saint-John's-wort." A man in his 20's said: "Usually, I cycle up the hill to my apartment in 12th gear. In the days after my shot, I ride it easily in 16th." A 40-year-old executive who took testosterone for bodybuilding purposes told me: "I walk into a business meeting now and I just exude self-confidence. I know there are lots of other reasons for this, but my company has just exploded since my treatment. I'm on a roll. I feel capable of almost anything."

When you hear comments like these, it's no big surprise that strutting peacocks with their extravagant tails and bright colors are supercharged with testosterone and that mousy little male sparrows aren't. "It turned my life around," another man said. "I felt stronger—and not just in a physical sense. It was a deep sense of being strong, almost spiritually strong." Testosterone's antidepressive power is only marginally understood. It doesn't act in the precise way other antidepressants do, and it probably helps alleviate gloominess primarily by propelling people into greater activity and restlessness, giving them less time to think and reflect. (This may be one reason women tend to suffer more from depression than men.) Like other drugs, T can also lose potency if overused. Men who inject excessive amounts may see their own production collapse and experience shrinkage of their testicles and liver damage.

Individual effects obviously vary, and a person's internal makeup is affected by countless other factors—physical, psychological and external. But in this complex human engine, testosterone is gasoline. It revs you up. A 1997 study took testosterone samples from 125 men and 128 women and selected the 12 with the lowest levels of testosterone and the 15 with the highest. They gave them beepers, asked them to keep diaries and paged them 20 times over a four-day period to check on their actions, feelings, thoughts and whereabouts. The differences were striking. High-testosterone people "experienced more arousal and tension than those low in testosterone," according to the study. "They spent more time thinking, especially about concrete problems in the immediate present. They wanted to get things done and felt frustrated when they could not. They mentioned friends more than family or lovers."

Unlike Popeye's spinach, however, testosterone is also, in humans at least, a relatively subtle agent. It is not some kind of on-off switch by which men are constantly turned on and women off. For one thing, we all start out with different base-line levels. Some women may have remarkably high genetic T levels, some men remarkably low, although the male-female differential is so great that no single woman's T level can exceed any single man's, unless she, or he, has some kind of significant hormonal imbalance. For another, and this is where the social and political ramifications get complicated, testosterone is highly susceptible to environment. T levels can rise and fall depending on external circumstances—short term and long term. Testosterone is usually elevated in response to confrontational situations—a street fight, a marital spat, a presidential debate—or in highly charged sexual environments, like a strip bar or a pornographic Web site. It can also be raised permanently in continuously combative environments, like war, although it can also be suddenly lowered by stress.

Because testosterone levels can be measured in saliva as well as in blood, researchers like Alan Booth, Allan Mazur, Richard Udry and particularly James M. Dabbs, whose book *Heroes, Rogues and Lovers* will be out in the fall of 2000, have compiled quite a database on these variations. A certain amount of caution is advisable in interpreting the results of these studies. There is some doubt about the validity of onetime samples to gauge underlying testosterone levels. And most of the studies of the psychological effects of testosterone take place in culturally saturated environments, so that the difference between cause and effect is often extremely hard to disentangle. Nevertheless, the sheer number and scale of the studies, especially in the last decade or so, and the strong behavioral correlations with high testosterone, suggest some conclusions about the social importance of testosterone that are increasingly hard to gainsay.

TESTOSTERONE IS CLEARLY CORRELATED in both men and women with psychological dominance, confident physicality and high self-esteem. In most combative, competitive environments, especially physical ones, the person with the most T wins. Put any two men in a room together and the one with more testosterone will tend to dominate the interaction. Working women have higher levels of testosterone than women who stay at home, and the daughters of working women have higher levels of testosterone than the daughters of housewives. A 1996 study found that in lesbian couples in which one partner assumes the male, or "butch," role and another assumes the female, or "femme," role, the "butch" woman has higher levels of testosterone than the "femme" woman. In naval medical tests, midshipmen have been shown to have higher average levels of testosterone than plebes. Actors tend to have more testosterone than ministers, according to a 1990 study. Among 700 male prison inmates in a 1995 study, those with the highest T levels tended to be those most likely to be in trouble with the prison authorities and to engage in unprovoked violence. This is true among women as well as among men, according to a 1997 study of 87 female inmates in a maximum security prison. Although high testosterone levels often correlate with dominance in interpersonal relationships, it does not guarantee more social power. Testosterone levels are higher among blue-collar workers, for example, than among white-collar workers, according to a study of more than 4,000 former military personnel conducted in 1992. A 1998 study found that trial lawyers—with their habituation to combat, conflict and swagger—have higher levels of T than other lawyers.

The salient question, of course, is: How much of this difference in aggression and dominance is related to environment? Are trial lawyers naturally more testosteroned, and does that lead them into their profession? Or does the experience of the courtroom raise their levels? Do working women have naturally higher T levels, or does the prestige of work and power elevate their testosterone? Because of the limits of researching such a question, it is hard to tell beyond a reasonable doubt. But the social context clearly matters. It is even possible to tell who has won a tennis match not by watching the game, but by monitoring testosterone-filled saliva samples throughout. Testosterone levels rise for both players before the match. The winner of any single game sees his T production rise; the loser sees it fall. The ultimate winner experiences a postgame testosterone surge, while the loser sees a collapse. This is true even for people watching sports matches. A 1998 study found that fans backing the winning side in a college basketball game and a World Cup soccer match saw their testosterone levels rise; fans rooting for the losing teams in both games saw their own T levels fall. There is, it seems, such a thing as vicarious testosterone.

One theory to explain this sensitivity to environment is that testosterone was originally favored in human evolution to enable successful hunting and combat. It kicks in, like adrenaline, in anticipation of combat, mental or physical, and helps you prevail. But a testosterone crash can be a killer too. Toward the end of my two-week cycle, I can almost feel my spirits dragging. In the event of a just-lost battle, as Matt Ridley points out in his book *The Red Queen,* there's a good reason for this to occur. If you lose a contest with prey or a rival, it makes sense not to pick another fight immediately. So your body wisely prompts you to withdraw, filling your brain with depression and self-doubt. But if you have made a successful kill or defeated a treacherous enemy, your hormones goad you into further conquest. And people wonder why professional football players get into postgame sexual escapades and violence. Or why successful businessmen and politicians often push their sexual luck.

Similarly, testosterone levels may respond to more long-term stimuli. Studies have shown that inner-city youths, often exposed to danger in high-crime neighborhoods, may generate higher testosterone levels than unthreatened, secluded suburbanites. And so high T levels may not merely be responses to a violent environment; they may subsequently add to it in what becomes an increasingly violent, sexualized cycle. (It may be no accident that testosterone-soaked ghettos foster both high levels of crime and high levels of illegitimacy.) In the same way, declines in violence and crime may allow T levels to drop among young inner-city males, generating a virtuous trend of further reductions in crime and birth rates. This may help to explain why crime can decline precipitously, rather than drift down slowly, over time. Studies have also shown that men in long-term marriages see their testosterone levels progressively fall and their sex drives subsequently decline. It is as if their wives successfully tame them, reducing their sexual energy to a level where it is more unlikely to seek extramarital outlets. A 1993 study showed that single men tended to have higher levels of testosterone than married men and that men with high levels of testosterone turned out to be more likely to have had a failed marriage. Of course, if you start out with higher T levels, you may be more likely to fail at marriage, stay in the sexual marketplace, see your testosterone increase in response to this and so on.

None of this means, as the scientists always caution, that testosterone is directly linked to romantic failure or violence. No study has found a simple correlation, for example, between testosterone levels and crime. But there may be a complex correlation. The male-prisoner study, for example, found no general above-normal testosterone levels among inmates. But murderers and armed robbers had higher testosterone levels than mere car thieves and burglars. Why is this not surprising? One of the most remarkable, but least commented on, social statistics available is the sex differential in crime. For decades, arrest rates have shown that an overwhelmingly disproportionate number of arrestees are male. Although the sex differential has narrowed since the chivalrous 1930's, when the male-female arrest ratio was 12 to 1, it remains almost 4 to 1, a close echo of the testosterone differential between men and women. In violent crime, men make up an even bigger proportion. In 1998, 89 percent of murders in the United States, for example, were committed by men. Of course, there's a nature-nurture issue here as well, and the fact that the sex differential in crime has decreased over this century suggests that environment has played a part. Yet despite the enormous social changes of the last century, the differential is still 4 to 1, which suggests that underlying attributes may also have a great deal to do with it.

This, then, is what it comes down to: testosterone is a facilitator of risk—physical, criminal, personal. Without the influence of testosterone, the cost of these risks might seem to far outweigh the benefits. But with testosterone charging through the brain, caution is thrown to the wind. The influence of testosterone may not always lead to raw physical confrontation. In men with many options it may influence the decision to invest money in a dubious enterprise, jump into an ill-advised sexual affair or tell an egregiously big whopper. At the time, all these decisions may make some sort of testosteroned sense. The White House, anyone?

THE EFFECTS OF TESTOSTERONE are not secret; neither is the fact that men have far more of it than women. But why? As we have seen, testosterone is not synonymous with gender; in some species, it is the female who has most of it. The relatively new science of evolutionary psychology offers perhaps the best explanation for why that's not the case in humans. For neo-Darwinians, the aggressive and sexual aspects of testosterone are related to the division of labor among hunter-gatherers in our ancient but formative evolutionary past. This division—men in general hunted, women in general gathered—favored differing levels of testosterone. Women need some testosterone—for self-defense, occasional risk-taking, strength—but not as much as men. Men use it to increase their potential to defeat rivals, respond to physical threats in strange environments, maximize their physical attractiveness, prompt them to spread their genes as widely as possible and defend their home if necessary.

But the picture, as most good evolutionary psychologists point out, is more complex than this. Men who are excessively testosteroned are not that attractive to most women. Although they have the genes that turn women on—strong jaws and pronounced cheekbones, for example, are correlated with high testosterone—they can also be precisely the unstable, highly sexed creatures that childbearing, stability-seeking women want to avoid. There are two ways, evolutionary psychologists hazard, that women have successfully squared this particular circle. One is to marry the sweet class nerd and have an affair with the college quarterback: that way you get the good genes, the good sex and the stable home. The other is to find a man with variable T levels, who can be both stable and nurturing when you want him to be and yet become a muscle-bound, bristly gladiator when the need arises. The latter strategy, as Emma Bovary realized, is sadly more easily said than done.

So over millennia, men with high but variable levels of testosterone were the ones most favored by women and therefore most likely to produce offspring, and eventually us. Most men today are highly testosteroned, but not rigidly so. We don't have to live at all times with the T levels required to face down a woolly mammoth or bed half the village's young women. We can adjust so that our testosterone levels make us more suitable for co-parenting or for simply sticking around our mates when the sexual spark has dimmed. Indeed, one researcher, John Wingfield, has found a suggestive correlation in bird species between adjustable testosterone levels and males that have an active role to play in rearing their young. Male birds with consistently high testosterone levels tend to be worse fathers; males with variable levels are better dads. So there's hope for the new man yet.

From the point of view of men, after all, constantly high testosterone is a real problem, as any 15-year-old boy trying to concentrate on his homework will tell you. I missed one deadline on this article because it came three days after a testosterone shot and I couldn't bring myself to sit still long enough. And from a purely genetic point of view, men don't merely have an interest in impregnating as many women as possible; they also have an interest in seeing that their offspring are brought up successfully and their genes perpetuated. So for the male, the conflict between sex and love is resolved, as it is for the female, by a compromise between the short-term thrill of promiscuity and the long-term rewards of nurturing children. Just as the female does, he optimizes his genetic outcome by a stable marriage and occasional extramarital affairs. He is just more likely to have these affairs than a woman. Testosterone is both cause and effect of this difference.

And the difference is a real one. This is so obvious a point that we sometimes miss it. But without that difference, it would be hard to justify separate sports leagues for men and women, just as it would be hard not to suspect judicial bias behind the fact that of the 98 people executed last year in the United States, 100 percent came from a group that composes a little less than 50 percent of the population; that is, men. When the discrepancy is racial, we wring our hands. That it is sexual raises no red flags. Similarly, it is not surprising that 55 percent of everyone arrested in 1998 was under the age of 25—the years when male testosterone levels are at their natural peak.

It is also controversial yet undeniable that elevating testosterone levels can be extremely beneficial for physical and mental performance. It depends, of course, on what you're performing in. If your job is to whack home runs, capture criminals or play the market, then testosterone is a huge advantage. If you're a professional conciliator, office manager or teacher, it is probably a handicap. Major League Baseball was embarrassed that Mark McGwire's 1998 season home-run record might have been influenced by his use of androstenedione, a legal supplement that helps increase the body's own production of testosterone. But its own study into andro's effects concluded that regular use of it clearly raises T levels and so improves muscle mass and physical strength, without serious side effects. Testosterone also accelerates the rate of recovery from physical injury. Does this help make sense of McGwire's achievement? More testosterone obviously didn't give him the skill to hit 70 home runs, but it almost certainly contributed to the physical and mental endurance that helped him do so.

Since most men have at least 10 times as much T as most women, it therefore makes sense not to have coed baseball leagues. Equally, it makes sense that women will be underrepresented in a high-testosterone environment like military combat or construction. When the skills required are more cerebral or more endurance-related, the male-female gap may shrink, or even reverse itself. But otherwise, gender inequality in these fields is primarily not a function of sexism, merely of common sense. This is a highly controversial position, but it really shouldn't be. Even more unsettling is the racial gap in testosterone. Several solid studies, published in publications like Journal of the National Cancer Institute, show that black men have on average 3 to 19 percent more testosterone than white men. This is something to consider when we're told that black men dominate certain sports because of white racism or economic class rather than black skill. This reality may, of course, feed stereotypes about blacks being physical but not intellectual. But there's no evidence of any trade-off between the two. To say that someone is physically gifted is to say nothing about his mental abilities, as even N.F.L. die-hards have come to realize. Indeed, as Jon Entine points out in his new book, *Taboo,* even the position of quarterback, which requires a deft mix of mental and physical strength and was once predominantly white, has slowly become less white as talent has been rewarded. The percentage of blacks among N.F.L. quarterbacks is now twice the percentage of blacks in the population as a whole.

BUT FEARS OF natural difference still haunt the debate about gender equality. Many feminists have made tenacious arguments about the lack of any substantive physical or mental differences between men and women as if the political equality of the sexes depended on it. But to rest the equality of women on the physical and psychological equivalence of the sexes is to rest it on sand. In the end, testosterone bites. This year, for example, Toys "Я" Us announced it was planning to redesign its toy stores to group products most likely to be bought by the same types of consumers: in marketing jargon, "logical adjacencies." The results? Almost total gender separation. "Girl's World" would feature Easy-Bake Ovens and Barbies; "Boy's World," trucks and action figures. Though Toys "Я" Us denied that there was any agenda behind this—its market research showed that gender differences start as young as 2 years old—such a public outcry ensued that the store canceled its plans. Meanwhile, Fox Family Channels is about to introduce two new, separate cable channels for boys and girls, boyzChannel and girlzChannel, to attract advertisers and consumers more efficiently. Fox executives told *The Wall Street Journal* that their move is simply a reflection of what Nielsen-related research tells them about the viewing habits of boys and girls: that, "in general terms, girls are more interested in entertainment that is relationship-oriented," while boys are "more action-oriented." T anyone? After more than two decades of relentless legal, cultural and ideological attempts to negate sexual difference between boys and girls, the market has turned around and shown that very little, after all, has changed.

Advocates of a purely environmental origin for this difference between the sexes counter that gender socialization begins very early and is picked up by subtle inferences from parental interaction and peer pressure, before being reinforced by the collective culture at large. Most parents observing toddlers choosing their own toys and play patterns can best judge for themselves how true this is. But as Matt Ridley has pointed out, there is also physiological evidence of very early mental differences between the sexes, most of it to the advantage of girls. Ninety-five percent of all hyperactive kids are boys; four times as many boys are dyslexic and learning-disabled as girls. There is a greater distinction between the right and left brain among boys than girls, and worse linguistic skills. In general, boys are better at spatial and abstract tasks, girls at communication. These are generalizations, of course. There are many, many boys who are great linguists and model students, and vice versa. Some boys even prefer, when left to their own devices, to play with dolls as well as trucks. But we are talking of generalities here, and the influence of womb-given testosterone on those generalities is undeniable.

Some of that influence is a handicap. We are so used to associating testosterone with strength, masculinity and patriarchal violence that it is easy to ignore that it also makes men weaker in some respects than women. It doesn't correlate with economic power: in fact, as we have seen, blue-collar workers have more of it than white-collar workers. It gets men into trouble. For reasons no one seems to understand, testosterone may also be an immune suppressant. High levels of it can correspond, as recent studies have shown, not only with baldness but also with heart disease and a greater susceptibility to infectious diseases. Higher levels of prostate cancer among blacks, some researchers believe, may well be related to blacks' higher testosterone levels. The aggression it can foster and the risks it encourages lead men into situations that often wound or kill them. And higher levels of testosterone-driven promiscuity make men more prone to sexually transmitted diseases. This is one reason that men live shorter lives on average than women. There is something, in other words, tragic about testosterone. It can lead to a certain kind of male glory; it may lead to valor or boldness or impulsive romanticism. But it also presages a uniquely male kind of doom. The cockerel with the brightest comb is often the most attractive and the most testosteroned, but it is also the most vulnerable to parasites. It is as if it has sacrificed quantity of life for intensity of experience, and this trade-off is a deeply male one.

So it is perhaps unsurprising that those professions in which this trade-off is most pronounced—the military, contact sports, hazardous exploration, venture capitalism, politics, gambling—tend to be disproportionately male. Politics is undoubtedly the most controversial because it is such a critical arena for the dispersal of power. But consider for a moment how politics is conducted in our society. It is saturated with combat, ego, conflict and risk. An entire career can be lost in a single gaffe or an unexpected shift in the national mood. This ego-driven roulette is almost as highly biased toward the testosteroned as wrestling. So it makes some sense that after almost a century of electorates made up by as many women as men, the number of female politicians remains pathetically small in most Western democracies. This may not be endemic to politics; it may have more to do with the way our culture constructs politics. And it is not to say that women are not good at government. Those qualities associated with low testosterone—patience, risk aversion, empathy—can all lead to excellent governance. They are just lousy qualities in the crapshoot of electoral politics.

IF YOU CARE ABOUT sexual equality, this is obviously a challenge, but it need not be as depressing as it sounds. The sports world offers one way out. Men and women do not compete directly against one another; they have separate tournaments and leagues. Their different styles of physical excellence can be appreciated in different ways. At some basic level, of course, men will always be better than women in many of these contests. Men run faster and throw harder. Women could compensate for this by injecting testosterone, but if they took enough to be truly competitive, they would become men, which would somewhat defeat the purpose.

The harder cases are in those areas in which physical strength is important but not always crucial, like military combat or manual labor. And here the compromise is more likely to be access but inequality in numbers. Finance? Business? Here, where the testosterone-driven differences may well be more subtly psychological, and where men may dominate by discrimination rather than merit, is the trickiest arena. Testosterone-induced impatience may lead to poor decision-making, but low-testosterone risk aversion may lead to an inability to seize business opportunities. Perhaps it is safest to say that unequal numbers of men and women in these spheres is not prima facie evidence of sexism. We should do everything we can to ensure equal access, but it is foolish to insist that numerical inequality is always a function of bias rather than biology. This doesn't mean we shouldn't worry about individual cases of injustice; just that we shouldn't be shocked if gender inequality endures. And we should recognize that affirmative action for women (and men) in all arenas is an inherently utopian project.

Then there is the medical option. A modest solution might be to give more women access to testosterone to improve their sex drives, aggression and risk affinity and to help redress their disadvantages in those areas as compared with men. This is already done for severely depressed women, or women with hormonal imbalances, or those lacking an adequate sex drive, especially after menopause. Why not for women who simply want to rev up their will to power? Its use needs to be carefully monitored because it can also lead to side effects, like greater susceptibility to cancer, but that's what doctors are there for. And since older men also suffer a slow drop-off in T levels, there's no reason they should be cold-shouldered either. If the natural disadvantages of gender should be countered, why not the natural disadvantages of age? In some ways, this is already happening. Among the most common drugs now available through Internet doctors and pharmacies, along with Viagra and Prozac, is testosterone. This summer, with the arrival of AndroGel, the testosterone gel created as a medical treatment for those four to five million men who suffer from low levels of testosterone, recreational demand may soar.

Or try this thought experiment: what if parents committed to gender equity opted to counteract the effect of testosterone on boys in the womb by complementing it with injections of artificial female hormones? That way, structural gender difference could be eradicated from the beginning. Such a policy would lead to "men and women with normal bodies but identical feminine brains," Matt Ridley posits. "War, rape, boxing, car racing, pornography and hamburgers and beer would soon be distant memories. A feminist paradise would have arrived." Today's conservative cultural critics might also be enraptured. Promiscuity would doubtless decline, fatherhood improve, crime drop, virtue spread. Even gay men might start behaving like lesbians, fleeing the gym and marrying for life. This is a fantasy, of course, but our increasing control and understanding of the scientific origins of our behavior, even of our culture, is fast making those fantasies things we will have to actively choose to forgo.

But fantasies also tell us something. After a feminist century, we may be in need of a new understanding of masculinity. The concepts of manliness, of gentlemanly behavior, of chivalry have been debunked. The New Age bonding of the men's movement has been outlived. What our increasing knowledge of testosterone suggests is a core understanding of what it is to be a man, for better and worse. It is about the ability to risk for good and bad; to act, to strut, to dare, to seize. It is about a kind of energy we often rue but would surely miss. It is about the foolishness that can lead to courage or destruction, the beauty that can be strength or vanity. To imagine a world without it is to see more clearly how our world is inseparable from it and how our current political pieties are too easily threatened by its reality.

And as our economy becomes less physical and more cerebral, as women slowly supplant men in many industries, as income inequalities grow and more highly testosteroned blue-collar men find themselves shunted to one side, we will have to find new ways of channeling what nature has bequeathed us. I don't think it's an accident that in the last decade there has been a growing focus on a muscular male physique in our popular culture, a boom in crass men's magazines, an explosion in violent computer games or a professional wrestler who has become governor. These are indications of a cultural displacement, of a world in which the power of testosterone is ignored or attacked, with the result that it re-emerges in cruder and less social forms. Our main task in the gender wars of the new century may not be how to bring women fully into our society, but how to keep men from seceding from it, how to reroute testosterone for constructive ends, rather than ignore it for political point-making.

For my part, I'll keep injecting the Big T. Apart from how great it makes me feel, I consider it no insult to anyone else's gender to celebrate the uniqueness of one's own. Diversity need not mean the equalization of difference. In fact, true diversity requires the acceptance of difference. A world without the unruly, vulnerable, pioneering force of testosterone would be a fairer and calmer, but far grayer and duller, place. It is certainly somewhere I would never want to live. Perhaps the fact that I write this two days after the injection of another 200 milligrams of testosterone into my bloodstream makes me more likely to settle for this colorful trade-off than others. But it seems to me no disrespect to womanhood to say that I am perfectly happy to be a man, to feel things no woman will ever feel to the degree that I feel them, to experience the world in a way no woman ever has. And to do so without apology or shame.

Malcolm Gladwell

John Rock's Error

FROM *THE NEW YORKER*

John Rock was christened in 1890 at the Church of the Immaculate Conception in Marlborough, Massachusetts, and married by Cardinal William O'Connell, of Boston. He had five children and nineteen grandchildren. A crucifix hung above his desk, and nearly every day of his adult life he attended the 7 A.M. Mass at St. Mary's in Brookline. Rock, his friends would say, was in love with his church. He was also one of the inventors of the birth-control pill, and it was his conviction that his faith and his vocation were perfectly compatible. To anyone who disagreed he would simply repeat the words spoken to him as a child by his home-town priest: "John, always stick to your conscience. Never let anyone else keep it for you. And I mean anyone else." Even when Monsignor Francis W. Carney, of Cleveland, called him a "moral rapist," and when Frederick Good, the longtime head of obstetrics at Boston City Hospital, went to Boston's Cardinal Richard Cushing to have Rock excommunicated, Rock was unmoved. "You should be afraid to meet your Maker," one angry woman wrote to him, soon after the Pill was approved. "My dear madam," Rock wrote back, "in my faith, we are taught that the Lord is with us always. When my time comes, there will be no need for introductions."

In the years immediately after the Pill was approved by the F.D.A., in 1960, Rock was everywhere. He appeared in interviews and documentaries on CBS and NBC, in *Time, Newsweek, Life, The Saturday Evening Post.* He toured the country tirelessly. He wrote a widely discussed book, *The Time Has Come: A Catholic Doctor's Proposals to End the Battle Over Birth Control,* which was translated into French, German, and Dutch. Rock was six feet three and rail-thin, with impeccable manners; he held doors open for his patients and addressed them as "Mrs." or "Miss." His mere association with the Pill helped make it seem respectable. "He was a man of great dignity," Dr. Sheldon J. Segal, of the Population Council, recalls. "Even if the occasion called for an open collar, you'd never find him without an ascot. He had the shock of white hair to go along with that. And posture, straight as an arrow, even to his last year." At Harvard Medical School, he was a giant, teaching obstetrics for more than three decades. He was a pioneer in in-vitro fertilization and the freezing of sperm cells, and was the first to extract an intact fertilized egg. The Pill was his crowning achievement. His two collaborators, Gregory Pincus and Min-Cheuh Chang, worked out the mechanism. He shepherded the drug through its clinical trials. "It was his name and his reputation that gave ultimate validity to the claims that the pill would protect women against unwanted pregnancy," Loretta McLaughlin writes in her marvellous 1982 biography of Rock. Not long before the Pill's approval, Rock travelled to Washington to testify before the F.D.A. about the drug's safety. The agency examiner, Pasquale DeFelice, was a Catholic obstetrician from Georgetown University, and at one point, the story goes, DeFelice suggested the unthinkable—that the Catholic Church would never approve of the birth-control pill. "I can still see Rock standing there, his face composed, his eyes riveted on DeFelice," a colleague recalled years later, "and then, in a voice that would congeal your soul, he said, 'Young man, don't you sell *my* church short.' "

In the end, of course, John Rock's church disappointed him. In 1968, in the encyclical "Humanae Vitae," Pope Paul VI outlawed oral contraceptives and all other "artificial" methods of birth control. The passion and urgency that animated the birth-control debates of the sixties are now a memory. John Rock still matters, though, for the simple reason that in the course of reconciling his church and his work he made an error. It was not a deliberate error. It became manifest only after his death, and through scientific advances he could not have anticipated. But because that mistake shaped the way he thought about the Pill—about what it was, and how it worked, and most of all what it meant—and because John Rock was one of those responsible for the way the Pill came into the world, his error has colored the way people have thought about contraception ever since.

John Rock believed that the Pill was a "natural" method of birth control. By that he didn't mean that it *felt* natural, because it obviously didn't for many women, particularly not in its earliest days, when the doses of hormone were many times as high as they are today. He meant that it worked by natural means. Women can get pregnant only during a certain interval each month, because after ovulation their bodies produce a surge of the hormone progesterone. Progesterone—one of a class of hormones known as progestin—prepares the uterus for implantation and stops the ovaries from releasing new eggs; it favors gestation. "It is progesterone, in the healthy woman, that prevents ovulation and establishes the pre- and post-menstrual 'safe' period," Rock wrote. When a woman is pregnant, her body produces a stream of progestin in part for the same reason, so that another egg can't be released and threaten the pregnancy already under way. Progestin, in other words, is nature's contraceptive. And what was the Pill? Progestin in tablet form. When a woman was on the Pill, of course, these hormones weren't coming in a sudden surge after ovulation and weren't limited to certain times in her cycle. They were being given in a steady dose, so that ovulation was permanently shut down. They were also being given with an additional dose of estrogen, which holds the endometrium together and—as we've come to learn—helps maintain other tissues as well. But to Rock, the timing and combination of hormones wasn't the issue. The key fact was that the Pill's ingredients duplicated what could be found in the body naturally. And in that naturalness he saw enormous theological significance.

In 1951, for example, Pope Pius XII had sanctioned the rhythm method for Catholics because he deemed it a "natural" method of regulating procreation: it didn't kill the sperm, like a spermicide, or frustrate the normal process of procreation, like a diaphragm, or mutilate the organs, like sterilization. Rock knew all about the rhythm method. In the nineteen-thirties, at the Free Hospital for Women, in Brookline, he had started the country's first rhythm clinic for educating Catholic couples in natural contraception. But how did the rhythm method work? It worked by limiting sex to the safe period that progestin created. And how did the Pill work? It worked by using progestin to extend the safe period to the entire month. It didn't mutilate the reproductive organs, or damage any natural process. "Indeed," Rock wrote, oral contraceptives "may be characterized as a 'pill-established safe period,' and would seem to carry the same moral implications" as the rhythm method. The Pill was, to Rock, no more than "an adjunct to nature."

In 1958, Pope Pius XII approved the Pill for Catholics, so long as its contraceptive effects were "indirect"—that is, so long as it was intended only as a remedy for conditions like painful menses or "a disease of the uterus." That ruling emboldened Rock still further. Short-term use of the Pill, he knew, could regulate the cycle of women whose periods had previously been unpredictable. Since a regular menstrual cycle was necessary for the successful use of the rhythm method—and since the rhythm method was sanctioned by the Church—shouldn't it be permissible for women with an irregular menstrual cycle to use the Pill in order to facilitate the use of rhythm? And if that was true why not take the logic one step further? As the federal judge John T. Noonan writes in *Contraception,* his history of the Catholic position on birth control:

> If it was lawful to suppress ovulation to achieve a regularity necessary for successfully sterile intercourse, why was it not lawful to suppress ovulation without appeal to rhythm? If pregnancy could be prevented by pill plus rhythm, why not by pill alone? In each case suppression of ovulation was used as a means. How was a moral difference made by the addition of rhythm?

These arguments, as arcane as they may seem, were central to the development of oral contraception. It was John Rock and Gregory Pincus who decided that the Pill ought to be taken over a four-week cycle—a woman would spend three weeks on the Pill and the fourth week off the drug (or on a placebo), to allow for menstruation. There was and is no medical reason for this. A typical woman of childbearing age has a menstrual cycle of around twenty-eight days, determined by the cascades of hormones released by her ovaries. As first estrogen and then a combination of estrogen and progestin flood the uterus, its lining becomes thick and swollen, preparing for the implantation of a fertilized egg. If the egg is not fertilized, hormone levels plunge and cause the lining—the endometrium—to be sloughed off in a menstrual bleed. When a woman is on the Pill, however, no egg is released, because the Pill suppresses ovulation. The fluxes of estrogen and progestin that cause the lining of the uterus to grow are dramatically reduced, because the Pill slows down the ovaries. Pincus and Rock knew that the effect of the Pill's hormones on the endometrium was so modest that women could conceivably go for months without having to menstruate. "In view of the ability of this compound to prevent menstrual bleeding as long as it is taken," Pincus acknowledged in 1958, "a cycle of any desired length could presumably be produced." But he and Rock decided to cut the hormones off after three weeks and trigger a menstrual period because they believed that women would find the continuation of their monthly bleeding reassuring. More to the point, if Rock wanted to demonstrate that the Pill was no more than a natural variant of the rhythm method, he couldn't very well do away with the monthly menses. Rhythm required "regularity," and so the Pill had to produce regularity as well.

It has often been said of the Pill that no other drug has ever been so instantly recognizable by its packaging: that small, round plastic dial pack. But what was the dial pack if not the physical embodiment of the twenty-eight-day cycle? It was, in the words of its inventor, meant to fit into a case "indistinguishable" from a woman's cosmetics compact, so that it might be carried "without giving a visual clue as to matters which are of no concern to others." Today, the Pill is still often sold in dial packs and taken in twenty-eight-day cycles. It remains, in other words, a drug shaped by the dictates of the Catholic Church—by John Rock's desire to make this new method of birth control seem as natural as possible. This was John Rock's error. He was consumed by the idea of the natural. But what he thought was natural wasn't so natural after all, and the Pill he ushered into the world turned out to be something other than what he thought it was. In John Rock's mind the dictates of religion and the principles of science got mixed up, and only now are we beginning to untangle them.

IN 1986, A YOUNG SCIENTIST named Beverly Strassmann travelled to Africa to live with the Dogon tribe of Mali. Her research site was the village of Sangui in the Sahel, about a hundred and twenty miles south of Timbuktu. The Sahel is thorn savannah, green in the rainy season and semi-arid the rest of the year. The Dogon grow millet, sorghum, and onions, raise livestock, and live in adobe houses on the Bandiagara escarpment. They use no contraception. Many of them have held on to their ancestral customs and religious beliefs. Dogon farmers, in many respects, live much as people of that region have lived since antiquity. Strassmann wanted to construct a precise reproductive profile of the women in the tribe, in order to understand what female biology might have been like in the millennia that preceded the modern age. In a way, Strassmann was trying to answer the same question about female biology that John Rock and the Catholic Church had struggled with in the early sixties: What is natural? Only, her sense of "natural" was not theological but evolutionary. In the era during which natural selection established the basic patterns of human biology—the natural history of our species—how often did women have children? How often did they menstruate? When did they reach puberty and menopause? What impact did breast-feeding have on ovulation? These questions had been studied before, but never so thoroughly that anthropologists felt they knew the answers with any certainty.

Strassmann, who teaches at the University of Michigan at Ann Arbor, is a slender, soft-spoken woman with red hair, and she recalls her time in Mali with a certain wry humor. The house she stayed in while in Sangui had been used as a shelter for sheep before she came and was turned into a pigsty after she left. A small brown snake lived in her latrine, and would curl up in a camouflaged coil on the seat she sat on while bathing. The villagers, she says, were of two minds: was it a deadly snake—*Kere me jongolo,* literally, "My bite cannot be healed"—or a harmless mouse snake? (It turned out to be the latter.) Once, one of her neighbors and best friends in the tribe roasted her a rat as a special treat. "I told him that white people aren't allowed to eat rat because rat is our totem," Strassmann says. "I can still see it. Bloated and charred. Stretched by its paws. Whiskers singed. To say nothing of the tail." Strassmann meant to live in Sangui for eighteen months, but her experiences there were so profound and exhilarating that she stayed for two and a half years. "I felt incredibly privileged," she says. "I just couldn't tear myself away."

Part of Strassmann's work focussed on the Dogon's practice of segregating menstruating women in special huts on the fringes of the village. In Sangui, there were two menstrual huts—dark, cramped, one-room adobe structures, with boards for beds. Each accommodated three women, and when the rooms were full, latecomers were forced to stay outside on the rocks. "It's not a place where people kick back and enjoy themselves," Strassmann says. "It's simply a nighttime hangout. They get there at dusk, and get up early in the morning and draw their water." Strassmann took urine samples from the women using the hut, to confirm that they were menstruating. Then she made a list of all the women in the village, and for her entire time in Mali—seven hundred and thirty-six consecutive nights—she kept track of everyone who visited the hut. Among the Dogon, she found, a woman, on average, has her first period at the age of sixteen and gives birth eight or nine times. From menarche, the onset of menstruation, to the age of twenty, she averages seven periods a year. Over the next decade and a half, from the age of twenty to the age of thirty-four, she spends so much time either pregnant or breast-feeding (which, among the Dogon, suppresses ovulation for an average of twenty months) that she averages only slightly more than one period per year. Then, from the age of thirty-five until menopause, at around fifty, as her fertility rapidly declines, she averages four menses a year. All told, Dogon women menstruate about a hundred times in their lives. (Those who survive early childhood typically live into their seventh or eighth decade.) By contrast, the average for contemporary Western women is somewhere between three hundred and fifty and four hundred times.

Strassmann's office is in the basement of a converted stable next to the Natural History Museum on the University of Michigan campus. Behind her desk is a row of battered filing cabinets, and as she was talking she turned and pulled out a series of yellowed charts. Each page listed, on the left, the first names and identification numbers of the Sangui women. Across the top was a time line, broken into thirty-day blocks. Every menses of every woman was marked with an X. In the village, Strassmann explained, there were two women who were sterile, and, because they couldn't get pregnant, they were regulars at the menstrual hut. She flipped through the pages until she found them. "Look, she had twenty-nine menses over two years, and the other had twenty-three." Next to each of their names was a solid line of X's. "Here's a woman approaching menopause," Strassmann went on, running her finger down the page. "She's cycling but is a little bit erratic. Here's another woman of prime childbearing age. Two periods. Then pregnant. I never saw her again at the menstrual hut. This woman here didn't go to the menstrual hut for twenty months after giving birth, because she was breast-feeding. Two periods. Got pregnant. Then she miscarried, had a few periods, then got pregnant again. This woman had three menses in the study period." There weren't a lot of X's on Strassmann's sheets. Most of the boxes were blank. She flipped back through her sheets to the two anomalous women who were menstruating every month. "If this were a menstrual chart of undergraduates here at the University of Michigan, all the rows would be like this."

Strassmann does not claim that her statistics apply to every preindustrial society. But she believes—and other anthropological work backs her up—that the number of lifetime menses isn't greatly affected by differences in diet or climate or method of subsistence (foraging versus agriculture, say). The more significant factors, Strassmann says, are things like the prevalence of wet-nursing or sterility. But over all she believes that the basic pattern of late menarche, many pregnancies, and long menstrual-free stretches caused by intensive breast-feeding was virtually universal up until the "demographic transition" of a hundred years ago from high to low fertility. In other words, what we think of as normal—frequent menses—is in evolutionary terms abnormal. "It's a pity that gynecologists think that women have to menstruate every month," Strassmann went on. "They just don't understand the real biology of menstruation."

To Strassmann and others in the field of evolutionary medicine, this shift from a hundred to four hundred lifetime menses is enormously significant. It means that women's bodies are being subjected to changes and stresses that they were not necessarily designed by evolution to handle. In a brilliant and provocative book, *Is Menstruation Obsolete?*, Drs. Elsimar Coutinho and Sheldon S. Segal, two of the world's most prominent contraceptive researchers, argue that this recent move to what they call "incessant ovulation" has become a serious problem for women's health. It doesn't mean that women are always better off the less they menstruate. There are times—particularly in the context of certain medical conditions—when women ought to be concerned if they aren't menstruating: In obese women, a failure to menstruate can signal an increased risk of uterine cancer. In female athletes, a failure to menstruate can signal an increased risk of osteoporosis. But for most women, Coutinho and Segal say, incessant ovulation serves no purpose except to increase the occurence of abdominal pain, mood shifts, migraines, endometriosis, fibroids, and anemia—the last of which, they point out, is "one of the most serious health problems in the world."

Most serious of all is the greatly increased risk of some cancers. Cancer, after all, occurs because as cells divide and reproduce they sometimes make mistakes that cripple the cells' defenses against runaway growth. That's one of the reasons that our risk of cancer generally increases as we age: our cells have more time to make mistakes. But this also means that *any* change promoting cell division has the potential to increase cancer risk, and ovulation appears to be one of those changes. Whenever a woman ovulates, an egg literally bursts through the walls of her ovaries. To heal that puncture, the cells of the ovary wall have to divide and reproduce. Every time a woman gets pregnant and bears a child, her lifetime risk of ovarian cancer drops ten per cent. Why? Possibly because, between nine months of pregnancy and the suppression of ovulation associated with breast-feeding, she stops ovulating for twelve months—and saves her ovarian walls from twelve bouts of cell division. The argument is similar for endometrial cancer. When a woman is menstruating, the estrogen that flows through her uterus stimulates the growth of the uterine lining, causing a flurry of potentially dangerous cell division. Women who do not menstruate frequently spare the endometrium that risk. Ovarian and endometrial cancer are characteristically modern diseases, consequences, in part, of a century in which women have come to menstruate four hundred times in a lifetime.

In this sense, the Pill really does have a "natural" effect. By blocking the release of new eggs, the progestin in oral contraceptives reduces the rounds of ovarian cell division. Progestin also counters the surges of estrogen in the endometrium, restraining cell division there. A woman who takes the Pill for ten years cuts her ovarian-cancer risk by around seventy per cent and her endometrial-cancer risk by around sixty per cent. But here "natural" means something different from what Rock meant. He assumed that the Pill was natural because it was an unobtrusive variant of the body's own processes. In fact, as more recent research suggests, the Pill is really only natural in so far as it's *radical*—rescuing the ovaries and endometrium from modernity. That Rock insisted on a twenty-eight-day cycle for his pill is evidence of just how deep his misunderstanding was: the real promise of the Pill was not that it could preserve the menstrual rhythms of the twentieth century but that it could disrupt them.

Today, a growing movement of reproductive specialists has begun to campaign loudly against the standard twenty-eight-day Pill regimen. The drug company Organon has come out with a new oral contraceptive, called Mircette, that cuts the seven-day placebo interval to two days. Patricia Sulak, a medical researcher at Texas A. & M. University, has shown that most women can probably stay on the Pill, straight through, for six to twelve weeks before they experience breakthrough bleeding or spotting. More recently, Sulak has documented precisely what the cost of the Pill's monthly "off" week is. In a paper in the February 2000 issue of the journal *Obstetrics and Gynecology,* she and her colleagues documented something that will come as no surprise to most women on the Pill: during the placebo week, the number of users experiencing pelvic pain, bloating, and swelling more than triples, breast tenderness more than doubles, and headaches increase by almost fifty per cent. In other words, some women on the Pill continue to experience the kinds of side effects associated with normal menstruation. Sulak's paper is a short, dry, academic work, of the sort intended for a narrow professional audience. But it is impossible to read it without being struck by the consequences of John Rock's desire to please his church. In the past forty years, millions of women around the world have been given the Pill in such a way as to maximize their pain and suffering. And to what end? To pretend that the Pill was no more than a pharmaceutical version of the rhythm method?

IN 1980 AND 1981, Malcolm Pike, a medical statistician at the University of Southern California, travelled to Japan for six months to study at the Atomic Bomb Casualties Commission. Pike wasn't interested in the effects of the bomb. He wanted to examine the medical records that the commission had been painstakingly assembling on the survivors of Hiroshima and Nagasaki. He was investigating a question that would ultimately do as much to complicate our understanding of the Pill as Strassmann's research would a decade later: why did Japanese women have breast-cancer rates six times lower than American women?

In the late forties, the World Health Organization began to collect and publish comparative health statistics from around the world, and the breast-cancer disparity between Japan and America had come to obsess cancer specialists. The obvious answer—that Japanese women were somehow genetically protected against breast cancer—didn't make sense, because once Japanese women moved to the United States they began to get breast cancer almost as often as American women did. As a result, many experts at the time assumed that the culprit had to be some unknown toxic chemical or virus unique to the West. Brian Henderson, a colleague of Pike's at U.S.C. and his regular collaborator, says that when he entered the field, in 1970, "the whole viral- and chemical-carcinogenesis idea was huge—it dominated the literature." As he recalls, "Breast cancer fell into this large, unknown box that said it was something to do with the environment—and that word 'environment' meant a lot of different things to a lot of different people. They might be talking about diet or smoking or pesticides."

Henderson and Pike, however, became fascinated by a number of statistical peculiarities. For one thing, the rate of increase in breast-cancer risk rises sharply throughout women's thirties and forties and then, at menopause, it starts to slow down. If a cancer is caused by some toxic outside agent, you'd expect that rate to rise steadily with each advancing year, as the number of mutations and genetic mistakes steadily accumulates. Breast cancer, by contrast, looked as if it were being driven by something specific to a woman's reproductive years. What was more, younger women who had had their ovaries removed had a markedly lower risk of breast cancer; when their bodies weren't producing estrogen and progestin every month, they got far fewer tumors. Pike and Henderson became convinced that breast cancer was linked to a process of cell division similar to that of ovarian and endometrial cancer. The female breast, after all, is just as sensitive to the level of hormones in a woman's body as the reproductive system. When the breast is exposed to estrogen, the cells of the terminal-duct lobular unit—where most breast cancer arises—undergo a flurry of division. And during the mid-to-late stage of the menstrual cycle, when the ovaries start producing large amounts of progestin, the pace of cell division in that region doubles.

It made intuitive sense, then, that a woman's risk of breast cancer would be linked to the amount of estrogen and progestin her breasts have been exposed to during her lifetime. How old a woman is at menarche should make a big difference, because the beginning of puberty results in a hormonal surge through a woman's body, and the breast cells of an adolescent appear to be highly susceptible to the errors that result in cancer. (For more complicated reasons, bearing children turns out to be protective against breast cancer, perhaps because in the last two trimesters of pregnancy the cells of the breast mature and become much more resistant to mutations.) How old a woman is at menopause should matter, and so should how much estrogen and progestin her ovaries actually produce, and even how much she weighs after menopause, because fat cells turn other hormones into estrogen.

Pike went to Hiroshima to test the cell-division theory. With other researchers at the medical archive, he looked first at the age when Japanese women got their period. A Japanese woman born at the turn of the century had her first period at sixteen and a half. American women born at the same time had their first period at fourteen. That difference alone, by their calculation, was sufficient to explain forty per cent of the gap between American and Japanese breast-cancer rates. "They had collected amazing records from the women of that area," Pike said. "You could follow precisely the change in age of menarche over the century. You could even see the effects of the Second World War. The age of menarche of Japanese girls went up right at that point because of poor nutrition and other hardships. And then it started to go back down after the war. That's what convinced me that the data were wonderful."

Pike, Henderson, and their colleagues then folded in the other risk factors. Age at menopause, age at first pregnancy, and number of children weren't sufficiently different between the two countries to matter. But weight was. The average post-menopausal Japanese woman weighed a hundred pounds; the average American woman weighed a hundred and forty-five pounds. That fact explained another twenty-five per cent of the difference. Finally, the researchers analyzed blood samples from women in rural Japan and China, and found that their ovaries—possibly because of their extremely low-fat diet—were producing about seventy-five per cent the amount of estrogen that American women were producing. Those three factors, added together, seemed to explain the breast-cancer gap. They also appeared to explain why the rates of breast cancer among Asian women began to increase when they came to America: on an American diet, they started to menstruate earlier, gained more weight, and produced more estrogen. The talk of chemicals and toxins and power lines and smog was set aside. "When people say that what we understand about breast cancer explains only a small amount of the problem, that it is somehow a mystery, it's absolute nonsense," Pike says flatly. He is a South African in his sixties, with graying hair and a salt-and-pepper beard. Along with Henderson, he is an eminent figure in cancer research, but no one would ever accuse him of being tentative in his pronouncements. "We understand breast cancer extraordinarily well. We understand it as well as we understand cigarettes and lung cancer."

What Pike discovered in Japan led him to think about the Pill, because a tablet that suppressed ovulation—and the monthly tides of estrogen and progestin that come with it—obviously had the potential to be a powerful anti-breast-cancer drug. But the breast was a little different from the reproductive organs. Progestin prevented ovarian cancer because it suppressed ovulation. It was good for preventing endometrial cancer because it countered the stimulating effects of estrogen. But in breast cells, Pike believed, progestin wasn't the solution; it was one of the hormones that *caused* cell division. This is one explanation for why, after years of studying the Pill, researchers have concluded that it has no effect one way or the other on breast cancer: whatever beneficial effect results from what the Pill does is cancelled out by how it does it. John Rock touted the fact that the Pill used progestin, because progestin was the body's own contraceptive. But Pike saw nothing "natural" about subjecting the breast to that heavy a dose of progestin. In his view, the amount of progestin and estrogen needed to make an effective contraceptive was much greater than the amount needed to keep the reproductive system healthy—and that excess was unnecessarily raising the risk of breast cancer. A truly natural Pill might be one that found a way to suppress ovulation *without* using progestin. Throughout the nineteen-eighties, Pike recalls, this was his obsession. "We were all trying to work out how the hell we could fix the Pill. We thought about it day and night."

PIKE'S PROPOSED SOLUTION is a class of drugs known as GnRHAs, which has been around for many years. GnRHAs disrupt the signals that the pituitary gland sends when it is attempting to order the manufacture of sex hormones. It's a circuit breaker. "We've got substantial experience with this drug," Pike says. Men suffering from prostate cancer are sometimes given a GnRHA to temporarily halt the production of testosterone, which can exacerbate their tumors. Girls suffering from what's called precocious puberty—puberty at seven or eight, or even younger—are sometimes given the drug to forestall sexual maturity. If you give GnRHA to women of childbearing age, it stops their ovaries from producing estrogen and progestin. If the conventional Pill works by convincing the body that it is, well, a little bit pregnant, Pike's pill would work by convincing the body that it was menopausal.

In the form Pike wants to use it, GnRHA will come in a clear glass bottle the size of a saltshaker, with a white plastic mister on top. It will be inhaled nasally. It breaks down in the body very quickly. A morning dose simply makes a woman menopausal for a while. Menopause, of course, has its risks. Women need estrogen to keep their hearts and bones strong. They also need progestin to keep the uterus healthy. So Pike intends to add back just enough of each hormone to solve these problems, but much less than women now receive on the Pill. Ideally, Pike says, the estrogen dose would be adjustable: women would try various levels until they found one that suited them. The progestin would come in four twelve-day stretches a year. When someone on Pike's regimen stopped the progestin, she would have one of four annual menses.

Pike and an oncologist named Darcy Spicer have joined forces with another oncologist, John Daniels, in a startup called Balance Pharmaceuticals. The firm operates out of a small white industrial strip mall next to the freeway in Santa Monica. One of the tenants is a paint store, another looks like some sort of export company. Balance's offices are housed in an oversized garage with a big overhead door and concrete floors. There is a tiny reception area, a little coffee table and a couch, and a warren of desks, bookshelves, filing cabinets, and computers. Balance is testing its formulation on a small group of women at high risk for breast cancer, and if the results continue to be encouraging, it will one day file for F.D.A. approval.

"When I met Darcy Spicer a couple of years ago," Pike said recently, as he sat at a conference table deep in the Balance garage, "he said, 'Why don't we just try it out? By taking mammograms, we should be able to see changes in the breasts of women on this drug, even if we add back a little estrogen to avoid side effects.' So we did a study, and we found that there were huge changes." Pike pulled out a paper he and Spicer had published in the *Journal of the National Cancer Institute,* showing breast X-rays of three young women. "These are the mammograms of the women before they start," he said. Amid the grainy black outlines of the breast were large white fibrous clumps—clumps that Pike and Spicer believe are indicators of the kind of relentless cell division that increases breast-cancer risk. Next to those X-rays were three mammograms of the same women taken after a year on the GnRHA regimen. The clumps were almost entirely gone. "This to us represents that we have actually stopped the activity inside the breasts," Pike went on. "White is a proxy for cell proliferation. We're slowing down the breast."

Pike stood up from the table and turned to a sketch pad on an easel behind him. He quickly wrote a series of numbers on the paper. "Suppose a woman reaches menarche at fifteen and menopause at fifty. That's thirty-five years of stimulating the breast. If you cut that time in half, you will change her risk not by half but by half raised to the power of 4.5." He was working with a statistical model he had developed to calculate breast-cancer risk. "That's one-twenty-third. Your risk of breast cancer will be one-twenty-third of what it would be otherwise. It won't be zero. You can't get to zero. If you use this for ten years, your risk will be cut by at least half. If you use it for five years, your risk will be cut by at least a third. It's as if your breast were to be five years younger, or ten years younger—*forever.*" The regimen, he says, should also provide protection against ovarian cancer.

Pike gave the sense that he had made this little speech many times before, to colleagues, to his family and friends—and to investors. He knew by now how strange and unbelievable what he was saying sounded. Here he was, in a cold, cramped garage in the industrial section of Santa Monica, arguing that he knew how to save the lives of hundreds of thousands of women around the world. And he wanted to do that by making young women menopausal through a chemical regimen sniffed every morning out of a bottle. This was, to say the least, a bold idea. Could he strike the right balance between the hormone levels women need to stay healthy and those that ultimately make them sick? Was progestin really so important in breast cancer? There are cancer specialists who remain skeptical. And, most of all, what would women think? John Rock, at least, had lent the cause of birth control his Old World manners and distinguished white hair and appeals from theology; he took pains to make the Pill seem like the least radical of interventions—nature's contraceptive, something that could be slipped inside a woman's purse and pass without notice. Pike was going to take the whole forty-year mythology of "natural" and sweep it aside. "Women are going to think, I'm being manipulated here. And it's a perfectly reasonable thing to think." Pike's South African accent gets a little stronger as he becomes more animated. "But the modern way of living represents an extraordinary change in female biology. Women are going out and becoming lawyers, doctors, presidents of countries. They need to understand that what we are trying to do isn't abnormal. It's just as normal as when someone hundreds of years ago had menarche at seventeen and had five babies and had three hundred fewer menstrual cycles than most women have today. The world is not the world it was. And some of the risks that go with the benefits of a woman getting educated and not getting pregnant all the time are breast cancer and ovarian cancer, and we need to deal with it. I have three daughters. The earliest grandchild I had was when one of them was thirty-one. That's the way many women are now. They ovulate from twelve or thirteen until their early thirties. Twenty years of uninterrupted ovulation before their first child! That's a brand-new phenomenon!"

JOHN ROCK'S LONG BATTLE on behalf of his birth-control pill forced the Church to take notice. In the spring of 1963, just after Rock's book was published, a meeting was held at the Vatican between high officials of the Catholic Church and Donald B. Straus, the chairman of Planned Parenthood. That summit was followed by another, on the campus of the University of Notre Dame. In the summer of 1964, on the eve of the feast of St. John the Baptist, Pope Paul VI announced that he would ask a committee of Church officials to reexamine the Vatican's position on contraception. The group met first at the Collegio San Jose, in Rome, and it was clear that a majority of the committee were in favor of approving the Pill. Committee reports leaked to the *National Catholic Register* confirmed that Rock's case appeared to be winning. Rock was elated. *Newsweek* put him on its cover, and ran a picture of the Pope inside. "Not since the Copernicans suggested in the sixteenth century that the sun was the center of the planetary system has the Roman Catholic Church found itself on such a perilous collision course with a new body of knowledge," the article concluded. Paul VI, however, was unmoved. He stalled, delaying a verdict for months, and then years. Some said he fell under the sway of conservative elements within the Vatican. In the interim, theologians began exposing the holes in Rock's arguments. The rhythm method " 'prevents' conception by abstinence, that is, by the non-performance of the conjugal act during the fertile period," the Catholic journal *America* concluded in a 1964 editorial. "The pill prevents conception by suppressing ovulation and by thus abolishing the fertile period. No amount of word juggling can make abstinence from sexual relations and the suppression of ovulation one and the same thing." On July 29, 1968, in the "Humanae Vitae" encyclical, the Pope broke his silence, declaring all "artificial" methods of contraception to be against the teachings of the Church.

In hindsight, it is possible to see the opportunity that Rock missed. If he had known what we know now and had talked about the Pill not as a contraceptive but as a cancer drug—not as a drug to prevent life but as one that would save life—the Church might well have said yes. Hadn't Pius XII already approved the Pill for therapeutic purposes? Rock would only have had to think of the Pill as Pike thinks of it: as a drug whose contraceptive aspects are merely a means of attracting users, of getting, as Pike put it, "people who are young to take a lot of stuff they wouldn't otherwise take."

But Rock did not live long enough to understand how things might have been. What he witnessed, instead, was the terrible time at the end of the sixties when the Pill suddenly stood accused—wrongly—of causing blood clots, strokes, and heart attacks. Between the mid-seventies and the early eighties, the number of women in the United States using the Pill fell by half. Harvard Medical School, meanwhile, took over Rock's Reproductive Clinic and pushed him out. His Harvard pension paid him only seventy-five dollars a year. He had almost no money in the bank and had to sell his house in Brookline. In 1971, Rock left Boston and retreated to a farmhouse in the hills of New Hampshire. He swam in the stream behind the house. He listened to John Philip Sousa marches. In the evening, he would sit in the living room with a pitcher of Martinis. In 1983, he gave his last public interview, and it was as if the memory of his achievements was now so painful that he had blotted it out.

He was asked what the most gratifying time of his life was. "Right now," the inventor of the Pill answered, incredibly. He was sitting by the fire in a crisp white shirt and tie, reading *The Origin,* Irving Stone's fictional account of the life of Darwin. "It frequently occurs to me, gosh, what a lucky guy I am. I have no responsibilities, and I have everything I want. I take a dose of equanimity every twenty minutes. I will not be disturbed about things."

Once, John Rock had gone to seven-o'clock Mass every morning and kept a crucifix above his desk. His interviewer, the writer Sara Davidson, moved her chair closer to his and asked him whether he still believed in an afterlife.

"Of course I don't," Rock answered abruptly. Though he didn't explain why, his reasons aren't hard to imagine. The Church could not square the requirements of its faith with the results of his science, and if the Church couldn't reconcile them how could Rock be expected to? John Rock always stuck to his conscience, and in the end his conscience forced him away from the thing he loved most. This was not John Rock's error. Nor was it his church's. It was the fault of the haphazard nature of science, which all too often produces progress in advance of understanding. If the order of events in the discovery of what was natural had been reversed, his world, and our world, too, would have been a different place.

"Heaven and Hell, Rome, all the Church stuff—that's for the solace of the multitude," Rock said. He had only a year to live. "I was an ardent practicing Catholic for a long time, and I really believed it all then, you see."

HELEN EPSTEIN

The Mystery of AIDS in South Africa

FROM *THE NEW YORK REVIEW OF BOOKS*

1.

One Sunday evening in early May, I went for a walk in one of Johannesburg's prosperous suburban neighborhoods. The whitewashed stucco houses on well-tended lawns with hissing sprinklers and swimming pools, the twittering birds, the leaning jacaranda trees lined up on quiet streets, resembled similar scenes in Los Angeles or Melbourne, Australia. That is, if it were not for the barbed wire curled above the gates, or the dogs that roared at me from behind each fence as I passed by. By the time I had walked half a block, it seemed as though all the dogs in Johannesburg were barking. I didn't go far. South Africa lives under a kind of self-imposed curfew. By sundown, the streets from the Cape to the Transvaal are eerily empty. Gates are bolted, alarms are set, car doors locked, and windows rolled up.

South Africa is one of the most dangerous countries in the world that is not at war. Everyone I met warned me to be careful. One acquaintance spent ten minutes listing all the people he knew who, in the past six years, had been shot, killed, raped, or who had been hijacked in their cars, robbed, thrown in the trunk, and then deposited, naked, by a roadside. Another South African told me that the bank in his ordinary, middle-class neighborhood had been robbed five times in six months. A Johannesburg taxi driver said that, in his company alone, a driver is murdered every month. There were more than 50,000 reported rapes in South Africa in 1999, and this number has recently been rising. Reported rapes are believed to represent only a fraction of the actual number committed, and according to some estimates, as many as a million rapes may have occurred in 1999. A doctor I met who has worked in black hospitals in the Eastern Cape for decades, through the worst years of apartheid, told me she now sees a growing caseload of gonorrhea and syphilis in children as young as two years of age, the result of an epidemic of child sexual abuse.

Crime in South Africa affects everyone, black, white, Asian, rich, and poor. Last year, someone walked off with an entire automatic teller machine that had been installed inside a police building in Johannesburg. In Cape Town, rapes and burglaries have been committed by members of Parliament, within the Parliament buildings themselves. The sense of suspicion and paranoia seemed to me to pervade even the fancy shopping malls, tourist beaches, and expensive hotels. It even informs the country's policies, including its response to the greatest health threat in its history.

I WENT TO SOUTH AFRICA for three weeks in May 2000 to write about the AIDS epidemic there. AIDS is caused by the HIV virus, which is passed from person to person through sexual fluids, blood, or blood products, or from mother to unborn child in the womb or through breast-feeding. The virus destroys the immune system that protects the body from infectious diseases. A person may live for ten years or more with HIV and have no symptoms, but eventually his immune system begins to disintegrate, and other viruses, bacteria, and fungi, which a healthy immune system would normally fight off, take hold. AIDS is the name given to the syndrome in which the patient slowly rots alive from these opportunistic infections.

The South African Ministry of Health estimates that 4.2 million South Africans carry the HIV virus, and about 1,700 more people are infected every day. Those most at risk are poor black people, particularly those who have been socially displaced, such as migrant workers, truck drivers, sex workers, and miners from rural areas, and their wives, girlfriends, and children back home.

I became particularly interested in the AIDS crisis in South Africa in the winter of 1999, when I heard that President Thabo Mbeki had begun to solicit the opinions of a murky group of California scientists and activists who believe that AIDS is caused not by HIV but by a vague collection of factors, including malnutrition, chemical pollution, recreational drugs, and by the very pharmaceutical drugs that are used to treat the disease. These "AIDS dissidents" may not agree among themselves about what the cause of AIDS actually is, but most of them seem to believe that the tens of thousands of scientists who work on HIV and AIDS are, largely unwittingly, part of a vast conspiracy cooked up by the pharmaceutical industry to justify the market in anti-AIDS drugs, such as AZT, worth billions of dollars a year. This conspiracy, the AIDS dissidents argue, relies on the demonization of HIV, a harmless virus in their eyes, and the promotion of wildly expensive, toxic drugs that have serious, even deadly, side effects. Mbeki is the only head of state known to have taken the views of the AIDS dissidents seriously, and many doctors and AIDS activists in South Africa and in the West have begun to wonder whether the President and his health minister have taken leave of their senses.

In the 1980s, the AIDS dissidents received considerable attention from journalists and even mainstream scientists who felt that the evidence that HIV caused AIDS was not strong enough. By now, however, the evidence that HIV is the cause of AIDS is very strong indeed,[1] and the implications for countries like South Africa are horrifying. During the past five years or so, the AIDS dissidents seemed to fade from the scene. They continued to publish their views on the Internet, but journalists and scientists paid little attention to them. Many people were surprised when President Mbeki began expressing interest in their ideas. When Mbeki invited a group of them, including Berkeley professor Peter Duesberg, to South Africa to present their views to a presidential panel on AIDS in Africa in May, observers were even more surprised. One scientist told me, "It's like the movie *Friday the 13th.* Just when you think you've finally killed the monster, it just keeps coming back to life." The interest that Mbeki has taken in the AIDS dissidents is more than an intellectual diversion. It is contributing to a public health disaster by distracting the Health Ministry and other official institutions from addressing the epidemic, and by failing to prevent HIV infection in South African children.

WESTERN PHARMACEUTICAL COMPANIES now sell a range of some fifteen drugs known as anti-retrovirals for people with HIV infection and AIDS. In 1998, researchers in Thailand and the US found that a cheap, short course of AZT taken around the time of childbirth can reduce by half the chances that a mother will transmit HIV to her baby. Many developing countries, including Thailand, Botswana, and Uganda, are now putting programs in place so that eventually every HIV-positive pregnant woman will be offered AZT, or another anti-retroviral drug called nevirapine, which has also proved effective in preventing transmission of HIV during childbirth.

In 1998, a number of maternity wards in South Africa's public hospitals were also getting ready to establish pilot projects to see how feasible it would be to offer AZT to every South African HIV-positive mother-to-be, no matter how poor. Almost immediately, then Minister of Health Nkosazana Zuma suspended public funds for these projects, because, she said, even these short courses of AZT were too costly.[2] AIDS activists and doctors mounted protests, but the Ministry of Health maintained its anti-AZT policy. In February 1999, after the President began to entertain the ideas of the AIDS dissidents, he stated in an address to Parliament that AZT was not only expensive, it was also alleged to be toxic, or so he had learned from some of the AIDS dissidents' websites. AZT would not, therefore, be administered to pregnant women attending public hospitals until it had been thoroughly investigated.[3]

Then in March, Parks Mankahlana, Mbeki's spokesman, shed new light on the President's thinking about AZT in an article in the *Business Day* newspaper. The drug was not only expensive and toxic, he said, but was part of a corporate conspiracy to rip off Africa's poor:

> Like the marauders of the military industrial complex who propagated fear to increase their profits, the profit-takers who are benefiting from the scourge of HIV/AIDS will disappear to the affluent beaches of the world to enjoy wealth accumulated from a humankind ravaged by a dreaded disease. . . .
>
> Sure, the shareholders of Glaxo Wellcome [the company that makes AZT] will rejoice to hear that the SA government has decided to supply AZT to pregnant women who are HIV-positive. The source of their joy will not be concern for those people's health, but about profits and shareholder value.[4]

Today, about two hundred babies are born in South Africa every day with the HIV virus. If all of their mothers had been given AZT around the time of delivery, as many as half of those babies might have been spared HIV infection and AIDS.[5] In 1997 Glaxo Wellcome had offered the drug at a discount to the Health Ministry for use in public maternity wards, but the Health Ministry turned the offer down.[6]

Have you ever seen a child dying of AIDS? I worked on AIDS research in

Uganda in the early 1990s, and what surprised me about the pediatric AIDS wards I visited there was how quiet they were. The thin, breathless children were too sick to cry. Before I left for South Africa, I had arranged to speak to some of South Africa's most important scientists and government officials about the government's anti-AZT policy, which did not make sense to me. What I found when I got there was even more disturbing than what I had expected.

2.

IN THE MID-1990S, scientists in the United States discovered that combinations, or "cocktails," of anti-retroviral drugs can help people with HIV live longer. The cocktails are expensive, must be taken according to complicated schedules, can have unpleasant side effects, and don't help everyone. However, for many AIDS patients, the right cocktails of anti-retrovirals can add years of health to their lives. Since these drugs were developed in the early 1990s, hospital AIDS wards have been closing across the US and Western Europe, and many patients once on the verge of death have gone home, to contemplate a future they never thought they'd have.

The annual cost of AIDS drug cocktails is more than $10,000 per patient, and these days the only way a poor HIV-positive South African can obtain them is by participating in a clinical trial sponsored by one of the large Western pharmaceutical companies that make the drugs. These companies test their drugs in Africa because there are so many HIV-positive people there who have never taken anti-retroviral drugs before. Such patients are unlikely to have developed resistance to any of the individual drugs in the cocktails, and this gives the cocktails a better chance of success, which in turn allows the companies to obtain clearer results. At present hundreds of such trials are underway across the country, and HIV-positive South Africans seem eager to sign up for them. "People are desperate," Florence Ngobeni of the Township AIDS Project in Soweto told me. "There is so much confusion and fear." A doctor who runs clinical trials of AIDS drugs at the University of Pretoria told me that she has no trouble finding participants. "They come to me," she said.

While I was in South Africa, a doctor named Costa Gazi, whom I was interviewing about the anti-AZT policy, told me about one clinical trial that seemed to have had disastrous results. I had read wire reports about the trial on the Internet in April.[7] Around five hundred HIV-positive people had been enrolled to compare two different cocktails of anti-retroviral drugs at sixteen hospital clinics around South Africa. Since the trial began in September 1999, five participants had died, two of them from liver damage possibly caused by one of the drugs in the cocktails. According to subsequent news reports in the South African press, a sixth patient died in April.[8] Gazi told me that five of the six deaths on the trial had occurred at a single hospital in Pretoria, called Kalafong, out of only forty-two patients enrolled there. In addition, six other patients from the same hospital had experienced side effects so severe that some of them had had to be hospitalized.

I wondered whether this could possibly be true. The reports that five out of five hundred patients died on the trial were disturbing, although perhaps they could be explained by the occurrence of unexpected side effects or cases of disease. But if, in fact, eleven people out of only forty-two had become very ill or died, something must have gone terribly wrong. I knew which drugs were in the cocktail being tested in the clinical trial. They are known to have side effects, which, in rare cases, can be fatal if patients aren't properly examined and monitored. But thousands of people with HIV in Europe and the US have taken these drugs for years, and such severe, widespread toxicity has never been seen before.

Since then, I have been trying to find out whether these eleven people could have become seriously ill or died from such strange symptoms at one hospital in so short a time. For the moment, the truth is buried under endless evasions, fantasies, and myths; but the obstacles to finding out what happened are themselves revealing.

SHORTLY AFTER I ARRIVED in South Africa, the health minister, Manto Tshabalala-Msimang, gave a speech at the opening of the presidential AIDS panel, which consisted of about thirty-three people, of whom around half were AIDS dissidents and half were researchers from various countries who believed that HIV causes AIDS. "[This is] not just a simple academic indulgence. . . ," she said. "We place on record our determination [to fight the AIDS epidemic] with every means at our disposal." Certainly this was arguable, since public hospitals were not providing AZT to pregnant women.

Other evidence suggested that the government's response to AIDS was incoherent and disorganized. In financial year 1999/2000, the AIDS directorate in the Ministry of Health failed to spend 40 percent of its funds. In February, the government appointed a national AIDS council that included an athlete, a TV producer, numerous politicians, and two traditional healers, but did not include South Africa's most important scientists, doctors, activists, and representatives of non-governmental AIDS organizations. Less than 60 percent of South Africa's health facilities are able to carry out AIDS testing and counseling, and few community health care clinics are equipped or staffed to manage AIDS patients. Some people with HIV have reported being turned away, even for complaints not related to AIDS.[9] Ineffective and expensive campaigns to promote public awareness of AIDS had been mounted, including "National Condom Week," during which free condoms that had unfortunately been stapled to a card were distributed. I had just visited a public hospital in the East Rand, where I had asked a doctor what he had to offer people with AIDS. "We have no [anti-retroviral] drugs here," he said. "Not even for needle-stick injuries."

This shocked me. Doctors and nurses frequently stick themselves with bloody needles by mistake, and in this way may expose themselves to HIV. If someone who is exposed to HIV through a needle stick takes a high dose of anti-retroviral drugs immediately afterward, he can reduce his chances of becoming infected by about 80 percent. These drugs, in kit form, are supposed to be available in all hospitals for health care workers at risk of needle-stick injuries. But according to this doctor, they were not, at least "not since the President started talking to [the AIDS dissidents]." I was later assured by Patricia Lambert, a lawyer who works with the health minister, that all hospitals were supposed to have anti-retroviral kits for needle-stick injuries, but I still wondered how even one hospital might fail to have them. Perhaps the Health Ministry was too preoccupied with the presidential panel and the AIDS dissidents to enforce regulations concerning these drugs, so that at least one hospital fell through the cracks.

Perhaps this distracted Health Ministry had also failed to implement adequate ethical and safety guidelines for clinical trials. Patients taking anti-retroviral drugs should be given regular blood tests to ensure that the drugs are not causing any side effects. Perhaps those patients who, according to Gazi, became ill or died at Kalafong hospital had been on a clinical trial that was poorly monitored. Maybe the patients got a bad batch of drugs from the pharmaceutical company, or were particularly sensitive to them for some reason, and no one recognized it in time. Of course there might have been other explanations for what seemed to have happened at Kalafong.

3.

About six miles west of Pretoria, the township of Atteridgeville blankets the tall grass and thorn bushes of the South African veldt. Like many of the townships on the outskirts of South Africa's cities, Atteridgeville was built to house Pretoria's black servants and factory workers.

As the grip of apartheid began to loosen, the population of Atteridgeville exploded, and the township spread over the brown hillsides. The gap between rich and poor is greater in South Africa today than in any other country except Brazil, and as you turn into Atteridgeville from the main road from Pretoria, the inequalities of the new South Africa unfold before you.[10] On the hill stand rows of new houses, many of them built since the elections in 1994 for South Africa's emerging black elite, mainly lawyers, civil servants, and businessmen. The hill is a modest version of the grander suburbs once reserved for whites. The houses are stucco, two-story structures, with tidy back yards and metal gates and signs indicating that the houses are alarmed and that trespassers will be met with an armed response. At the foot of the hill lies the neat grid of streets and tiny brick houses built for the workers during the apartheid years, and just beyond this stretches the vast plain of Atteridgeville's squatter camp.

In the squatter camp, the listing shacks, made of sheet metal, wood, and heavy cardboard, are occupied by the overflow from the overcrowded township itself and by migrants from Zimbabwe, Mozambique, and other countries to the north. There are no municipal services here. Garbage piles up by the roadside, and the squatters string wires from power lines to steal electricity. Water is collected from communal taps and sewage is ad hoc.

A tin shack I visited in the squatter camp was dark and cool and orderly. Its dirt floor was swept, and bright plastic jugs were lined up beside a small portable stove. The few pieces of furniture were draped with patterned cloth, and an ornamental mask hung on the wall. Andrew, the shack's resident, wore a sweater and button-down shirt, and looked very young, like a teenager about to have his yearbook photograph taken. He lost his job as a schoolteacher in Zimbabwe in the early 1990s, when the government imposed sharp cuts in the education budget to satisfy requirements for receiving a World Bank loan.[11] After the change of government in South Africa, Andrew and his wife came here to look for jobs, but without a South African degree, he had little hope of finding work as a teacher. Instead he picks up odd jobs as a plasterer or bricklayer on some of the new building sites on the hill.

In March of 1998, Andrew's wife gave birth to a daughter. The baby did not thrive. When she was about a year old, she became feverish and ill and her mother took her to Kalafong, the large, state-run hospital that serves the black population of Atteridgeville. The child was tested for HIV and found to be positive.[12] Then Andrew and his wife were tested as well. When they were both found to be positive, they were referred to a Dr. Ingrid Steenkamp, who runs the HIV clinic at Kalafong. Dr. Steenkamp is also affiliated with the University of Pretoria, under whose auspices she conducts clinical research on AIDS patients.[13]

THE UNIVERSITY OF PRETORIA had recently been chosen by an American pharmaceutical company to be one of the sites for their clinical trial of a new anti-retroviral drug cocktail. The university in turn had chosen Dr. Steenkamp to recruit HIV-positive people to participate in the trial. She would conduct blood tests to determine who was eligible to participate, give participants the trial drugs, counsel them about how to take them, and conduct follow-up examinations and lab tests to determine how well the drugs were working and to detect any side effects. Because Dr. Steenkamp worked at Kalafong, she had access to a large and growing population of HIV-positive people, most of whom were very poor.

Dr. Steenkamp invited Andrew and his wife to sign up for the trial of this new drug cocktail that would fight the virus in their blood and extend their lives. They would receive four bottles of pills that had to be taken at precise times. Andrew and his wife agreed to sign up.

To be eligible for the trial, the participants had to be HIV-positive, but they also had to be healthy and not suffering from AIDS. Andrew had never been sick before, he told me, and it was only after his daughter's illness and death that he discovered he was HIV-positive. However, almost as soon as he began taking the drugs, he felt very weak. He mentioned this to Dr. Steenkamp, but "she said it was not the tablets that made me sick, it was the HIV, and that I should keep taking the medicine." Within a month, Andrew had to be admitted to Kalafong hospital, with vomiting, rash, fever, and painful, bleeding sores that covered his entire body.[14] Andrew stopped taking the trial drugs, and his wife, who did not experience any of these symptoms, also dropped out of the trial.

Molly also lives in Atteridgeville. She is about thirty years old, and has been HIV-positive since 1995. She was being treated for tuberculosis when, she says, Dr. Steenkamp enrolled her on a clinical trial at Kalafong. The TB treatment had been working, and Molly felt well enough to go to her job as a counselor at an HIV support center, and to care for her two children. However, soon after she started taking the drugs that Dr. Steenkamp had given her, her health deteriorated. She came down with pneumonia, severe constipation, and cramps, and her throat became very sore. Most frightening of all, she said, soon after she started to take the drugs, her sight began to fade and she nearly went blind. "Dr. Steenkamp said, 'This is caused by the HIV, and you won't see for the rest of your life,' " Molly told me. But as soon as Molly stopped taking the drugs that Dr. Steenkamp had given her, her sight returned. When I met her, Molly was sitting on the edge of a bed in the TB hospital, before a plate of French fries and sausage that she seemed too weak to eat. She was extremely thin, found it difficult to talk, and would lean over at intervals and cough violently into a heap of tissue paper on the bed.

"They say it's TB," she said, "but I don't believe them. The treatment was working before. I don't know why it's not working now."

4.

COSTA GAZI, THE DOCTOR, who had first alerted me to the possible problems on the Kalafong trial, is a member of Parliament for the Pan-Africanist Congress of Azania, or PAC. The PAC split off from the African National Congress in the late 1950s, and its first leader, Robert Mangaliso Sobukwe, led the protest against the notorious Pass Laws that erupted into the Sharpeville massacre in 1960. Famous for their slogan "One Boer, One Bullet," the PAC took a harder line against whites than the ANC did. Today, the PAC calls for racial reconciliation and for the redress of such lingering roots of inequity as corruption, crime, and poor health care and education, which it claims the ruling African National Congress is not dealing with effectively. Some PAC policies seem eccentric, such as calling for a repeal of the ANC's ban on cigarette advertising in order to preserve the tobacco industry and advertising jobs, or for the establishment of an office for traditional African kings inside the president's office. The PAC is today a beleaguered party, having won only three seats in the 1999 elections. Nevertheless, what remains of the PAC is very outspoken, and often controversial. In April 1999, Gazi himself accused Nkosazana Zuma, who was then health minister, of manslaughter for not providing AZT to pregnant HIV-positive women. He was later fined R1000 (about $160) for "bringing Zuma into disrepute." He is contesting the charge.[15]

Confirming what Dr. Gazi had told me about what had happened at Kalafong proved extremely hard to do. He put me in touch with another PAC MP named Patricia de Lille, whom I met in Cape Town. She told me what she had heard about the Kalafong trial, and she explained that she felt monitoring procedures on clinical trials were inadequate, and that the government was wasting time with the dissidents. De Lille then referred me to a lapsed Anglican priest in Johannesburg named Johan Viljoen.

During the winter of 1999, Viljoen spent five months working with the Motivation and Educational Trust, an AIDS organization that runs a support center for HIV-positive people based at Kalafong hospital. Around Christmas, a few months after he started working for the Motivation and Educational Trust, Viljoen began to hear disturbing reports from some of its clients who had been participating in the AIDS drug trial run by Dr. Steenkamp at Kalafong.

"Many of the people on the trial knew each other. I got involved when some of them started coming to me and telling me about this one that died, and then that one. People were really worried. There was this spate of four or five deaths right before Christmas." According to what the patients told Viljoen, four women who went on the trial in October were dead by Christmas, and a further five women and one man, including Andrew and Molly, became very ill after taking the pills they were given. Then, in April, another woman died.

Viljoen gave me copies of signed testimonies that he had collected from the six patients who said they had become ill on the trial, and from a man who said his mother had died. All these patients, including Andrew and Molly, testified that they were HIV-positive; they all named Dr. Steenkamp and all said that she had put them on a clinical trial. Only one of the testimonies specifically named the American pharmaceutical company trial, however. The symptoms the patients described were very strange. "I developed severe headaches and fevers, I felt extreme stress, depression, and anger," testified one woman. Another woman said, "I began to experience muscle cramps, spasms, and fits. I also feel stress, to such an extent I feel I might get a stroke." Another wrote, "In February 2000, I began to have a serious rash, all over my body."

VILJOEN BECAME AN Anglican priest in the mid-1980s. At the time, South Africa was practically at war with its own people. In the townships and homelands, the anti-apartheid struggle erupted into a nearly constant series of battles between factions of black freedom fighters and the state security forces, which were using their most devious and cruel tactics to keep their grip on the country. There were raids, firebomb and tear gas attacks, torture, detentions, and murders, and this was exactly where Viljoen wanted to be. "I wanted to really live and work with people in the townships," he told me, "and the clergy were the only white people who were allowed to do that. There was always something going on, there were riots practically every week, and all the sermons I gave were about politics and how we had to do something about the injustices of apartheid." After the trouble began to die down, Viljoen quit the priesthood, which he found bureaucratic, and went to work with a Catholic refugee organization in Mozambique, Angola, and South Africa.

"I'm a real Afrikaner," Viljoen told me. "My ancestors came over from Europe in the 1600s, they were in the Great Trek, the Boer War, the whole thing." Viljoen's father was a diplomat for the apartheid government, and the family moved around a lot, to Switzerland, Italy, Israel. While I was in South Africa, I met a number of young whites who told me how hard it was for them to understand how their parents, who might have been perfectly nice people, could have lived under apartheid without horror, and could have tolerated its injustices. I wondered how Viljoen felt about that, and whether it had influenced his decision to work with disadvantaged people. "I didn't decide to do good works because I felt I had to make reparations or anything," he said, "but I guess I had seen both sides of South Africa, and I wanted to work to make it a better place."

In the fall of 1999, Viljoen met Father Barry Hughes-Gibbs, the Anglican priest who runs the Motivation and Educational Trust. "Father Barry called me up and told me that he and his wife had had a vision that I would quit my job working with refugees to go to work with him at the Motivation and Educational Trust." Viljoen is himself HIV-positive. "I guess that did influence my decision to take the job. I felt I had something to offer those people."

There was something about Viljoen's account that did not make sense. The patients' complaints were so varied and so strange, and many of them did not correspond to any known side effects of anti-retroviral drugs. In retrospect, I should have wondered more about the situations they described in their testimonies. It was certainly possible that these patients were not, after all, on a trial of anti-retroviral drugs sponsored by an American company. Perhaps they were entirely mistaken about being on any sort of trial, or perhaps they were on a different trial altogether. No proof of any wrong-doing by any person involved in the trials emerged, but much remains unclear. In any case, I wanted to know whether the government, or any other authority, was doing anything to find out what had happened.

5.

I HAD HEARD that a report on the American AIDS drug trial had been drafted within the Health Ministry, but that it had not been made public. I very much wanted to speak to someone who had read it. In addition, I wanted to discuss the other controversies surrounding AIDS and its treatment in South Africa, such as AZT for pregnant women and Mbeki's relationship with the AIDS dissidents. But one by one, most of my appointments were canceled.

One senior government scientist changed our appointment six times and then left on a trip to the United States. Many others never returned phone calls, e-mails, or faxes. Dr. Nono Simelela, the head of the AIDS directorate in the Ministry of Health, agreed to meet me in Pretoria on the Thursday after I arrived. When I turned up at her office, her assistant told me that Dr. Simelela had been called away suddenly, just fifteen minutes previously, to the Director General's office. I was introduced instead to Cornelius Lebeloe, who is in charge of counseling, testing, and support of AIDS patients. He told me he was not allowed to talk about the AIDS dissidents, the government's position on AZT for pregnant women, or the trial at Kalafong. When Dr. Simelela appeared an hour later, she said she was too busy to talk to me.

"How about tomorrow?" I asked.

Dr. Simelela opened her diary.

"No."

"Saturday?"

"No."

"Sunday?"

"No."

"Monday?"

"No."

"Tuesday?"

"OK, call me at ten in the morning on Tuesday."

When I phoned Dr. Simelela the following Tuesday, I was told that she was out. Her assistant suggested I call again on Friday morning at nine. On Friday morning, Dr. Simelela was not in the office. I called several times the following week, but Dr. Simelela was either out of town, in a meeting, or otherwise unavailable. After I returned to New York I sent an e-mail, requesting an interview by phone. I have yet to receive a reply.

I also arranged to meet Dr. William Makgoba, the head of the South

African Medical Research Council. Dr. Makgoba had attended the presidential AIDS panel where the AIDS dissidents had made their presentations. He was an outspoken critic of the President's policies on HIV and of the anti-AZT policy, and he considered the AIDS dissidents "pseudo-scientists." However, Makgoba is also an ideological ally of Mbeki's. He recently edited a book of essays called *African Renaissance*,[16] which includes a preface by the President. In South Africa, discussion of the anti-AZT policy and the President's relationship with the AIDS dissidents usually involves speculation about the President's psychology. What could he possibly be thinking? Mbeki takes the African Renaissance very seriously, and I thought Makgoba, if anyone, might have some insight into how it could be influencing his thinking on AIDS.

THE AFRICAN RENAISSANCE was conceived by black intellectuals in order to supply Africa with a founding ideology that would inform development strategies that draw on African, rather than Western, models, and take account of African realities. It may, for example, provide an alternative to the kind of Western corporate exploitation that Mankahlana described in his article about AIDS in *Business Day.*

Observers often find the African Renaissance perplexing. How can anyone talk about an African Renaissance in the shadow of the Rwanda genocide and the wars that have broken out across nearly all of Central Africa, from Sierra Leone to Angola, to the Democratic Republic of the Congo, to Ethiopia, conflicts that seem to be pulling the rest of the continent down with them? Not to mention the floods in Mozambique, the drought in Ethiopia, the resurgence of malaria, and the AIDS epidemic?

One of Mbeki's most famous speeches, given at the adoption of South Africa's new constitution in May 1996, movingly addressed this paradox:

> I am an African. I owe my being to the hills and the valleys, the mountains and the glades, the rivers, the deserts, the trees, the flowers, the seas and the ever-changing seasons that define the face of our native land. . . . I owe my being to the Khoi and the San whose desolate souls haunt the great expanses of the beautiful Cape—they who fell victim to the most merciless genocide our native land has ever seen. . . . I am formed of the migrants who left Europe to find a new home on our native land. Whatever their actions, they remain still part of me. . . . I am the grandchild of the warrior men and women that Hintsa and Sekhukhune led. . . .

> I have seen our country torn asunder as these, all of whom are my people, engaged one another in a titanic battle. . . . The pain of the violent conflict that the peoples of Liberia, Somalia, the Sudan, Burundi and Algeria experience is a pain I also bear. The dismal shame of poverty, suffering and human degradation of my continent is a blight that we share. . . . This thing that we have done today says that Africa reaffirms that she is continuing her rise from the ashes. Whatever the setbacks of the moment, nothing can stop us now! . . . However much we have been caught by the fashion of cynicism and loss of faith in the capacity of the people, let us say today: Nothing can stop us now!

Thus Mbeki expresses his faith in an Africa whose suffering is its strength, and is finally optimistic about a continent whose future can't possibly be worse than its past.

In the book *African Renaissance,* one of the contributors, Bernard Makhosezwe Magubane, praises Barbara Tuchman's idea, derived from William McNeill's book *Plagues and Peoples,* that Europe's Renaissance was partly made possible by the calamities of the Middle Ages such as plague, inequality, and war. Magubane speculates that Africa's Renaissance may also emerge from Africa's current political, ecological, and health crises.

I was particularly interested in the scientific elements of the African Renaissance movement, which emphasizes the exploration of indigenous knowledge systems, such as traditional medicine. I wondered if Dr. Makgoba knew whether any South African scientists were looking for a cure for AIDS in the herbal pharmacopeia of Africa's indigenous medicine men and women. I was very eager to speak to Dr. Makgoba about this because, if such research were going on, it would not be the first time the government had been involved in unconventional AIDS research.

IN 1997, A GROUP OF researchers at the University of Pretoria approached Mbeki, who was then deputy president under Nelson Mandela, and Nkosazana Zuma, who was health minister at the time. The researchers claimed to have a cure for AIDS, which they called Virodene. Zuma invited them to address the Cabinet on their research, a highly unusual move in any country. Virodene consisted of an impure solution of dimethylformamide, an industrial chemical used in dry cleaning, among other things. It is very unlikely to be in any way beneficial to AIDS patients, and may well be harmful, since it is known to cause liver damage and skin rashes and may even activate HIV, possibly accelerating the course of AIDS.[17]

Clinical trials of Virodene on HIV-positive people were carried out without the authorization of the Medicines Control Council, the state body that regulates research and approval of drugs,[18] and Zuma, Mbeki, and the Virodene researchers were criticized for their behavior concerning the drug by numerous South African doctors and scientists and by the South African Medical Research Council, the Democratic Party, and the Medicines Control Council.[19] In an open letter in the South African press, Mbeki wrote an indignant defense: "How alien these goings-on seem to be to the noble pursuits of medical research! In our strange world, those who seek the good for all humanity have become the villains of our time." He defended his decision to try to "facilitate the carrying out of the critical clinical trials that would test the efficacy of Virodene."[20] Shortly after the Medicines Control Council tried to block further trials of Virodene in human subjects, its chairman and several other officials were fired.

Dubious AIDS cures are nothing new in Africa, or anywhere else, but the strange thing about Virodene was that it seemed to attract interest in such high places. The owners of the company that manufactured Virodene, called CPT, were engaged in a court battle over ownership of the company, and during the proceedings, a memorandum from the company surfaced, which stated that "the ANC is to receive six percent shares in the CPT."[21] Zuma and Mbeki denied any knowledge of this document, and there is no evidence that they had been aware of the offer. The Virodene story made headlines in South Africa and elsewhere in 1998, but since then the furor has subsided, and no one I spoke to in South Africa seemed to know whether any further research on it is being conducted.

Virodene was not an indigenous African cure for AIDS, but it did emerge from an African university, and I wondered whether Mbeki, who was so hostile to AIDS drugs made by Western pharmaceutical companies, might not have been interested in it simply because he wanted to give support to drugs developed in Africa.

The Virodene episode, Mbeki's skepticism of Western approaches to the AIDS epidemic, and his support for the African Renaissance reminded me of an essay I had read in *The Sunday Times* of South Africa by Mark Gevisser, Thabo Mbeki's biographer.[22] Gevisser describes a correspondence between Mbeki and his friend Rhianon Gooding that took place when Mbeki was studying in Moscow in 1969. In his letters, Mbeki discusses his admiration for

Coriolanus, who fought to save Rome, but was too high-minded to live in it, and ended up making war on his own people. Gevisser does not quote Coriolanus' most famous lines from the play, but I immediately thought of them as I read the essay. When the Senate threatens to banish Coriolanus from Rome because he refuses to accept the mantle of a Roman hero, Coriolanus replies, "I banish you! . . . There is a world elsewhere." Like Coriolanus, Mbeki has stubbornly decided to debate AIDS on his own terms, whether by defending the Virodene researchers or by denying fifteen years of research on HIV and AIDS. As he does so, his pride may well destroy his own people.

DR. MAKGOBA ARRIVED an hour late for our meeting, and had to leave almost immediately to catch a plane. As I waited in his office, his secretary told me that she had reminded him of our meeting that very morning, and he had said he would be able to make it. But then he had gone to see his dietitian, and as far as she knew, he had been detained there. This seemed odd. Why suddenly consult a dietitian? And then be detained there? I began to feel as though I had come to a land in a fairy tale, where everybody is evasive and ignores appointments.

When Dr. Makgoba finally appeared, he didn't want to talk about the AIDS dissidents, the African Renaissance, or Coriolanus. He did tell me that research on traditional African medicine for AIDS was underway in South Africa, but when I asked for details, he said he didn't want to talk about it. After fifteen minutes he asked me to leave.

6.

BACK IN JOHANNESBURG, I finally found someone who would talk to me about the Kalafong trial. Professor Geoffrey Falkson is a retired oncologist, and he also runs the Ethics Committee at the University of Pretoria, which approved Dr. Steenkamp's application to conduct the clinical trial on behalf of the American pharmaceutical company. As we sat in Falkson's small, sunny office in the academic hospital he said, in answer to my questions, that five patients out of forty-two had died on the trial at Kalafong. "Isn't that rather a lot?" I asked.

"They had full-blown AIDS!" he said.

"No, they didn't."

He took the protocol book off his shelf and turned to the page where the inclusion criteria for the trial were listed. Clearly, patients with AIDS were not allowed on the trial.

Professor Falkson seemed confused about many of the details of the trial, and he didn't seem to appreciate the gravity of what he was telling me had happened, that five healthy HIV-positive people suddenly died after receiving medications that were supposed to extend their lives. He talked about the difficulty of conducting trials in South Africa, the high ethical standards that prevail there, the selflessness of investigators like Dr. Steenkamp. Finally, he wrote down a web address for me, where, it turned out, some of his oil paintings could be viewed.

Finally I asked Dr. Falkson to tell me who was ultimately responsible for examining and caring for people like Andrew and Molly, who might have been harmed on a clinical trial, and for all the people who died.

"Clindipharm are the ones to talk to about that," he said. Clindipharm is a private company based in Pretoria that serves as an intermediary between pharmaceutical companies wishing to conduct clinical trials of their drugs using South African patients and South African doctors and scientists wishing to carry out the work itself. Clindipharm packages the drugs, distributes them to the doctors, and manages the data that come in from the laboratories and doctors' offices. The medical director of Clindipharm refused to speak to me, but he did tell me that all adverse events on clinical trials are reported immediately to the Medicines Control Council, at which point they cease to be Clindipharm's responsibility.

DURING MY LAST WEEK in Johannesburg, I tried without success to reach Dr. Helen Rees, the new head of the Medicines Control Council. By now, of course, I was not surprised when Dr. Rees did not return my calls. However, just as I was leaving South Africa, I spotted Dr. Rees in the departure lounge of the airport in Johannesburg.

"You have your facts wrong," she said, when I told her I had heard that five people died on the Kalafong trial.

"Well, how many people did die at Kalafong, then?" She did not remember. I asked Dr. Rees who was responsible for following up adverse events on clinical trials, and examining patients like Molly and Andrew who might have been harmed by experimental drugs.

"We refer the matter to the local Ethics Committee. We take these things very seriously," she said.

I said I thought that sounded strange, because Professor Falkson, the head of the Ethics Committee at the University of Pretoria, had told me that Clindipharm was responsible for following up adverse events. Clindipharm, in turn, had told me that they referred all reports of adverse events to the Medicines Control Council, and now the head of the Medicines Control Council was telling me she refers them back to the Ethics Committee. The responsibility seemed to have gone full circle.

Dr. Rees told me that new legislation governing the ethics of clinical trials was now in draft form, and had been under negotiation for more than two years. It would be presented to Parliament soon, she said. I wondered whether the legislation had not come too late for the patients at Kalafong. Clearly it would be important to have such legislation in place before clinical trials of drugs that the government itself alleges can be toxic should be allowed to proceed. As it turns out, the ethical guidelines currently in force were drafted in the 1960s, a time when high ethical standards did not prevail in South Africa. This only deepened my suspicion that the Ministry of Health might have been distracted by the AIDS dissidents and was failing to expedite far more important issues.

On the airplane, I found myself sitting next to a young white South African who worked as a salesman for a Pretoria company that sold CAT scanning machines. He followed the controversies over AIDS pretty closely in the South African newspapers. "It's all political," he said. "Everything is political in South Africa."

I told the CAT scanner salesman about my baffling interview with Dr. Rees, and the extreme evasiveness of so many people in government and other scientific institutions. "You see," he said, "these people aren't used to journalists asking questions. During the struggle, they only had to speak to reporters on their own side. Now when a reporter challenges them they think 'she's on the wrong side' and they'll do anything to avoid you." I realized that so much of my information about the trial, and about AIDS in South Africa in general, had come from one side only, that of activists like Gazi and de Lille in the PAC, and the lapsed priest, Viljoen. This was why I had not been able, in the three weeks I was in South Africa, to figure out with any certainty what had happened at Kalafong. I had been trapped in a hall of mirrors, and many of them reflected only a particular political agenda.

7.

When I returned to New York, I called the American pharmaceutical company that had sponsored the clinical trial in South Africa and asked them to tell me how many patients died at the Kalafong site. I was surprised to learn that, contrary to what I had heard in South Africa, only one patient on the trial had died at Kalafong. Viljoen had given me the names of the patients whose families said they had died after being enrolled on the trial by Dr. Steenkamp. I called Dr. Steenkamp in South Africa, and she told me that none of these patients were, in fact, on the trial. She would say nothing about the patients who were sick but still alive because, she said, that would violate their privacy. When I told Dr. Steenkamp that all of these patients seemed to think they were on a trial, according to signed testimonies given to me by Viljoen, and all of them claimed to have come regularly to receive drugs from her, Dr. Steenkamp told me she had certainly been treating some of those patients, but she would not say whether they had been on a trial. What on earth was going on?

My first thought was that some of the patients might have been screened for the trial, but because they were already sick with AIDS, they were not eligible. Dr. Steenkamp may have given them some medication, but the patients may have misunderstood and thought they were actually on the trial. Perhaps these people became much sicker right after their appointment with Dr. Steenkamp, simply because the disease was taking its course. When they died, their families blamed the doctor.

The HIV/AIDS epidemic came late to South Africa, but since 1993, the virus has been spreading very fast. Today, 15 percent of South African adults are HIV-positive, but most people were infected recently, in the last seven years. People may live perfectly healthy lives with HIV for a decade, so the great wave of death from AIDS in South Africa is now beginning to break. Perhaps these five deaths were an early swell of this great wave. Perhaps five people died of AIDS and another six got sick. Could it be that the sick ones and the relatives of the dead ones could not accept what was happening, and were looking for a culprit? Perhaps whispers circulated in the townships about a doctor at Kalafong. Hysteria may have ensued, perhaps fueled by reports of the controversy in the President's office over whether AZT and other antiretroviral drugs were part of a Western plot and might even be the cause of AIDS. In the eyes of the surviving patients and the families of the dead, Dr. Steenkamp may have been seen as some sort of witch.

BUT DETAILS OF THE STORY that just didn't fit together still perplexed me. Why would six patients and the families of five others think they were all on a clinical trial if they had not been? Why did Professor Falkson, who was admittedly weak on details, tell me that five patients had died on the Kalafong trial if in fact there was only one death there? Why did one of the patients tell Viljoen that after she and the others had received their trial drugs from Dr. Steenkamp, they all sat around together in the lunchroom at the Motivation and Educational Trust, unwrapping their parcels of pills, only to discover that not everybody got the same kinds of pills? Some people got two bottles, some four, and the pills looked different. The American trial was placebo-controlled and blinded, which means everybody should have received exactly the same thing. Why did one of the patients who received drugs from Dr. Steenkamp write in her testimony, "[Steenkamp] . . . told me not to talk to anybody else about [the trial] because it was our secret"? When I called Dr. Steenkamp from New York, she insisted that she had not been conducting any other clinical trials with AIDS patients at the time she was working on the American drug trial.

Why did the patients who became ill report such strange symptoms? When Andrew described his symptoms to me, they sounded a lot like Stevens-Johnson syndrome, a potentially fatal, but rare, side effect of one of the anti-retroviral drugs on the trial sponsored by the American pharmaceutical company. But when I spoke to the pharmaceutical company representative, he told me that no cases of Stevens-Johnson syndrome at the Kalafong site had been reported to him. If any cases had occurred there, he assured me he would have known. So if Andrew's symptoms were not drug-associated Stevens-Johnson syndrome, what were they?

Molly's testimony said she was put on a "drug trial," but she had tuberculosis, which would have excluded her from the American trial. Why, after Molly got sick, was she prevented from looking at her own medical file by the authorities at Kalafong? A friend of Viljoen's had sneaked into the hospital disguised as a nun and had stolen the file, but when it was opened, Viljoen told me, the most recent three months of notes were missing. And why did Molly say she went temporarily blind? I have spoken to a number of AIDS researchers who have long experience with anti-retroviral drugs, and temporary blindness is not a known side effect of any of them. Blindness does occur in AIDS patients, but it does not get better.

Clinical trials are often surrounded by secrecy. South African politics and evasion obscure matters even further. In early June, I was leafing through old copies of *The Lancet* in the library and I came across an editorial criticizing Mbeki's anti-AZT policy.[23] "Is there some other agenda," the author speculated, "such as the promotion of locally developed drugs?" Perhaps South Africa's paranoia was beginning to affect me too, but I thought of the unauthorized clinical trials of Virodene, and of Mbeki's interest in the African Renaissance and in finding African solutions to African problems. I also thought of Dr. Makgoba's reluctance to tell me about South African research into traditional African medicine for AIDS. I began to wonder whether some accident might not have occurred at Kalafong. This seemed unlikely; but what did happen remains unexplained.

8.

PRESIDENT MBEKI ARRIVED in the United States on the same day I did. On Tuesday, May 23, he was interviewed on *The NewsHour with Jim Lehrer,* and he was asked specifically about his controversial AIDS policies. His replies were a series of evasions. He claimed that he had never said that HIV was not the cause of AIDS, but he did not deny that he had questioned the link between them. Nor did he explain why, if he did not question this link, he had invited Duesberg and his followers to present their views to the presidential AIDS panel. When he was asked why he refused to make AZT available in public maternity wards for the prevention of HIV transmission from mother to newborn child, he answered that the state could not afford to make anti-retroviral drugs available for life to all HIV-positive people in South Africa. But he was not answering the question that had been put to him. He had not been asked why he didn't provide all HIV-positive South Africans with anti-retroviral therapy for life. He had been asked why he didn't provide it only to HIV-positive women for a month, or perhaps only a week, around the time they give birth. This would cost about $50 million a year, which the Health Ministry could well afford, especially in view of the fact that it will cost many times more to treat the children who might have been spared HIV infection when they eventually come down with AIDS.

Nevertheless, Mbeki and his government continue to argue that providing these women with anti-retroviral therapy is too expensive and the drugs are too toxic. His critics, meanwhile, continue to argue that the drugs are not too expensive, and that their prices are falling anyway, and that they are very unlikely to be toxic in the short doses used to prevent infection during childbirth.

THIS ISSUE MAY BE clarified at the upcoming Thirteenth International AIDS Conference in Durban in early July of 2000. A group of researchers working at Chris Hani-Baragwanath Hospital in Soweto have been testing a cheaper anti-retroviral drug called nevirapine, to see how well it protects babies born to HIV-positive women from HIV. There is already evidence that it does.[24] The government has quietly approved nevirapine pilot projects in public maternity wards, and there is hope that the results of the Baragwanath trials will be encouraging enough for these projects to be approved by the Medicines Control Council.

The science and politics of AIDS can seem complicated, and even illusory, to those who aren't familiar with them. Like some mystical Hebrew text, it may seem as though they can be interpreted in many different ways. As I watched Mbeki on television dodging every direct question, I felt that he wasn't being straight with his American audience, or with his own people. He seemed to be hiding something.

The anti-apartheid struggle never quite erupted into a real war, but there is nevertheless a vaguely postwar atmosphere in South Africa, in the self-imposed curfew, the corruption and crime, the drug shortages, the shifty-eyed officials, and in the sense that something terrible has happened there, in a nation haunted by death and ruined lives, with a visionary ruler who has not adequately confronted the most deadly threat facing his country.

NOTES

1. Interested readers can find a discussion of the evidence that HIV is the sole cause of AIDS on the website of the National Institutes of Health (www. niaid.nih.gov/publications/hivaids/all.htm). See also Richard Horton, "Truth and Heresy about AIDS," *The New York Review*, May 23, 1996.

2. See Howard Barrell and Stuart Hess, "Zuma Defends AZT Policy," *Mail and Guardian* (Johannesburg), October 16, 1998. An AZT pilot project is underway in Khayelitsha Township in Western Cape Province, with funding from the provincial government and Doctors Without Borders. The project has been allowed to continue, and is likely to be expanded, because Western Cape Province is under the control of the National Party, which has been critical of the anti-AZT policy. South Africa's other provinces have suspended AZT pilot projects. They are all under the control of the ANC, except for KwaZulu Natal, which is controlled by the Inkatha Freedom Party.

3. See "AIDS Exists. Let's Fight It Together," *Mail and Guardian*, February 11, 2000. AZT and other anti-retroviral drugs are unlikely to be toxic to unborn children in the doses needed to prevent HIV transmission. The Centers for Disease Control has followed

the cases of more than 20,000 children born to women in the US who took AZT during pregnancy, and found no evidence of long-term toxic effects. In France, two children died from what seemed to be toxic exposure to AZT and another anti-retroviral drug, 3TC. These children were exposed to higher doses of the drugs than would be administered in South Africa. See "Timeline of events related to follow-up of children exposed to anti-retrovirals perinatally," Centers for Disease Control, 2000.

4. Parks Mankahlama, "Buying Anti-AIDS Drugs Benefits the Rich," *Business Day*, March 20, 2000.

5. For logistical reasons, the actual proportion of babies that might be spared infection with HIV if a universal AZT program were implemented in maternity wards would be lower than 50 percent, at least at first.

6. See "Ministry Refuses Anti-HIV Drug Discount," *Mail and Guardian*, May 7, 1999.

7. "Triangle Pharmaceuticals Announces Clinical Hold on Study FTC-302," *PRNewswire*, April 7, 2000.

8. See "AIDS Trial Woman Dies," *The Citizen*, April 24, 2000.

9. "A Duty of Care Means Good Money Shouldn't Lie Idle," *Financial Mail*, March 17, 2000; Ivor Powell, "Uproar over AIDS Council," *Mail and Guardian*, January 28, 2000; Judith Soal, "HIV Patient Was Denied Treatment," *Cape Times*, May 11, 2000; N. Lamati, testimony given at the Health Portfolio Committee hearings, Parliament of South Africa, Cape Town, May 10, 2000; see also www.tac.org.za/parlhear.txt.

10. This gap has actually widened since the African National Congress came to power in 1994. The wealth gap between blacks and whites has narrowed, but inequalities within the black population have widened considerably. (R. W. Johnson, Helen Suzman Foundation, unpublished.)

11. During the 1980s and 1990s, the World Bank established the Structural Adjustment Program, which provided loans to the poorest countries that had such bad credit they were unable to obtain loans from other banks. One of the conditions for receiving a structural adjustment loan was that countries had to reduce public spending. For the Zimbabwe government, this meant cutting the education budget and laying off teachers.

12. HIV testing may not be conducted in South Africa, or in most other countries, without consent, or, in the case of a child, without the consent of the parents. Andrew maintains that neither he nor his wife gave permission for the child to be tested, and if this is true, then the hospital committed a serious ethical violation.

13. The names of the trial participants and Dr. Steenkamp have been changed.

14. At the beginning of the trial Andrew had been given a form with a telephone number on it to call if he wished to know his rights as a participant on the trial. The phone number was in fact a fax number, of little use to someone feeling sick in the middle of an impoverished squatter camp.

15. P. Dickson, "AIDS Activists Set Up Watchdog Body," *Mail and Guardian*, February 4, 2000.

16. Cape Town: Mafube/Tafelberg, 1999.

17. S. J. Klebanoff et al., "Activation of the HIV type 1 long terminal repeat and viral replication by dimethylsulfoxide and related solvents," *AIDS Research and Human Retroviruses*, Vol. 13 (September 20, 1997), pp. 1221–1227.

18. "Virodene Is Still Being Sold: SAPS," *The Citizen*, March 6, 1998.

19. D. C. Spencer, "Virodene—support misguided," B. W. Hugo, "Virodene—support misguided," F. Mahomed, "Virodene—support misguided," *South African Medical Journal*, Vol. 87, No. 3 (March 1997), pp. 613–614; Peter I. Folb et al., "Virodene—support misguided," *South African Medical Journal*, Vol. 87, No. 4 (April 1997), pp. 613–614; W. J. Kalk et al., "Virodene—support misguided," *South African Medical Journal*, Vol. 87, No. 6 (June 1997), pp. 775–776; Pat Sidley, "South African Research into AIDS 'Cure' Severely Criticised," *BMJ*, Vol. 314 (March 15, 1997), p. 771; Mike Ellis, "Virodene: What the DP Is Saying," *The Citizen*, March 4, 1998.

20. Thabo Mbeki, " 'Alien Goings-On' Mar Virodene Fight," *The Citizen*, March 8, 1998; see also Prakash Naidoo and Pippa Green, "Mbeki Slams Doctors over AIDS Drug," *The Citizen*, March 8, 1998; Chris Barron, "Government Swallowed the Wrong AIDS Pill," *The Citizen*, March 1998; Jack Lundin, "Tortuous Tale of AIDS Drug Unfolds," *Financial Mail*, April 10, 1998; Brian Stuart, "Zuma Seeks Testing Despite Virodene 'Toxic' Tag," *The Citizen*, March 5, 1998.

21. Brian Stuart, "Mbeki and Zuma in Virodene Row," *The Citizen*, March 3, 1998.

22. Mark Gevisser, "Thabo Mbeki: The Chief," *The Sunday Times* (South Africa), June 20, 1999.

23. "Politicisation of Debate on HIV Care in South Africa," *The Lancet*, April 29, 2000, p. 1473.

24. Laura A. Guay et al., "Intrapartum and neonatal single dose nevirapine compared with zidovudine for prevention of mother-to-child transmission of HIV-1 in Kampala, Uganda," *The Lancet*, September 4, 1999, pp. 795–802.

DEBBIE BOOKCHIN and JIM SCHUMACHER

The Virus and the Vaccine

FROM *THE ATLANTIC MONTHLY*

Harvey Pass, the chief of thoracic surgery at the National Cancer Institute, in Bethesda, Maryland, was sitting in his laboratory one spring afternoon in 1993 when Michele Carbone, a wiry young Italian pathologist who was working as a researcher at the NCI, strode in with an unusual request. Pass had never before met Carbone, and had talked to him for the first time, on the telephone, only a few hours before. Now Carbone was asking Pass for his help in proving a controversial theory he had developed about the origins of mesothelioma, a deadly cancer that afflicts the mesothelial cells in the lining of the chest and the lung. Mesothelioma was virtually unheard of prior to 1950, but the incidence of the disease has risen steadily since then. Though it is considered rare—accounting for the deaths of about 3,000 Americans a year, or about one half of one percent of all domestic cancer deaths—the disease is particularly pernicious. Most patients die within eighteen months of diagnosis.

Pass, one of the world's leading mesothelioma surgeons, knew, like other scientists, that the disease was caused by asbestos exposure. But Carbone had a hunch he wanted to explore. He told Pass that he wondered if the cancer might also be caused by a virus—a monkey virus, known as simian virus 40, or SV40, that had widely contaminated early doses of the polio vaccine, but that had long been presumed to be harmless to people.

Pass listened as Carbone described for him the history of the early polio vaccine. A breakthrough in the war against polio had come in the early 1950s, when Jonas Salk took advantage of a new discovery: monkey kidneys could be used to culture the abundant quantities of polio virus necessary to mass-produce a vaccine. But there were problems with the monkey kidneys. In 1960 Bernice Eddy, a government researcher, discovered that when she injected hamsters with the kidney mixture on which the vaccine was cultured, they developed tumors. Eddy's superiors tried to keep the discovery quiet, but Eddy presented her data at a cancer conference in New York. She was eventually demoted, and lost her laboratory. The cancer-causing virus was soon isolated by other scientists and dubbed SV40, because it was the fortieth simian virus discovered. Alarm spread through the scientific community as researchers realized that nearly every dose of the vaccine had been contaminated. In 1961 federal health officials ordered vaccine manufacturers to screen for the virus and eliminate it from the vaccine. Worried about creating a panic, they kept the discovery of SV40 under wraps and never recalled existing stocks. For two more years millions of additional people were needlessly exposed—bringing the total to 98 million Americans from 1955 to 1963. But after a flurry of quick studies, health officials decided that the virus, thankfully, did not cause cancer in human beings.

After that the story of SV40 ceased to be anything more than a medical curiosity. Even though the virus became a widely used cancer-research tool, because it caused a variety of tumors so easily in laboratory animals, for the better part of four decades there was virtually no research on what SV40 might do to people.

Carbone had reviewed some old research papers on the contamination and some of the early tests on SV40. He had even reviewed the notes from a crucial 1963 epidemiological study, by Joseph Fraumeni, an NCI researcher, which had concluded that children inoculated with contaminated vaccine did not show increased mortality rates. The studies did not impress Carbone: no one had systematically searched for evidence of the virus in tumors, and, as Fraumeni himself noted, the epidemiological study was too short to have detected certain slow-developing cancers. (Mesothelioma can take twenty to forty years to develop.)

Carbone had just finished a series of experiments in which he had injected the virus into dozens of hamsters. Every one of them developed mesothelioma and died within three to seven months. The results made Carbone wonder if SV40 might also play a role in human mesothelioma. He had come to see Pass because he had heard that the senior surgeon had meticulously saved tumor tissue from every one of the dozens of mesothelioma surgeries he had performed, and now had one of the largest collections of mesothelioma biopsies in the world. Carbone asked Pass if he could look for SV40 DNA in Pass's tumor-tissue samples, using a sophisticated molecular technique, known as polymerase chain reaction, or PCR, to extract tiny fragments of DNA from the frozen tissue and then amplify and characterize them.

As they talked, Pass became more and more impressed with Carbone. The young scientist was energetic and extremely self-confident—something Pass attributed to Carbone's surgical patrimony. (Carbone's father is a well-known orthopedic surgeon in Italy.) When Carbone had finished describing his proposed experiment, Pass realized that the implications were potentially significant. Only a handful of viruses have been directly associated with human cancers, and none of them are simian in origin. If SV40 was linked to mesothelioma in people, might it also cause bone and brain cancers in human beings, as it had done in hamsters? What if the monkey virus could spread from person to person? And if the virus was cancer-causing, or oncogenic, what was one to make of the fact that millions of Americans had been exposed to it as part of a government-sponsored vaccination program?

"I thought to myself, He's got this wild-assed idea," Pass recalls. "If it's true, it's unbelievable. Even if it's not, I'm going to get a hell of an education in state-of-the-art molecular biology."

Others at the National Institutes of Health—including some of the scientists who had been around at the time of the contamination scare—were less receptive to the novel theory. They told Carbone that the last thing anyone wanted to hear was that the exalted polio vaccine was linked to cancer. Too much was at stake. Implicating a vaccine contaminant in cancer—even if the contamination occurred some forty years ago—might easily shake public confidence in vaccines in general. And besides, everyone knew that asbestos was the cause of mesothelioma.

Carbone sought the advice of two renowned pathologists, Umberto Saffiotti, the chief of the NCI's Laboratory of Experimental Pathology, and Harold L. Stewart, a former director of pathology at the NCI who was once the head of the American Association for Cancer Research. Both urged Carbone to follow his intuition. "Forget what people tell you," Stewart told Carbone. "They told me I was wrong all my life. If you want to do it, you should, or you will regret it." That spring afternoon in 1993, with Pass's mesothelioma samples in hand, Carbone called an old friend, Antonio Procopio, a professor of experimental pathology in Italy who had worked for three years at the NIH. "I asked him if he was willing to do this crazy project with me," Carbone says.

"I told him I could not pay him or his expenses." A month later Procopio arrived in Bethesda. "We had no money," Carbone recalls. "He slept in my house for six months, and we worked day and night."

It turned out that Pass's samples were loaded with the monkey virus: 60 percent of the mesothelioma samples contained SV40 DNA; the nontumor tissues used as controls were negative. Moreover, Carbone found that in most of the positive samples he tested, the monkey virus was active, producing proteins—suggesting to Carbone that the SV40 was not just an opportunistic "passenger virus" that had found a convenient hiding place in the malignant cells but was likely to have been involved in causing the cancer.

In 1994 Carbone, Pass, and Procopio published the results of their experiment in *Oncogene,* one of the world's leading cancer-research journals. They proposed SV40 as a possible co-carcinogen in human mesothelioma. It was the first time researchers had put forward hard evidence that the all-but-forgotten vaccine contaminant might cause cancer in human beings.

A Solution to an Enigma

MICHELE CARBONE is almost stereotypically Italian: generous with his emotions, outspoken, and jovial. He is strikingly handsome, with large brown eyes and shoulder-length brown hair. Carbone grew up in a cultured home in Calabria, on the shores of the Mediterranean in southern Italy. As a youth he often spent hours poring over medical texts, some of them 300 years old, in the voluminous library started by the first of the seven generations of Carbone physicians to date. If his father gave him science, from his mother he may have inherited the strong intuition that is his distinguishing characteristic as a researcher. She is an accomplished artist whose work is exhibited widely in Europe.

Carbone graduated in 1984, at the top of his class, from the University of Rome Medical School, one of the largest in the world, and quickly won a coveted NIH doctoral fellowship. In 1993 he received a Ph.D. in human pathology. In less than a decade he has risen to the top of his profession. Today he is internationally recognized as an expert in mesothelioma.

Since 1994 Carbone has written more than twenty studies and reviews investigating SV40's link to human cancer. "There is no doubt that SV40 is a human carcinogen," he says. "SV40 is definitely something you don't want in your body." Carbone suggests that the virus works in tandem with asbestos or by itself to transform healthy mesothelial cells into cancerous ones.

Since he published his first study, scientists at seventeen major laboratories—in the United States, Great Britain, France, Belgium, Italy, and New Zealand—have confirmed Carbone's research with respect to the presence of SV40 in human mesothelioma. Their results point to a solution to an enigma that long puzzled researchers. At least 20 percent of mesothelioma victims report no asbestos exposure, and only 10 percent of people who have had heavy exposure to asbestos ever develop mesothelioma. The experiments suggest that SV40 may be another factor at work in the tumors.

Two very recent studies, from Finland and Turkey, found no SV40 in domestic mesothelioma samples but did find it, respectively, in American and Italian samples. The authors observe that their negative findings lend support to the theory that contaminated polio vaccine is associated with the disease: neither Turkey nor Finland used SV40-contaminated vaccines. Today Finland has one of the lowest rates of mesothelioma in the Western world.

The virus has also been located in other kinds of tumors. More than a dozen laboratories have found SV40 in various kinds of rare brain and bone tumors. In 1996 Carbone reported that he had found SV40 in a third of the osteosarcomas (bone cancers of a type that afflicts about 900 Americans a year) and nearly half of the other bone tumors he tested—research that has since been confirmed by numerous laboratories. The virus has also been detected in pituitary and thyroid tumors.

The possibility of a link between SV40 and brain tumors is particularly intriguing. Like mesothelioma, brain tumors have become dramatically more common in recent years. Brain tumors will be diagnosed in about 3,000 children in the United States alone this year. In 1995 Janet Butel, the chairman of the department of molecular virology and microbiology at the Baylor College of Medicine, in Texas, and her chief collaborator, John Lednicky, also a Baylor virologist, reported that they had found SV40 in a number of children's brain tumors. Butel and Lednicky reported that DNA sequencing revealed that the virus was not a hybrid but rather authentic SV40—the same as the SV40 found in monkeys. In the fall of 1996 an Italian research team, led by Mauro Tognon, of the University of Ferrara, announced that it had found SV40 DNA in a large percentage of brain and neurological tumors, including glioblastomas, astrocytomas, ependymomas, and papillomas of the choroid plexus. The researchers suggested that SV40 may be a "viral cofactor" involved in the sharp rise in human brain tumors. In 1999 an extensive study undertaken in China reinforced those results. The study examined sixty-five brain tumors, finding SV40 in each of the eight ependymomas and two choroid-plexus papillomas, common brain tumors among children. It also found the virus in 33 to 90 percent of five other kinds of brain tumor examined. The authors, writing in the November, 1999, issue of *Cancer,* noted that the virus was actively expressing proteins.

Recent research also indicates that SV40 has gained a secure foothold in the human species. In 1996 Tognon and his collaborators reported that they had also found the virus in 45 percent of the sperm samples and 23 percent of the blood samples they tested from healthy people, suggesting that the monkey virus could spread through sexual contact or unscreened blood products. In 1998 the presence of SV40 antibodies in human blood samples was reported by Butel, who tested several hundred American blood samples and found antibodies to SV40 in about 10 percent of them. Butel's laboratory also tested samples from children born from 1980 to 1995—decades after the contaminated vaccine was removed from the market. A surprising six percent tested positive—offering evidence that the virus may now be spreading from person to person, including from mother to child.

THE PRESENCE OF SV40 in human tumors has been reported on in more than forty independent research papers. But one molecular study that has had an enormous impact on the direction of SV40 research and funding was performed not by a virologist, like Butel, or a molecular pathologist, like Carbone, but by an epidemiologist named Howard Strickler. Strickler served as a senior clinical investigator in the NCI's Viral Epidemiology Branch for many years before he joined the Albert Einstein College of Medicine, in New York, in January 1999. He has been persistently skeptical of any association between the vaccine contaminant and tumors. Though he is no longer at the NCI, he remains instrumental in the government response.

In June of 1996 Strickler published a paper with Keerti Shah, of the School of Public Health at Johns Hopkins University, in Baltimore, in the journal *Cancer Epidemiology, Biomarkers and Prevention.* Strickler and Shah reported that they had come up empty-handed in their search for SV40 in fifty mesothelioma samples. Their study and a 1999 British study are the only two published SV40 studies with negative results. These two papers, particularly Strickler's, are cited again and again by federal health officials as proof that the dozens of peer-reviewed papers reporting SV40's presence in human tumors are unpersuasive and that a major research effort on SV40 is unnecessary.

Strickler acknowledges that he has never done PCR himself (Shah was responsible for the PCR work for their 1996 collaboration), but he challenges the work of other labs that have found SV40 in human tumors. "I feel that the data are mixed regarding the detection of SV40 DNA in human tissues," Strickler says, citing his own negative study and the British study. Strickler also points out that when SV40 is found in tumor cells, it often occurs only at very low levels. Whereas human papilloma virus (HPV), which causes cervical cancer, can be detected at rates of fifty viruses per cancer cell, SV40 is sometimes found at a rate of one virus per cell. "I find it curious that even the laboratories that detect SV40 in the cancers report that the virus is present at such extremely low levels," Strickler says. John Lednicky, of Baylor, counters that HPV is very different from SV40. Strickler "is comparing an apple with an orange," he says. "SV40 is known to be far more tumorigenic than HPV in animals. One copy of SV40 per cell is enough to transform a cell."

Several SV40 researchers have criticized Strickler's 1996 study and the more recent British one, saying that they treated specimens in a manner that would not result in the efficient extraction of SV40 DNA. Bharat Jasani, the director of the molecular diagnostic unit at the University of Wales, in Cardiff, has found SV40 in British mesothelioma samples. He recently wrote a lengthy critique of the two studies that has not yet been published. In this critique Jasani concludes that the negative results "are explainable by the paucity of the diagnostic biopsy material used and/or insufficient sensitivity of the overall PCR methodology used." Jasani says that Strickler's PCR technique would have missed low levels of SV40.

Federal health officials are understandably concerned that any link between SV40 and human cancers could frighten people away from the polio vaccine and vaccination in general. They stress that before SV40 in the polio vaccine can be linked definitively to cancer, the proposition must clear important scientific hurdles. Carbone and others must prove that the SV40 they have found is not a laboratory contaminant. They must demonstrate that SV40 is responsible for the cellular damage that leads to cancer and is not just a benign "passenger" in human tumors. And they must show that it was introduced into human beings through the polio vaccine.

In assessing the research to date, Strickler is perplexed that the virus has been found in so many kinds of tumors. In addition to the confirmed research reporting the virus in more than a half dozen kinds of brain tumors and a similar number of bone tumors, researchers in new, isolated studies have reported finding the virus in Wilms' tumors, which afflict the kidney, and adenosarcomas, rare cancers of the uterus. "It's not likely that a single virus causes ten thousand different diseases," Strickler says. "That's not how it works."

These anomalies have fueled Strickler's suspicion that many of the SV40 findings in human tumors may really be false positives resulting from laboratory contamination. He points out that SV40 is used for cancer research in so many laboratories around the world that almost any lab involved with tumor assays could conceivably harbor it. "Is it possible that SV40 is in human tumors and that SV40 is at some level circulating in the human population?" Strickler asks. "Could it be true? I can't exclude the possibility, but the studies to demonstrate it haven't really been done, and the data in our hands have been negative." Strickler's former boss, James Goedert, the chief of the NCI's Viral Epidemiology Branch, agrees. Although he says he has an open mind about SV40, he believes that contamination may lie behind the findings of Carbone, Butel, and others.

In 1997, largely in response to Strickler's study, the International Mesothelioma Interest Group set out to determine once and for all if the virus was present in human mesothelioma samples. The organization asked an internationally known molecular geneticist, Joseph R. Testa, the director of the Human Genetics Program at the Fox Chase Cancer Center, in Philadelphia, to oversee a study. Testa, who specializes in mesothelioma research, confesses that initially he doubted the idea that SV40 could be found in human mesotheliomas, because he believed it was well established that asbestos was the cause of the disease. "I'm a very careful person," Testa says. "I had a fair amount of skepticism about it." But the results of the investigation he led changed his mind. Four laboratories participated in the tightly controlled study, including Carbone's. All four found SV40 in at least nine out of the twelve mesothelioma samples they tested. Each laboratory's control samples tested negative, suggesting that the positive SV40 samples were not the result of laboratory contamination. The results were published in the journal *Cancer Research* in 1998.

Strickler believes that Testa's study "did not really move the ball forward" in determining whether contamination lies behind findings of SV40 in human tumors. He questions Testa's conclusions. "They are trying to make a large point out of the fact that results were reproduced," he says. But according to Strickler, that such a high percentage of tumors tested positive actually casts doubt on the study's reliability and raises the possibility that the labs merely exchanged contaminated samples. "The prevalence [of SV40-positive samples] was so high . . . that you have no way to make the distinction between [contamination] and a true positive result," he says.

Carbone and some of the other scientists we have interviewed say that

Strickler's contamination theory is a red herring. "We've documented that it is the case that this virus is present and is expressed in these tumors," Testa says. "I think the onus is on [federal health officials] to take this new research into consideration." Carbone, not surprisingly, is even more adamant. "The idea that these tumor samples, tested in laboratories all over the world, were all contaminated, while all the controls remained negative, is ridiculous," he says. "There is no scientific evidence in support of contamination, and plenty of evidence to the contrary. Moreover, many labs have demonstrated SV40 using techniques other than PCR."

Recently we asked several prominent scientists to evaluate the SV40 studies. George Klein, at the Karolinska Institute, in Stockholm, who chaired the Nobel Assembly, and is a longtime expert on SV40, read Testa's study. His conclusion was different from Strickler's. According to Klein, the Testa study is "quite convincing concerning the association between SV40 and mesothelioma," and "the evidence suggests that SV40 may contribute to the genesis of some human tumors, mesothelioma in particular."

Carlo Croce, the editor of *Cancer Research* and a member of the National Academy of Sciences, agreed. Not only is it indisputable that SV40 is present in human tumor samples, he told us, but "it looks like the presence of the virus contributes to the cause of mesothelioma."

Janet Rowley, the editor of the journal *Genes, Chromosomes and Cancer* and a professor of molecular genetics and cell biology at the University of Chicago, is a pioneer in the study of chromosome abnormalities in cancer. Rowley's groundbreaking research was itself called into question for years. "People didn't believe that chromosome abnormalities had anything to do with leukemia," she recalled. "It took a long time to break down that prejudice." She told us that Carbone had faced the same kind of doubts that first greeted her. "Everybody had assumed that mesothelioma was associated with asbestos. One of the important things in medicine is not to let your assumptions and those generally accepted paradigms obscure the fact that maybe there's more." Rowley believes that Carbone and Testa's work strongly implicates SV40 as a causal factor in some mesotheliomas.

"Like Somebody Set Off a Bomb"

CARBONE'S OFFICE is tucked into a quiet second-floor corner of the glass-and-concrete Cardinal Bernardin Cancer Center, at Loyola University, in Maywood, Illinois. The center is just a few miles west of Chicago and about ten minutes by car from Oak Park, where Carbone lives in a stately Frank Lloyd Wright house, with his wife and two daughters. Carbone came to Loyola in 1996 after a two-year stint at the University of Chicago. Now an associate professor of pathology, he works with Paola Rizzo, his senior scientist and closest collaborator, and a handful of post-docs and lab assistants in a tidy laboratory just down the hall from his office.

The lab is lively. Carbone has recruited compatriots as some of his research assistants, and the whir of high-tech machinery is punctuated by good-natured banter in Italian. This afternoon Carbone is examining an SV40-infected cell-culture plate under a microscope. He speaks almost fondly of the virus he has studied for most of the past decade. SV40 is "the smallest perfect war machine ever," Carbone murmurs. "He's so small. But he's got everything he needs."

Magnified 50,000 times under an electron microscope, SV40 doesn't seem particularly menacing. It looks almost pretty—bluish snowflakes, against a field of white. The virus consists of six proteins, three of which make up the twenty-sided triangular scaffolding that is the virus's protein skin. But one of the remaining proteins, called large T-antigen (for "tumor antigen"), is, according to Carbone, the most oncogenic protein ever discovered. It is unique, he says, in its ability to cause cancer when it is set loose inside a cell.

In 1997, in *Nature Medicine,* Carbone published the first in a series of papers that outlined how large T-antigen blocks crucial tumor-suppressor pathways in human mesothelial cells. Whenever a cell begins to divide, in the process known as mitosis, a small army of quality-control agents goes to work. Running up and down the cell's DNA, these genes and proteins work together to scrutinize the DNA's integrity. If at any stage of cell division they detect DNA abnormalities that cannot be repaired, mitosis is halted and the cell undergoes apoptosis, or cellular suicide. The principal in this elaborate regulatory dance is a gene called p53. Arnold Levine, the president of The Rockefeller University, in New York City, and the discoverer of p53, says that 60 percent of all cancers involve some sort of p53 damage, mutation, or inactivation. "The p53 gene is central to human cancers," he says, describing it as "the first line of defense against cancer formation."

Carbone's experiments have shown that in human mesotheliomas large T-antigen attacks p53, binding to it so that it cannot function properly. Large T-antigen also strangles a series of proteins called Rbs, which together serve as some of the final gatekeepers in cellular division.

No other cancer-causing virus uses just one protein to knock out two different regulatory pathways simultaneously. For example, human papilloma virus must produce two proteins, E6 and E7, to inactivate p53 and the Rbs respectively; SV40 does its damage in one stroke. Levine calls large T-antigen "a remarkable protein."

Large T-antigen's cancer-causing havoc isn't limited to disabling a cell's most important tumor suppressors. It can also damage chromosomes by adding or deleting whole sections of DNA or reshuffling the genes. Once the virus is finished with a cell, Joseph Testa says, "it looks like somebody set off a bomb inside the cell's nucleus, because of all these chromosome rearrangements." Carbone says that because SV40 binds to tumor-suppressor genes and also causes genetic damage, it "is one of the strongest carcinogens we know of."

Yet he emphasizes that most people who carry SV40 in their cells won't develop cancer, because a healthy immune system generally seeks out and destroys invading viruses. He points out that large T-antigen normally provokes a particularly strong immune response, unless a person has been exposed to asbestos, a known immunosuppressant. "Human beings," Carbone says, "have devised many mechanisms to defend themselves against cancer. This is one of the reasons that human beings live so long compared with other animals. Human cancer is usually the result of a number of unfortunate events that together cause a malignant cell to emerge."

But SV40 may have evolved other strategies to elude the immune system. In a recently published article Carbone writes that sometimes SV40 produces such small amounts of large T-antigen that the virus escapes detection. Paradoxically, in this hypothesis small amounts of the virus are even more dangerous than large amounts.

Other scientists suspect that SV40 can inflict damage and then disappear completely, in what is described as a "hit-and-run" attack. This analogy is lent credence by a recent German study in which rat cells were infected with SV40 and transformed into cancer cells. When scientists searched for large T-antigen, it was no longer present in some of the cells. Further, these cells appeared to be even more malignant than those that were still expressing the protein, because the immune system could no longer recognize them as a threat.

The new theory may explain how SV40 and perhaps other viruses can induce cancer and yet not be readily detectable once tumors start proliferating rapidly. But that notion runs counter to traditional scientific thinking about cancer. "As a geneticist, I would like to see every single cell have evidence of the virus," Testa says, noting that the hit-and-run theory must still be proved. But, Testa observes, "This is an area that's going to perhaps establish a new paradigm."

Although Carbone's T-antigen research has bolstered his contention that the SV40 found in human tumors is not simply a passenger virus, until recently he had no answer to a criticism commonly voiced by those skeptical that the polio vaccine could be linked to cancer: some of the SV40 he and others have isolated in human tumors has a crucial genetic difference from the virus that contaminated the polio vaccine. The SV40 that its discoverers isolated from the polio vaccine in 1960 had a genetic feature that allowed it to replicate more quickly than the SV40 subsequently found in human bone and brain cancers and in most monkeys. That led some to question the idea that the SV40 that researchers were finding in these tumors was related to the SV40 in the polio vaccine.

To settle the issue Carbone sought to examine old vaccine stocks. He was told by government and drug-company officials that they had thrown out all the old lots. Then, two years ago, Carbone found an elderly Chicago-area physician who had an unopened case of polio vaccine from 1955, which he had stored in his refrigerator for more than forty years. "I would have gone all the way to Alaska to find this stuff, and here it was three miles away," Carbone says. Last summer Carbone finally completed tests on the vintage vaccine. He found that the tiny vials contained SV40 genetically identical to the strains found in human bone and brain tumors and in monkeys. "This proves that the SV40 that was present in the polio vaccine is identical to the SV40 we are finding in these human tumors," he says. Why was the SV40 isolated from the 1960 vaccine the faster-growing version? Because, Carbone says, both kinds occurred in the monkey kidneys used to grow the vaccine. Carbone and Janet Butel say that the SV40 that grew more quickly might have had an advantage in cell cultures—perhaps explaining why it was the strain originally isolated from the vaccine. However, the slower-growing virus would almost certainly have an advantage in tumor formation, because it would be less likely to be detected by the immune system.

Because he believed that the slower-growing SV40 was more likely to induce tumors, Carbone wanted to see if federally mandated vaccine-screening tests for viruses were adequate to detect it. Vaccine manufacturers are not required to use state-of-the-art molecular techniques—PCR, for example—for virus detection. Instead they rely on ordinary light-microscope examination to look for evidence of cellular damage by viral contaminants after fourteen-day cycles in tissue culture. Although the current screening protocols—themselves forty years old—are, according to Carbone, more than adequate to detect the faster-growing form of SV40, his tests found that the slower-growing SV40 took at least nineteen days to grow out, and thus wouldn't be detected in the fourteen-day screening cycles. Carbone says his experiments suggest that any slow-growing SV40 present in the vaccine after the early 1960s could have gone undetected.

Carbone recently tested six vials of polio vaccine manufactured in 1996, and found that they were negative for SV40. He concludes that the colonies of monkeys used today must be free of the virus, because if slow-growing strains were present, the tests used for routine screening would not detect them. (Today's injected vaccine is produced on monkey cell lines, and is therefore free of any viral contaminants, whereas the oral vaccine is still produced on actual kidneys. Under Centers for Disease Control regulations that went into effect last month, American children should now receive only injected vaccine.) In a paper on his tests of vaccines Carbone recommends conducting extensive molecular testing of polio-vaccine stocks from the 1960s, 1970s, and 1980s to look for the slower-growing SV40. The issue is more than academic: the results would help to establish whether SV40 is present in young children today as a result of continued exposure to contaminated vaccine or as a result of human-to-human transmission based on the original, 1955–1963 exposure.

A Study Plagued by Strife

DESPITE THE ACCUMULATING EVIDENCE of SV40's association with human tumors, the NCI has been preoccupied with determining whether the virus is even present in human tumors. For more than two years the NCI's chief focus with respect to SV40 was the design and administration of a multi-laboratory study whose stated purpose was to assess whether PCR was a reliable tool for identifying the presence of SV40 in human tissue. Critics of the study, including scientists at some of the participating labs, worried that other agendas were involved. The study was directed by Howard Strickler and overseen by James Goedert. Nine labs participated in the study, including those of Keerti Shah, at Johns Hopkins; Bharat Jasani, at the University of Wales; and Janet Butel, at Baylor, but not Carbone's. The study, which was planned and administered by the NCI's Viral Epidemiology Branch, had a fairly unusual design. Instead of just seeing whether different labs could replicate one another's work, as is usually done, the labs were asked to prove that they could replicate their own work. Each lab was given a variety of samples from unidentified human mesothelioma tissues and asked to see if it could find SV40 DNA. Then it was asked to find SV40 DNA again in masked samples from the same tumor tissue.

We asked Richard Klausner, the director of the NCI, about his views on

SV40 and about the design of the experiment. Klausner said that the research to date hadn't quelled his doubts that SV40 is present in human tumor tissue, and he questioned the reliability of the techniques that Carbone and others have been using. "These sorts of molecular technologies are wonderful tools but very complicated and sometimes misleading to use," Klausner said. "I think there is very good reason to question whether there has been the development of adequate standards or probes, PCR probes," for detecting the virus.

Like Strickler and James Goedert, Klausner raised the possibility of contamination to explain the positive findings of dozens of laboratories. "I do not see any compelling molecular data" to support the association of SV40 with human tumors, he told us. "In the absence of compelling clinical or epidemiological data, it's very difficult to say this looks like a pressing problem." We asked him about the many molecular studies, from numerous independent laboratories around the world, that had identified SV40 in human tumors. "There's too much irreproducibility and too many good explanations for artifact," he said. Klausner told us that the NCI has taken "an open approach but a critical one" to the notion that SV40 is associated with human tumors, and he insisted that it is seriously studying the issue. Michele Carbone's work, for instance, has been funded by the NCI. (Carbone is also funded by the American Cancer Society.)

We asked Klausner to explain why the Viral Epidemiology Branch had directed the multi-laboratory molecular-biology study, especially given that neither Strickler nor the head of the branch, Goedert, has a strong background in the field. Why hadn't he tapped an NCI division with more expertise in DNA extraction, sequencing, and characterization? "Their expertise in viruses and virus-associated disease makes [the Viral Epidemiology Branch] really the right place to do it. . . . As an expert in doing this sort of work, I feel that I can make that decision and I feel very comfortable with the decision," Klausner said. "What we are trying to do is establish some agreed-upon probes and standards that independent laboratories could utilize to provide ways of either validating or not validating molecular findings."

On another issue, Klausner referred to an epidemiological study that Strickler had done to determine whether SV40 was linked to human cancer. That study appeared in 1998 in the *Journal of the American Medical Association*, and received extensive publicity upon its release. It concluded that the NCI's database on cancer incidence shows no statistically significant correlation between exposure to SV40-contaminated vaccine and rates of cancer, including rarer cancers such as mesotheliomas, ependymomas, and osteosarcomas.

Strickler did find elevated cancer rates among those exposed to SV40, including a threefold increase in mesothelioma. Susan Fisher, an associate professor of epidemiology and biostatistics at Loyola, says that although the correlation Strickler found did not achieve statistical significance, it was at least "scientifically interesting." Strickler's study was "technically correct," Fisher says, but "it's hard to look at these numbers and turn around and say there is no evidence to suggest an association."

Moreover, Fisher says, standard epidemiological techniques may be useless in determining whether SV40 exposure is linked to higher cancer rates. If the research of Janet Butel and others is correct and SV40 is now spreading among human beings, it may be impossible to assemble an experimental group that has never been exposed to SV40.

The multi-lab NCI study concluded with six of the nine laboratories detecting SV40 in some samples. However, only two of the labs got the same positive results on samples from the same tissues. Although the multi-lab study was completed at the end of 1998, at the time this article was written it had yet to be submitted for publication.

Memos sent to Strickler by some of the participating laboratories show that from its inception the study was plagued by considerable internal strife. (Participating laboratories we approached declined to share the memos or discuss them. We obtained them independently.) Two laboratories suggested that poor DNA-extraction techniques by the outside laboratory Strickler had chosen to provide the DNA samples were to blame for the largely negative results obtained. Their concerns were heightened when it was learned that the contractor had contaminated some of the negative controls.

They also complained that Strickler was wrongly using the study to imply that previous positive findings were caused by contamination. "It cannot be that all of these laboratories are contaminated and that contamination always happens in mesotheliomas, osteosarcomas and brain tumors, while the negative controls are always negative," a scientist from one of the laboratories wrote Strickler. "Contamination is a random event. . . . [The] flaws and unresolved scientific issues . . . have become so cumulative as to outweigh any positive scientific benefit which might be derived from the publication of this study." From another laboratory came this objection: "We feel that our comments about data interpretation are being dismissed and ignored. Your intransigence about the interpretation of the data and the conclusions of the study have forced us to admit that the collegiality and the scientific collaboration that was the basis of this study is very strained." Both laboratories maintained that Strickler's draft manuscript summarizing the study results was wrong in asserting that contamination was the cause of previous SV40 findings.

An unlikely ally in the laboratories' cause has been William Egan, the acting head of the Food and Drug Administration's Office of Vaccines Research and Review. Egan believes there is no strong epidemiological proof that SV40 is associated with human cancers and emphasizes that the current polio vaccine is free of SV40. However, he says, there is evidence that the virus may well be present in some tumor samples. After he had reviewed Strickler's draft manuscript, last February, Egan wrote a lengthy letter to Strickler criticizing it. "I think that this paragraph, and the following paragraph, imply, unintentionally so, that the positive results [of SV40 in tumors] that have been reported are due to laboratory contamination; I do not think that this should be implied." Strickler responded, "This study would not have been conducted if there was not some doubt. That point must be made and made clearly."

Later Egan chided Strickler about another section of his draft, which stated, "This multi-institutional study failed to demonstrate the reproducible detection of SV40 in human mesotheliomas." Egan wrote,

> More exactly, it failed to demonstrate SV40 sequences in *this set* of mesotheliomas. This is not inconsistent with SV40 being found by others previously. Indeed, the fact that laboratories that previously found SV40 in their samples do not now find SV40 in these samples (and get the study controls correct) only lends credence to their previous findings. . . . These laboratories are able to find SV40 when it is there, and do not find it when it is not there.

Frustrated by continuing objections, Goedert and Strickler considered publishing the study without the approval of the dissenting labs, but that plan was dropped. In September 1999 an independent arbitrator was called in to rewrite Strickler's manuscript. The dissenters apparently gained some ground. The arbitrator made major changes in its tone and conclusions. The study now states that "laboratory contamination was unlikely to have been the source of SV40 DNA" found in human tumors in previous experiments (by Butel, Jasani, and the other participating labs).

The Search for a Vaccine

THIRTY MILES NORTH of Venice, in the seaside resort town of Lignano Sabbiadoro, 200 clinicians and researchers are gathered at the international Conference on Malignant Pleural Mesothelioma. At a similar conference in Paris in 1995 Carbone startled his audience when he presented his first SV40 paper.

Today a significant portion of the conference is devoted to SV40's association with mesothelioma—testament to a sea change among researchers regarding the simian virus. Brooke Mossman, the director of the environmental-pathology program at the University of Vermont, was the first scientist to tease out the complex molecular pathways by which asbestos disrupts cellular regulatory mechanisms and causes mesothelioma. She has been impressed by Carbone's work. At Lignano she and Carbone are co-chairing a panel on the molecular pathways employed by asbestos and SV40 which lead to tumor development. In another presentation Luciano Mutti, a researcher at the Salvatore Maugeri Foundation's Institute for Research and Care, in Pavia, will report that mesothelioma patients who test positive for SV40 have a shorter life-span than those who test negative.

At the moment the floor belongs to David Schrump, the new chief of thoracic surgery at the NCI. Schrump matter-of-factly announces the results of a series of experiments he has just completed. When he "turned off" SV40 large T-antigen, he says, human-mesothelioma cell cultures that contained the virus stopped proliferating and started to die. Schrump explains that he undertook the experiment partly because he was skeptical of SV40's role in the development of mesothelioma. He and his team assembled human mesotheliomas that tested positive for SV40 and then devised a genetic bullet, a strand of RNA called an "antisense," which would prevent the expression of SV40 large T-antigen.

Within days after the antisense was administered to the cancer cultures, Schrump found, the growth rates of mesotheliomas with SV40 in them dropped dramatically; the negative controls were unaffected. One important finding was that even very low levels of SV40 appeared to be biologically important—a discovery that speaks to Strickler's objection about the low levels of SV40 often found in tumor tissue. Schrump's study was published in 1999 in *Cancer Research.*

Another study in that same issue also supports the idea that SV40 is actively involved in mesothelioma. Adi Gazdar is a professor of pathology and the deputy director of the Hamon Cancer Center, at the University of Texas Southwestern Medical Center. He originally doubted Carbone's work on SV40. "Here's a monkey virus suddenly popping up in a rare tumor—I was skeptical of the data," he told us. So Gazdar devised an experiment that could determine at one stroke whether the SV40 found in tumors was a lab contaminant and whether the virus is involved in tumor formation. Gazdar used a technique called laser microdissection to separate cancerous cells from nearby noncancerous ones. He found SV40 in more than half of the mesothelioma tumors. He also found the virus in some adjacent pre-cancerous cells. Significantly, 98 percent of nearby noncancerous cells tested negative for SV40. "That rules out any contamination," Gazdar says, "because if a specimen was contaminated, the SV40 would be in all parts of the specimen—it wouldn't be localized to the mesothelium alone." Moreover, Gazdar says, his study "suggests that the virus is in the right type of cells many years before they become malignant"—evidence that SV40 contributes to the development of cancer. Gazdar says of Carbone's work, "I feel everything he's said, I've been able to confirm, and more."

Gazdar and other scientists believe that the time has come for a major federal funding initiative on SV40 to better understand who is infected, how the virus works, and what might be done to prevent disease. "There's still a lot we don't know about the basic biology of this virus in human infections, including what tissues it infects, how it is transmitted, and when people become infected with it," Janet Butel says. "Until more studies are done, we don't know if we're looking at the only types of cancers that have an association with SV40," she says of the lung, bone, and brain cancers with which SV40 has been associated most often. "Maybe these are just the ones we've recognized so far. There may be others people haven't run across." Gazdar says, "It's such a crucial issue. Possibly millions of people are sitting with this virus in their mesothelium or other tissues and are at risk for developing cancers." Cancers that were once rare "may suddenly become not so rare," he says. "I think it's an enormous potential health problem."

Arnold Levine, of The Rockefeller University, is not convinced that the virus causes cancer in human beings, but he, too, believes that the discovery of SV40 in human tumors warrants a serious federal response. "If it's part of the cause of a disease," he says, "it has a significance in public health and I think we ought to find that out. That's a good reason to spend taxpayers' money: to do science to find out whether the public health is really monitored here properly. I think that maybe there's enough evidence in the literature now that the National Cancer Institute ought to put out an RFA." The reference is to a Request for Applications, the formal process by which the federal agency identifies a major health-research initiative and invites scientists to apply for research funds. "That would stimulate people to come in and design experiments and replicate these things." Carbone made the same suggestion to federal health officials in 1997 but was rebuffed.

Like the NCI, the Atlanta-based Centers for Disease Control maintains a stance of neutrality with undertones of skepticism. In a four-page fact sheet called "Questions and Answers on Simian Virus 40 (SV40) and Polio Vaccine" the CDC notes that SV40 has been found in some tumors and adds that "more research is needed" to confirm a causative link with human disease. It also raises the possibility of contamination as an explanation. It cites Strickler's work by name but not that of Carbone, Butel, or Testa.

Some researchers plan to conduct screenings for the virus. Joseph Testa hopes to initiate a screening program at Fox Chase's new cancer-prevention pavilion that focuses on asbestos exposure. He is collaborating with officials from the Asbestos Workers Local 14, in Philadelphia, to identify people who are particularly at risk. Carbone applauds that effort. "If you test positive for this virus, you should not be anywhere near asbestos," he says. Bharat Jasani, who has found SV40 DNA in a high percentage of the British mesotheliomas he examined, has begun testing British and Canadian mesothelioma patients, at their request. He hopes they may be candidates for future SV40-targeted therapy.

In 1999 scientists reported that a vaccine they had developed targeting large T-antigen appeared to help prevent and reverse tumors expressing large T-antigen in mice. Carbone and Harvey Pass, who is now the chief of thoracic oncology at the Karmanos Cancer Institute, at Wayne State University, in Detroit, are collaborating with Martin Sanda and Michael Imperiale, of the University of Michigan at Ann Arbor, who are among the vaccine's developers. They hope soon to bring the experimental vaccine to Phase I clinical trials, in which it will be tested for its safety in human beings, though not yet for whether it works. Even if the vaccine eventually proves effective in human beings, years may well pass before it is widely available.

In an age of uncontrolled AIDS in Asia and Africa, rampant tuberculosis in Russia, and antibiotic-resistant microbes in American hospitals, does SV40 really warrant a significant public-health response? There is no doubt, Carbone says, that the virus is linked to some cancers. What's more, millions of Americans now have been exposed to the virus. Studying SV40 may teach us something about the dangers of cross-species infection at a time when the use of animal tissue for medical purposes is gaining acceptance.

Good science is ultimately about the exchange of ideas unfettered by presuppositions. Sometimes great breakthroughs come out of theories that at first seemed heretical or even nonsensical. "Can you think of anything more different on earth than asbestos and a monkey virus?" Carbone says. "Yet you stick them together and they work together to be more deadly than either one of them is alone." He goes on, "This research is important in so many different ways. It's not just about SV40 and mesothelioma. It helps us understand the whole picture of how viruses interact with environmental carcinogens. This research can help us understand how completely unrelated carcinogens can work together in causing disease—a mystery we have barely begun to unravel."

Stephen Jay Gould

Syphilis and the Shepherd of Atlantis

FROM *NATURAL HISTORY*

We usually manage to confine our appetite for mutual recrimination to merely petty or mildly amusing taunts. Among English speakers, unannounced departures (especially with bills left unpaid) or military absences without permission go by the epithet of "taking French leave." But a Frenchman calls the same, presumably universal, human tendency *filer à l'anglaise,* or "taking English leave." I learned, during an undergraduate year in England, that the condoms I had bought (for no realized purpose, alas) were "French letters" to my fellow students. In France that summer, my fellow students of another nation called the same item a *chapeau anglais,* or "English hat."

But this form of pettiness can escalate to danger. Names and symbols inflame us, and wars have been fought over flags and soccer matches. Thus, when syphilis first began to ravage Europe in the 1480s or 1490s (the distinction, as we shall see, becomes crucial), a debate erupted about naming rights for this novel plague—that is, the right to name the disease for your enemies. The first major outbreak had occurred in Naples in the mid-1490s, so the plague became, for some, the Italian or the Neapolitan disease. According to one popular theory (still under debate, in fact), syphilis had arrived from the New World, brought back by Columbus's sailors, who had pursued the usual activities in novel places—hence "the Spanish disease." The plague had been sufficiently acute a bit northeast of Columbus's site of return—hence "the German disease." In the most popular moniker of all, for this nation maintained an impressive supply of enemies, syphilis became "the French disease" (*morbus Gallicus* in medical treatises, then usually published in Latin), with blame cast upon the troops of the young French king, Charles VIII, who had conquered Naples, where the disease first reached epidemic proportions, in 1495. Supporters of this theory then blamed the spread through the rest of Europe on the activities of Charles's large corps of mercenary soldiers, who, upon demobilization, fanned out to their homes all over the continent.

I first encountered this debate in a succinct summary written by Ludovico Moscardo, who described potential herbal remedies in the catalog of his museum, published in 1672: "*Ne sapendo, a chi dar la colpa, li Spagnuoli lo chiamorono mal Francese, li Francesi male Napolitano, e li Tedeschi, mal Spagnuolo*" (not knowing whom to blame, the Spaniards call it the French disease, the French the Neapolitan disease, and the Germans the Spanish disease). Moscardo then added that other people attribute the origins of syphilis to bad airs generated by a conjunction of the three most distant planets—Mars, Jupiter, and Saturn—in the night sky.

How, then, did the new plague receive its modern name of syphilis, and what does "syphilis" mean, anyway? The peculiar and fascinating tale of the naming of syphilis can help us to understand two key principles of scholarship that may seem contradictory at first but that must be amalgamated into a coherent picture if we hope to appreciate both the theories of our forebears and the power of science to overcome past error: first, that the apparently foolish concepts of early scientists made sense in their times and can therefore teach us to respect their struggles, and second, that these older beliefs were truly erroneous and that science both progresses, in any meaningful sense of the term, and holds immense promise for human benefit through correction of error and discovery of genuine natural truths.

"Syphilis," the proper name of a fictional shepherd, entered our language in a long poem composed in 1,300 verses of elegant Latin hexameter and published in 1530 by the greatest physician of his generation (and my second favorite character of the time, after Leonardo da Vinci)—a gentleman from Verona (also the home of Romeo and Juliet), Girolamo Fracastoro (1478–1553). Fracastoro dabbled in astronomy (he became friendly with Copernicus when both studied medicine at Padua), made some crucial geological observations about the nature of fossils, wrote dense philosophical treatises and long classical poems, and held high status as the most celebrated physician of his time (in his role as papal doctor, for example, he supervised the transfer of the Council of Trent to Bologna in 1547, both to honor his holiness's political preferences and to avoid a threatened epidemic). In short, a Renaissance man of the Renaissance itself.

My inspiration for this essay flowed from the stark contrast between Fracastoro's christening of syphilis in 1530 and the style and substance of a 1998 paper on the genome of the bacterium that truly causes syphilis. Fracastoro could not resolve the origins of syphilis and didn't even recognize its venereal mode of transmission. So he wrote a poem and devised a myth, naming syphilis to honor a fictional shepherd of his own invention. In greatest contrast, the sober paper published by thirty-three coauthors in *Science* magazine (July 17, 1998) resolves the 1,138,006 base pairs—arranged in a sequence of 1,041 genes—in the genome of *Treponema pallidum*, the undoubted biological cause of syphilis.

Fracastoro's shepherd may have ended an acrimonious debate by donating his neutral name, but Fracastoro himself, as a Veronese patriot, made his own allegiances clear in the full title of his epic poem: *Syphilis sive morbus Gallicus* (Syphilis, or the French disease).

To epitomize some horrendous complexities of local politics: Verona had long been controlled by the more powerful neighboring city of Venice. Italy did not yet exist as a nation, and the separate kingdom of Naples maintained no formal ties to Venice. But commonalities of language and interest led the citizens of Verona to side with Naples against the invading French forces of Charles VIII, while general French designs on Italian territory prompted nearly a half century of war and strong Italian enmity, especially following Charles's temporary occupation of Naples.

Meanwhile, Maximilian I, the Hapsburg Holy Roman Emperor (who ruled an Austrian-dominated confederation in western and central Europe, despite the name), added Spain to his extensive holdings by marrying both a son and a daughter to Spanish rulers. He also allied himself with the Pope, Venice, and Spain to drive Charles VIII out of Italy. A decade later, given the shifting alliances of realpolitik, Maximilian had made peace with France and even sought its aid to wage war on Venice. His successful campaign split Venetian holdings, and Maximilian occupied Fracastoro's city of Verona from 1509 until 1517, when control reverted to Venice by treaty.

Fracastoro had fled the territory to escape Maximilian's war with Venice. But he returned in 1509 and began to prosper both immediately and mightily, so I assume that his allegiances lay with Maximilian. But to shorten the tale and come to the relevant point, Maximilian (at least most of the time) controlled Spain and regarded France as his major enemy. Fracastoro, as a Veronese patriot and supporter of Maximilian, also despised the French presence and pretensions. Fracastoro's interest therefore lay with absolving Spain for the European spread of syphilis by denying the popular theory that Columbus's men had inadvertently imported "the Spanish disease" with their other spoils from the New World. Hence, for Fracastoro, his newly christened syphilis would be called *morbus Gallicus.*

I can't boast nearly enough Latin to appreciate Fracastoro's literary nuances, but experts then and now have heaped praise upon his Virgilian style. Joseph Scaliger, perhaps the greatest scholar of the generation after Fracastoro's, lauded the work as "a divine poem," and Geoffrey Etough, the major translator of our time, writes that "even Fracastoro's rivals acclaimed him second only to Virgil." In this essay, I will use Nahum Tate's English version of 1686, the first complete translation ever made into any other language and a highly influential work in its own right (despite the clunkiness of Tate's heroic couplets in utterly unrelieved iambic pentameter). This version remained a standard source for English readers for more than two centuries. Tate, one of England's least celebrated poets laureate (or is it poet laureates, or even poets laureates?), wrote the libretto for Henry Purcell's short operatic jewel *Dido and Aeneas.* A few devout choristers may also know his texts for "While Shepherds Watched" or "As Pants the Hart." We shall pass by his once-popular adaptation of *King Lear,* with its happy ending in Cordelia's marriage to Edgar.

Syphilis sive morbus Gallicus includes three parts, each with its own form and purpose. Part 1 discusses origins and causes, while parts 2 and 3 narrate myths in closely parallel structure, devised to illustrate the two most popular (though, in retrospect, not particularly effective) cures. Fracastoro begins by defending his choice of *morbus Gallicus* as a name for the disease:

> To Naples first it came
> From France, and justly took from France his name
> Companion from the war. . . .

He then considers the theory of New World transmission on Spanish ships and admits the tragic irony, if true:

> If then by Traffick thence this plague was brought
> How dearly dearly was that Traffick bought!

But Spanish shipping cannot be blamed, Fracastoro holds, because the disease appeared too quickly and in too many places, including areas that never received products from the New World, to validate a single point of origin:

> To whom all Indian Traffick is unknown
> Nor could th'infection from the Western Clime
> Seize distant nations at the self same time.

Spain must therefore be absolved:

> Nor can th'infection first be charged on Spain
> That sought new worlds beyond the Western main.
> Since from Pyrene's foot, to Italy
> It shed its bane on France, while Spain was free.
> . . . .
> From whence 'tis plain this Pest must be assigned
> To some more pow'rful cause and hard to find.

The remainder of part 1 presents Fracastoro's general view of nature as complex and puzzling but intelligible—thereby exemplifying Renaissance humanism, an attitude that tried to break through the strictures of Scholastic logical analysis to recover the presumed wisdom of classical times ("renaissance" means "rebirth") but that did not yet include the belief in the primacy of empirical documentation that would characterize the rise of modern science more than a century later. Fracastoro tells us that we must not view syphilis as divine retribution for human malfeasance (a popular theory at the time)—a plague that must be corrected but cannot, as a departure from nature's usual course, be comprehended.

Rather, syphilis originated by natural causes that can, in principle, be understood. But nature is far more complex and unattuned to human sensibilities than we had been willing to admit, and explanation will not come easily—for nature works in strange ways and at scales far from our easy perception. For example, Fracastoro argues, syphilis probably had no simple point of origin followed by later spread (thus absolving Spain once again). Its particles of contagion (whatever they may be) must be carried by air but may remain latent for centuries before breaking out. Thus, the plague of any moment may emerge for reasons set long before. Moreover, certain potent causes—planetary conjunctions, for example, that may send poisonous emanations to Earth—remain far from our potential observation or understanding. In any case, and on a note of hope, Fracastoro depicts plagues as comprehensible phenomena of complex nature. And just as they ravage us with sudden and unanticipated fury, the fostering conditions will change in time, and our distress shall lift:

> Since nature's then so liable to change
> Why should we think this late contagion strange?
> . . . .
> The offices of nature to define
> And to each cause a true effect assign
> Must be a task both hard and doubtful too.
> . . . .
> [But] nature always to herself is true.

Part 2 continues the central theme of natural causation and potential alleviation, but in a very different manner. Following the traditions of Latin epic poetry, Fracastoro now constructs a myth to illustrate both the dangers of human hubris and the power of salvation through knowledge. He begins by giving the usual sage advice about alleviation via good living: lots of vigorous exercise, healthy and frugal diet, and no sex. (This regimen, addressed to males alone, proscribes sex only as a drain upon bodily energy, not as a source of infection—for Fracastoro did not yet understand the venereal transmission of syphilis.) But cure also requires pharmacological aid. Fracastoro upheld the traditional Galenic theory of humors and regarded all disease, including syphilis, as an imbalance among essential components that must be corrected by such measures as bleeding, sweating, and purging:

> At first approach of Spring, I would advise,
> Or ev'n in Autumn months if strength suffice,
> To bleed your patient in the regal vein,
> And by degrees th'infected current drain.

Part 2 then extols the virtues of mercury as a cure in this context. Mercury can, in fact, retard the spread of the syphilis spirochete, but Fracastoro interpreted its benefits only in terms of humoral rebalancing and the purging of poisons—for mercury plasters induced sweating, while ingestion encouraged copious spitting. The treatment, he admitted, may be unpleasant in the extreme, but ever so preferable to the dementia, paralysis, and death imposed by syphilis in the final stages of worst cases:

> Nor let the foulness of the course displease.
> Obscene indeed, but less than your disease.
> . . . .
> The mass of humors now dissolved within,
> To purge themselves by spittle shall begin,
> Till you with wonder at your feet shall see,
> A tide of filth, and bless the remedy.

Finally, Fracastoro spins his myth about human hubris, repentance, and the discovery of mercury. A hunter named Ilceus kills one of Diana's sacred deer. Apollo, Diana's twin brother, becomes royally infuriated and inflicts the pox of syphilis upon poor Ilceus. But the contrite hunter prays mightily and sincerely for relief, and the goddess Callirhoe, feeling pity, carries Ilceus underground, far from the reach of the sun god's continuing wrath. There in the realms of mineralogy, Ilceus discovers the curative power of mercury.

Fracastoro wrote these first two parts in the early 1510s and apparently intended to publish them by themselves. But by the 1520s, a new (and ultimately ineffective) "wonder cure" had emerged, and Fracastoro therefore added a third part to describe the new remedy in the mythic form previously applied to mercury—the same basic plot, but this time with a shepherd named Syphilis in place of the hunter Ilceus. And thus, with thanks to readers for their patience, we finally come to Fracastoro's reason and motives for naming syphilis. (An excellent article by R. A. Anselment supplied these details of Fracastoro's composition: "Fracastoro's *Syphilis:* Nahum Tate and the Realms of Apollo," *Bulletin of the John Rylands University Library of Manchester* 73, 1991.)

Fracastoro's derivation of the name has never been fully resolved, but most scholars regard Syphilis (often spelled Syphilus) as a medieval form of Sipylus, a son of Niobe in Ovid's *Metamorphoses*—a classical source that would have appealed both to Fracastoro's Renaissance concern for ancient wisdom and to his abiding interest in natural change.

In part 3 of Fracastoro's epic, the sailors of a noble leader (unnamed, but presumably Columbus) find great riches in a new world but incur the wrath of the sun god by killing his sacred parrots (just as Ilceus had angered the same personage by slaying Diana's deer). Apollo promises horrible retribution in the form of a foul disease. But just as the sailors fall to their knees to beg the sun god's forgiveness, a group of natives arrives—"a race with human shape, but black as jet," in Tate's translation. They, too, suffer from syphilis and have come to the parrots' grove to perform an annual rite that recalls the origin of their misfortune and also permits them to use the curative powers of local botany.

These people, we learn, are the degraded descendants of the race that inhabited the lost isle of Atlantis. They had already suffered enough in losing their ancestral lands and flocks. But a horrendous heat wave then parched their new island and fell with special fury on the king's shepherd:

> A shepherd once (distrust not ancient fame)
> Possessed these downs, and Syphilus his name.
> A thousand heifers in these vales he fed,
> A thousand ewes to those fair rivers led.
> . . . .
> This drought our Syphilus beheld with pain,
> Nor could the sufferings of his flock sustain,
> But to the noonday sun with upcast eyes,
> In rage threw these reproaching blasphemies.

Syphilus cursed the sun, destroyed Apollo's altars, and then decided to start a new religion based on direct worship of his local king, Alcithous. The king, in turn, heartily approved this new arrangement:

> Th'aspiring prince with godlike rites o'erjoyed,
> Commands all altars else to be destroyed,
> Proclaims himself in earth's low sphere to be
> The only and sufficient deity.

Apollo becomes even angrier than before (for Ilceus alone had inspired his wrath in part 2), and he now inflicts the disease upon everyone—but first upon Syphilus, who thus gains eternal notoriety as name bearer:

> Th'all-seeing sun no longer could sustain
> These practices, but with enraged disdain
> Darts forth such pestilent malignant beams,
> As shed infection on air, earth and streams;

From whence this malady its birth received,
And first th'offending Syphilus was grieved.
. . . .
He first wore buboes dreadful to the sight,
First felt strange pains and sleepless passed the night;
From him the malady received its name,
The neighboring shepherds caught the spreading flame:
At last in city and in court 'twas known,
And seized t'ambitious monarch on his throne.

A shepherd or two could be spared, but the suffering of kings demands surcease. The high priest therefore suggests a human sacrifice to assuage the wrath of Apollo (now given his Greek name of Phoebus), and guess whom they choose? But fortunately, the goddess Juno decides to spare the unfortunate shepherd and to make a substitution, in obvious parallel to the Biblical tale of Abraham and Isaac:

On Syphilus the dreadful lot did fall,
Who now was placed before the altar bound
His head with sacrificial garlands crowned,
His throat laid open to the lifted knife,
But interceding Juno spared his life,
Commands them in his stead a heifer slay,
For Phoebus's rage was now removed away.

Ever since then, these natives, the former inhabitants of Atlantis, perform an annual rite of sacrifice to memorialize the hubris of Syphilus and the salvation of the people by repentance. The natives still suffer from syphilis, but their annual rites of sacrifice please Juno, who in return allows a wondrous cure, the guaiacum tree, to grow on their isle alone. The Spanish sailors, now also infected with the disease, learn about the new cure and bring guaiacum back to Europe.

Thus, the imprecation heaped upon Spain by calling syphilis the Spanish disease becomes doubly unfair. Not only should the Spaniards be absolved for importation (because the disease struck Europe all at once, and from a latent contagion that originated well before any ships reached the New World), but the same Spanish sailors, encountering a longer history of infection and treatment in the New World, had discovered a truly beneficent remedy.

Many people know about the former use of mercury in treating syphilis, for the substance had some benefit and the remedy endured for centuries. But the guaiacum cure has faded to a historical footnote because, in a word, this magical New World potion flopped completely. (By 1530, the year of Fracastoro's publication, Paracelsus himself had branded guaiacum as useless.) But Fracastoro devised his myth of Syphilus during the short period of euphoria about the power of the new nostrum. The treatment failed, but the name stuck.

We should not be surprised to learn that Fracastoro's attraction to guaiacum owed as much to politics as to scientific hope. The powerful Fugger family, the great German bankers, had lent vast sums to Maximilian's grandson Charles V in his successful bid to swing election as Holy Roman Emperor over his (and Fracastoro's) archenemy, Francis I of France. As partial repayment for Charles's debt, the Fuggers received a royal monopoly for importing guaiacum to Europe. (The Hapsburg Charles V also controlled Spain and, consequently, all shipping to and from Hispaniola, where the guaiacum tree grew.) In fact, the Fuggers built a chain of hospitals for the treatment of syphilis with guaiacum. Fracastoro's allegiances, for reasons previously discussed, lay with Charles V and the Spanish connection, so his tale of the shepherd Syphilus and the discovery of guaiacum suited his larger concerns as well. (Guaiacum, also known as *lignum vitae* or *lignum sanctum* ["wood of life" or "holy wood"], has some medicinal worth, although not for treating syphilis. As an extremely hard wood, of the quality of ebony, guaiacum also has value in building and decoration.)

Fracastoro did proceed beyond his politically motivated poetry to learn more about syphilis. In the later work that secured his enduring fame (but largely for the wrong reason)—his *De Contagione et Contagiosis Morbis et Curatione* (On Contagion and Contagious Diseases and Their Cure) of 1546—Fracastoro finally recognized the venereal nature of syphilis, writing that infection occurs "*verum non ex omni contactu, neque prompte, sed tum solum, quum duo corpora contactu mutuo plurimum incalvissent, quod praecipue in coitu eveniebat*" (truly not from all contact, nor easily, but only when two bodies join in most intense mutual contact, as primarily occurs in coitus). Fracastoro also recognized that infected mothers can pass the disease to their children, either at birth or through suckling.

Treating himself diplomatically and in the third person, Fracastoro admitted and excused the follies of his previous poem, written "*quum iuniores essemus*" (when we were younger). In this later prose work, Fracastoro accurately describes both the modes of transmission and the three temporal stages of symptoms—the small, untroublesome (and often overlooked) genital sore of the primary stage; the secondary stage of lesions and aches, occurring several months later; and the dreaded tertiary stage, developing months to years later and leading to death by destruction of the heart or brain (called paresis, or paralysis accompanied by dementia) in the worst cases.

In the hagiographical tradition still all too common in textbook accounts of the history of science, Fracastoro has been called the father of the germ theory of disease for his sensitive and accurate characterization, in this work, of three styles of contagion: by direct contact (as for syphilis), by transmission from contaminated objects, and at a distance through transport by air. Fracastoro discusses particles of contagion called *semina* (seeds), but this term, taken from ancient Greek medicine, carries no connotation of an organic nature or origin. Fracastoro does offer many speculations about the nature of contagious *semina,* but he never mentions microorganisms, a hypothesis that could scarcely be imagined more than fifty years before the invention of the microscope.

In fact, Fracastoro continues to argue that the infecting *semina* of syphilis may arise from poisonous emanations sparked by planetary conjunctions. He even invokes a linguistic parallel between transmission of syphilis by sexual contact *(coitus)* and the production of bad seeds by planetary overlap in the sky, for he describes the astronomical phenomenon with the same word, as *"coitum et conventum syderum"* (the coitus and conjunction of stars), particularly *"nostra trium superiorum, Saturni, Iovis et Martis"* (our three most distant bodies: Saturn, Jupiter, and Mars).

Nonetheless, we seem to need heroes, defined as courageous iconoclasts who discerned germs of modern truth through strictures of ancient superstition—and Fracastoro therefore wins false accolades under our cultural myth of prescience ("ahead of his time"), followed by rejection and later rediscovery, long after death and well beyond hope of earthly reward. For example, the *Encyclopaedia Britannica* entry on Fracastoro ends by proclaiming:

> Fracastoro's theory was the first scientific statement of the true nature of contagion, infection, disease germs, and modes of disease transmission. Fracastoro's theory was widely praised during his time, but its influence was soon obscured by the mystical doctrines of the Renaissance physician Paracelsus, and it fell into general disrepute until it was proved by Koch and Pasteur.

But Fracastoro deserves our warmest praise for his brilliance and compassion *within* the beliefs of his own time. We can appreciate his genius only when we understand the features of his work that strike us as most odd by current reckonings—particularly his choice of Latin epic poetry as a medium for describing syphilis and his christening of the disease for a mythical shepherd whose suffering also reflected Fracastoro's political needs and beliefs. In his article on Fracastoro for the *Dictionary of Scientific Biography,* Bruno Zanobio gives a far more accurate description, properly rooted in sixteenth-century knowledge, of Fracastoro's concept of contagious seeds:

> They are distinct imperceptible particles, composed of various elements. Spontaneously generated in the course of certain types of putrefaction, they present particular characteristics and faculties, such as increasing themselves, having their own motion, propagating quickly, enduring for a long time, even far from their focus of origin, [and] exerting specific contagious activity. . . .

A good description to be sure, but not buttressed by any hint that these *semina* might be living microorganisms. "Undoubtedly," Zanobio continues, "the *seminaria* derive from Democritean atomism via the *semina* of Lucretius and the gnostic and Neoplatonic speculations renewed by St. Augustine and St. Bonaventura." Fracastoro, in short, remained true to his Renaissance conviction that answers must be sought in the wisdom of classical antiquity.

Fracastoro surely probed the limits of his time, but medicine, in general, made very little progress in controlling syphilis until the twentieth century. Guaiacum failed, and mercury remained both minimally effective and maximally miserable. (We need only recall Erasmus's sardonic quip that in exchange for a night with Venus, one must spend a month with Mercury.) Moreover, since more than 50 percent of people infected with the spirochete never develop symptoms of the dreaded third stage, the disease, if left untreated, effectively "cures" itself in a majority of cases (although spirochetes remain in the body). Thus one can argue that traditional medicine usually did far more harm than good—a common situation, recalling Benjamin Franklin's remark that although Dr. Mesmer was surely a fraud, his ministrations should be regarded as benevolent because people who followed his "cures" by inducing "animal magnetism" didn't visit "real" physicians, thereby sparing themselves such useless and harmful remedies as bleeding and purging.

No truly effective treatment for syphilis existed until 1909, when Paul Ehrlich introduced preparation 606 (Salvarsan). Genuine (and gratifyingly easy) cures only became available in 1943, with the discovery and development of penicillin. Identification in the first stage, followed by one course of penicillin, can control syphilis, but infections that proceed to later stages may still be intractable.

I make no apologies for science's long record of failure in treating syphilis—a history that includes both persistent, straightforward error (the poisoning and suffering of millions with ineffective remedies based upon false theories) and, on occasion, morally indefensible practices as well (most notoriously, in American history, the Tuskegee study that purposely left a group of infected black males untreated as "controls" for testing the efficacy of treatments on another group; in a moving ceremony, President Clinton recently apologized for this national disgrace to the few remaining survivors of the untreated group). But syphilis can now be controlled and may even be a good candidate for total elimination (as we have done with smallpox), at least in the United States, if not in the entire world. We owe this blessing, after so much pain, to knowledge won by science. There is no other way.

And so, while science must own its shame (along with every other institution managed by that infuriating and mercurial creature known as *Homo sapiens*), science can also find cures, or at least discover some means of relief, for human miseries caused by external agents that must remain beyond our control until their factual nature and modes of operation become known. The sequential character of this duality—failures as necessary preludes to success, given the stepwise nature of progress in scientific knowledge—leads me to contrast Fracastoro's Latin hexameter with the stodgy prose of the 1998 *Science* article on the genome of *Treponema pallidum,* the syphilis spirochete.

The recent work boasts none of Fracastoro's grace or charm (even in Tate's heroic couplets)—no lovely tales about mythical shepherds who displease sun gods and no intricate pattern of dactyls and spondees. In fact, I can't imagine a duller prose ending than the last sentence of the 1998 *Science* article, with its impersonal subject and its entirely conventional plea for forging onward to further knowledge: "A more complete understanding of the biochemistry of this organism derived from genome analysis may provide a foundation for the development of a culture medium for *T. pallidum,* which opens up the possibility of future genetic studies." Any decent English teacher would run a big blue pencil through these words.

But consider the principal and ever so much more important difference between Fracastoro's efforts and our own. In an article written to accompany the genomic presentation, M. E. St. Louis and J. N. Wasserheit, of the Centers for Disease Control and Prevention in Atlanta, write:

> Syphilis meets all of the basic requirements for a disease susceptible to elimination. There is no animal reservoir; humans are the only host. The incubation period is usually several weeks, allowing for interruption of transmission with rapid prophylactic treatment of contacts, whereas infectiousness is limited to less than twelve months even if untreated. [Tertiary syphilis may be both dreadful and deadly, but the disease is not passed to others at this stage—S. J. G.] It can be diagnosed with inexpensive and widely available blood tests. In its infectious stage, it is treatable with a single dose of antibiotics. Antimicrobial resistance has not yet emerged.

Interestingly, Fracastoro knew that syphilis infected only humans, but he regarded this observation as a puzzle under his theory of poisonous airborne particles that might, in principle, harm all life. He discusses this anomaly at length in part 1 of *Syphilis sive morbus Gallicus:*

> Sometimes th'infected air hurts trees alone,
> To grass and tender flowers pernicious known.
> . . . .
> When earth yields store, yet oft some strange disease
> Shall fall and only on poor cattle seize.
> . . . . . .
> Since then by dear [in the British sense of "costly"] experiment we find
> Diseases various in their rise and kind
> Of this contagion let us take a view
> More terrible for being strange and new.

Thus, the very phenomenon that so puzzled Fracastoro for its anomalous nature under his concept of disease becomes an important clue under the microbial theory.

Similarly, the deciphering of a genome guarantees no automatic or rapid panacea, but what better source of information could we desire for a reservoir of factual hope? Already, several features of this base-level knowledge (base-level, that is, in both the literal and the figurative sense) indicate potentially fruitful directions of research. To cite just three items that caught my attention as I read the technical article on the decipherment of *T. pallidum*'s genome:

1. Several genes that promote motility—and that may help us to understand why these spirochetes become so invasive in so many tis-

sues—have been identified and found to be virtually identical to known genes in *B. burgdorferi,* the spirochete that causes Lyme disease.

2. The *T. pallidum* genome includes only a few genes coding for integral membrane proteins. This fact may help us to explain why the syphilis spirochete can be so successful in evading the human immune response. For if our antibodies can't detect *T. pallidum* because the invader, so to speak, presents too "smooth" an outer surface, then our natural defenses can become crippled. But if these proteins, even though few, can be identified and characterized, then we may be able to develop specific remedies or potentiators for our own immunity.
3. *T. pallidum*'s genome includes a large family of duplicated genes for membrane proteins that act as porins and adhesins—in other words, as good attachers and invaders. Again, genes that can be located and characterized become targets for study and candidates for demobilization.

Science may have needed nearly 500 years to reach our current state of hope, but we should look on the bright side of the differences between then and now. Fracastoro wrote verse and invented shepherds because he knew effectively nothing about the causes of a frightening plague whose effects could be specified and described in moving detail well suited for poetic treatment. The thirty-three modern authors, in maximal contrast, have obtained the goods for doing good. We may judge their prose as uninspired, but the greatest "poetry" ever composed about syphilis lies not in Fracastoro's hexameter of 1530 but in the intricate and healing details of a schematic map of 1,041 genes made of 1,138,006 base pairs, forming the genome of *Treponema pallidum* and published with the 1998 article—the adamantine beauty of genuine and gloriously complex factuality, full of lifesaving potential. Fracastoro did his best for his time; may he be forever honored in the annals of human achievement. But the modern map embodies far more beauty, both for its factuality and utility and as Fracastoro's finest legacy in the history of increasing knowledge—a truly epic tale that we must not shy from labeling by its right and noble name of progress.

TRACY KIDDER

The Good Doctor

FROM *THE NEW YORKER*

On maps of Haiti, National Highway 3 looks like a major thoroughfare. And, indeed, it is the *gwo wout la,* the biggest road across the Central Plateau, a dirt track where trucks of various sizes, overfilled with passengers, sway in and out of giant potholes, raising clouds of dust, their engines whining in low gear. A more numerous traffic plods along on donkeys and on foot, including a procession of the sick. They are headed for the village of Cange and the medical complex called Zanmi Lasante, Creole for Partners in Health. In an all but treeless landscape, it stands out like a fortress on a hillside, a large collection of concrete buildings half covered by tropical greenery. Now and then on the road, a bed moves slowly toward it, a bearer at each corner, a patient on the mattress.

Zanmi Lasante is famous in the Central Plateau, in part for its medical director, Dr. Paul Farmer, known as Doktè Paul, or Polo, or, occasionally, Blan Paul. The women in Zanmi Lasante's kitchen call him *ti blan mwen*—"my little white guy." Peasant farmers like to remember how, during the violent years of the coup that deposed President Aristide, the unarmed Doktè Paul faced down a soldier who tried to enter the complex carrying a gun. One peasant told me, "God gives everyone a gift, and his gift is healing." A former patient once declared, "I believe he is a god." It was also said, in whispers, "He works with both hands"—that is, both with science and with the magic necessary to remove ensorcellments, to many Haitians the deep cause of illnesses. Most of the encomiums seem to embarrass and amuse Farmer. But this last has a painful side. The Haitian belief in illness sent by sorcery thrives on deprivation, on the long absence of effective medicine. Farmer has dozens of voodoo priests among his patients.

On an evening in January 2000, Farmer sat in his office at Zanmi Lasante, dressed in his usual Haiti clothes, black pants and a T-shirt. He was holding aloft a large white plastic bottle. It contained indinavir, one of the new protease inhibitors for treating AIDS—the kind of magic he believes in. A sad-faced young man sat in the chair beside him. Patients never sat on the other side of his desk. He seemed bound to get as close to them as possible.

Farmer is an inch or two over six feet and thin, unusually long-legged and long-armed, and he has an agile way of folding himself into a chair and arranging himself around a patient he is examining that made me think of a grasshopper. He is about forty. There is a vigorous quality about his thinness. He has a narrow face and a delicate nose, which comes almost to a point. He peered at his patient through the little round lenses of wire-rimmed glasses.

The young man was looking at his feet. He wore ragged sneakers. They were probably Kennedys. Back in the nineteen-sixties, Farmer explained to me, J.F.K. had sponsored a program that sent industrial-grade oil to Haiti. The Haitians considered it of inferior quality, and the President's name ever since has been synonymous with shoddy or hand-me-down goods. The young man had AIDS. Farmer had been treating him with antibacterials, but his condition had worsened. The young man said he was ashamed.

"Anybody can catch this—I told you that already," Farmer said in Creole. He shook the bottle, and the pills inside rattled. He asked the young man if he'd heard of this drug and the other new ones for AIDS. The man hadn't.

Well, Farmer said, the drugs didn't cure AIDS, but they would take away his symptoms and, if he was lucky, let him live for many years as if he'd never caught the virus. Farmer would begin treating him soon. He had only to promise that he would never miss a dose. The young man was still looking at his shoes. Farmer leaned closer to him. "I don't want you to be discouraged."

The young man looked up. "Just talking to you makes me feel better. Now I know I'll sleep tonight." Clearly, he wanted to speak to Farmer some more, and just as clearly he was welcome to do so. Farmer likes to tell medical students that to be a good clinician you must never let a patient know that you have problems or that you're in a hurry. "And the rewards are so great for just those simple things!" Of course, this means that some patients wait most of a day to see him, and that he rarely leaves his office before stars shiver in the louvred windows. There is a price for everything, especially virtue.

"My situation is so bad," the young man said. "I keep injuring my head, because I'm living in such a crowded house. We have only one bed, and I let my children sleep on it, so I have to sleep under the bed, and I forget, and I hit my head when I sit up." He went on, "I don't forget what you did for me, Doktè Paul. When I was sick and no one would touch me, you used to sit on my bed with your hand on my head. I would like to give you a chicken or a pig."

When Farmer is relaxed, his skin is pale, with a suggestion of freckles underneath. Now it reddened instantly, from the base of his neck to his forehead. "You've already given me a lot. Stop it!"

The young man was smiling. "I am going to sleep well tonight."

"O.K., *neg pa*" ("my man").

Farmer put the bottle of pills back in his desk drawer. No one else is treating impoverished Haitians with the new anti-retroviral drugs. Even some of his allies in the Haitian medical establishment think he's crazy to try. The drugs could cost as much as eighteen thousand dollars a year per patient. But the fact that the poor are dying of illnesses for which effective treatments exist is, like many global facts of life, unacceptable to Farmer. Indeed, to him it is a sin.

In the fall of 1999, he gave a speech to a group in Massachusetts called Cambridge Cares About AIDS and said, "Cambridge cares about AIDS, but not nearly enough." He wondered if he'd gone too far, but afterward, at his suggestion, health-care workers in the audience and people with AIDS collected a bunch of unused drugs, and he ended up with enough to begin triple therapy for several of his Haitian patients. He is working on grant proposals to obtain a larger, more reliable supply. He doesn't seem to think there is a chance he'll fail. In his experience, when he begs for medicines someone always comes through. Begging of one sort or another is the main way in which he and others managed to create Zanmi Lasante. They didn't borrow, but he did a little stealing—the first microscope in Cange was one he had appropriated from Harvard Medical School.

Farmer graduated from Duke in 1982, summa cum laude and with a major in anthropology. He started coming to Haiti the following spring. On an early visit, he met an Episcopal priest named Fritz Lafontant, who became, and remains, the patriarch of Zanmi Lasante. In 1984, Farmer enrolled at Harvard Medical School, and two years later he enrolled in Harvard's graduate program in anthropology. He received both degrees simultaneously, in 1990. He worked hard at his studies, but often far away from Harvard, while helping to create, piece by piece, the medical complex that would become Zanmi Lasante and serving as an unlicensed doctor in Cange. By the time he got his M.D., he had dealt with more varieties of illness than most American physicians see in a lifetime. With several American friends, he had also founded Partners in Health, Zanmi Lasante's sponsoring organization, with headquarters in Cambridge.

Farmer had chosen to work in one of Haiti's poorest regions. His idea was to bring Boston medicine to the Central Plateau, and in some respects he has succeeded. About a million peasant farmers rely on the medical complex now. About a hundred thousand live in its catchment area—the area for which the organization provides community health workers. On many nights, a hundred people camp out in the courtyard beside the ambulatory clinic; by morning, three hundred, sometimes more, are waiting for treatment. Unlike almost every other hospital in Haiti, Zanmi Lasante charges only nominal fees, and women and children and the seriously ill pay nothing. Partners in Health pays the bills, which are remarkably small. My local hospital, in Massachusetts, which treats about a hundred and seventy-five thousand patients a year, has an annual operating budget of sixty million dollars. Zanmi Lasante spends only about one and a half million dollars to treat forty thousand patients a year. (Farmer spends about two hundred dollars to cure an uncomplicated case of TB in Haiti. The same cure in the United States costs between fifteen and twenty thousand dollars.)

Sometimes the pharmacy muddles a prescription or runs out of a drug. Now and then, the lab technicians lose a specimen. Seven doctors work at the complex, not all of them fully competent—Haitian medical training is mediocre at best. But Zanmi Lasante has built schools and communal water systems for most of the villages in its catchment area. A few years back, when Haiti suffered an outbreak of typhoid resistant to the drugs usually used to treat it, Partners in Health imported an effective but expensive antibiotic, cleaned up the water supply, and stopped the outbreak in the Central Plateau. The medical complex has launched programs in its catchment area for both the prevention and the treatment of AIDS, and has reduced the vertical transmission rate (from mothers to babies) to four per cent—about half the current rate in the United States. In Haiti, tuberculosis kills more adults than any other disease, but no one from the catchment area has died from it since 1988.

Farmer now has these titles, among others: associate professor in two different departments at Harvard Medical School; member of the senior staff in Infectious Disease at the Brigham and Women's Hospital, in Boston; chief consultant on tuberculosis in Russian prisons for the World Bank (unpaid, at his insistence—he deplores some of the World Bank's policies); and founding director of Partners in Health, which has outposts not only in Cange but in Mexico, Cambodia, Peru, and Roxbury, Massachusetts. The organization is perennially overextended, perennially just getting by financially. It raised about three million dollars last year, from grants and private donations—the largest from one of the founders, a Boston developer named Tom White, who has donated millions over the years. Farmer contributed, too, though he didn't know exactly how much.

In 1993, the MacArthur Foundation gave Farmer one of its so-called genius grants. He had the entire sum sent to Partners in Health—in this case, some two hundred and twenty thousand dollars. During his medical school years, Farmer camped out in Roxbury, in a garret in the rectory of St. Mary of the Angels. Later, during sojourns in Boston, he stayed in the basement of Partners in Health headquarters, and he went on staying there after he got married—to Didi Bertrand, the daughter of the schoolmaster in Cange, and "the most beautiful woman in Cange," people at Zanmi Lasante say. When a daughter was born, in 1998, Farmer saw no reason to change their Boston digs, but his wife did. Now they have an apartment in Eliot House, at Harvard. He never sees his paychecks from Harvard and the Brigham. The bookkeeper at Partners in Health cashes them, pays his family's bills, and puts the rest away in the treasury. One day not long ago, Farmer tried to use his credit card and was told he'd reached his limit, so he called the bookkeeper. She told him, "Honey, you are the hardest-working broke man I know."

By any standard, Farmer's life is complicated. Didi, who is thirty-one, and their daughter spend the academic year in Paris, where Didi is finishing her own studies in anthropology. Several friends have told Farmer that he should visit his family more often. "But I don't have any patients in Paris," he says forlornly. In theory, he works four months in Boston and the rest of the year in Haiti. In fact, those periods are all chopped up. Years ago, he got a letter from American Airlines welcoming him to its million-mile club. He has travelled at least two million miles since. I spent a month with Farmer: a little more than two weeks in Haiti, with a short trip to South Carolina wedged in; five days in Cuba, at a conference on AIDS; and the rest in Moscow, on TB business. He called this "a light month for travel." It had a certain roundness. A church group was, in effect, paying for his flight to South Carolina; the Cuban government covered his travel to Havana; and the Soros Foundation financed his trip to Moscow. "Capitalists, Commies, and Jesus Christers are paying," he said.

WHEN PAUL FARMER goes on a long trip, he carries medicines and carrousels of slides and gifts of Haitian art for his hosts, and ends up with room for only three shirts. He owns one suit, which is black, so that he can, for example, wipe the fuzz off the tip of his pen on his pant leg while writing up orders at the Brigham, catch a night flight to Lima or Moscow, and still look presentable when he arrives. En route to South Carolina, the zipper on his suit pants came apart. "Oh well, I'll button my coat," he said before his speech. "That's what a gentleman does when his zipper is broken."

He addressed a meeting of the Anderson County Medical Society. Some of its members visit Cange every year to treat patients free. Some in this particular audience were also church people—the Episcopal Diocese of Upper South Carolina has been making donations to Zanmi Lasante for almost two decades. It was a jacket-and-tie crowd. Farmer gave them his Haiti talk, a compendium of harsh statistics (per-capita incomes of about two hundred and thirty dollars a year and consequent burdens of preventable, treatable illnesses, which kill twenty-five per cent of Haitians before the age of forty) and cheerful photographs that showed what contributions from places like South Carolina could do. Here was a photograph of a girl who had come to Zanmi Lasante with extrapulmonary TB. She was bald, her limbs were wasted. And here was the same girl, with a full head of hair and chubby cheeks, smiling at the camera. Cries of surprise from the crowd, followed by applause. No matter who the audience, that pair of photographs always had the same effect. Farmer had felt jubilant, too, when he treated the girl, but the fact was that the "before" picture more nearly represented the Haitian norm. When the applause died away, Farmer made a little grin. "It's *almost* as if she had a treatable infectious disease."

Throughout the world, the poor stand by far the greatest chances of contracting treatable diseases and of dying from them. Not just simple poverty but even relative poverty in affluent countries is associated with large burdens of disease and untimely death. Medicine can address only some symptoms of poverty. Farmer likes to say that he and his colleagues will make common cause with anyone who is sincerely trying to change the "political economies" of countries like Haiti. In the meantime, though, the poor are suffering. They are "dying like smelt," as Farmer puts it. Partners in Health believes in providing services to them—directly, now. "We call it pragmatic solidarity," he told the audience in South Carolina. "It's probably a goofy term, but we mean it."

Farmer showed a slide with a quotation from the World Health Organization: "In developing countries, people with multi-drug-resistant TB usually die, because effective treatment is often impossible in poor settings." He asked, "Why is it impossible? Because adequate resources have not been brought to bear in places like Haiti and Peru." He showed another photograph of a child—a child who "did not want to be declared cost-ineffective." He said, "Cost-effectiveness analysis is good if you use it in order not to waste money. But what's so great about reducing health expenditures?" He told the audience, "We need to oppose this push for lower standards of care for the poor. We are physicians. I don't mean we should do bone-marrow transplants in Cange, but proven therapies. Equity is the only acceptable goal."

"He always kind of holds your feet to the fire," the MC said when Farmer was done, and, indeed, the applause sounded only slightly more than polite.

Like much of the audience, Farmer is himself religious. He subscribes to the Catholic doctrine called liberation theology, and to its central imperative—to provide "a preferential option for the poor." But, he told me, "I hang on to my Catholicism by a tiny thread. I'm still looking for something in the sacred texts that prohibits using condoms." Some of his beliefs, ones he hadn't openly expressed that night—for example, "I think there should be a massive redistribution of wealth to places like Haiti"—would have seemed extreme to this sedate audience. Yet he liked these people a great deal.

Farmer's politics are complex. He has problems with groups that on the surface would seem to be allies. With, for example, what he calls "W.L.s"—"white liberals," some of whose most influential spokespeople are black. "I love W.L.s, love 'em to death. They're on our side," Farmer once said. "But W.L.s think all the world's problems can be fixed without any cost to themselves. We don't believe that. There's a lot to be said for sacrifice, remorse, even pity. It's what separates us from roaches." As often as not, he prefers religious groups and what he calls "church ladies."

We stayed at the house of a church lady that night, an impeccably genteel Southerner. She lived in a retirement community. When we arrived, one of her neighbors, a retired dentist, was repainting the movable flags on all the residents' mailboxes. Of our hostess, Farmer had said, "She is a very good person. I'll take her over a Harvard smart-ass any day. I love her, actually." I was a little puzzled. This woman wasn't a person you'd suspect of threatening the world order. An hour before dawn the next morning, we climbed into her big new car and she turned on her headlights. They lit up her garage, which was filled to the rafters with boxes and crates—all the equipment Farmer had requested for a new ophthalmology clinic in Cange.

As we flew back toward Haiti, via Miami, Farmer worked on thank-you notes to patrons of Partners in Health. During the descent into Miami, Farmer said he had a fantasy that one day he'd look out at the skyline and at the count of ten all the buildings erected with drug money would collapse. He glanced out the window, disappointed once again. He had other Miami rituals. Depending on its length, a layover at the airport was either "a Miami day" or "a Miami day plus," and included a haircut from his favorite Cuban barber (they'd chat in Spanish) and a thorough reading of *People,* which he called the Journal of Popular Studies, or the J.P.S. (it took him fifty-five minutes, "about as long as Mass in the States"). And then it was up to the Admirals' Club, which he was in the habit of calling "Amirales." There he'd take a hot shower and then stake out a section of lounge (this was "making a cave" or "getting cavaceous at Amirales") and answer E-mail. He had a message from one of the staff in Cange:

> Dear Polo, we are so glad we will see you in a mere matter of hours. We miss you. We miss you as the cracked, dry earth misses the rain.

"After thirty-six hours?" Farmer said to his computer. "Haitians, man. They're totally over the top. My kind of people."

Days and nights ran together. He has a small house in Cange, the closest thing in his life to a home, perched on a cliff across the road from the medical complex. It's a modified *ti kay,* the better sort of peasant house, with a metal roof and concrete floors, and is exceptional in that it has a bathroom, albeit without hot water. Farmer told me that he slept about five hours a night, but, many times when I looked inside his house, his bed seemed unused. Once he told me, "I can't sleep. There's always somebody not getting treatment. I can't stand that." I suppose he slept some nights. His days usually began around dawn. He'd spend an hour or so among the people who had camped out in the lower courtyard, to make sure the staff hadn't missed someone critically ill, and another hour gobbling a little breakfast while answering E-mail, from Peru and other Partners in Health outposts, and Harvard students, and colleagues at the Brigham, and the various warring factions involved in the effort to stop the Russian TB epidemic. Then he saw patients in his office.

Most of the patients were indeed the poor and the maimed and the halt and the blind. For consolation, there was the man he called Lazarus, who had first arrived on a stretcher, wasted by AIDS and TB to about ninety pounds, and now weighed a hundred and fifty. There was a healthy-looking young woman with AIDS whose father only a month before had been saving for her coffin. But there was also a tiny old woman whose backbone had been eaten by TB bacilli, and who hobbled around with her torso at right angles to her legs. A sixteen-year-old boy who weighed only sixty pounds. ("His body has got used to starvation. We're gonna buff him up.") A lovely young woman being treated for drug-resistant TB, now in the midst of a sickle-cell crisis and moaning in pain. ("O.K., *doudou.* O.K., *cheri,*" Farmer cooed. He gave her morphine.) An elderly man with drug-resistant TB who was totally blind. (He'd wanted a pair of glasses anyway; Farmer had found him a pair.) He called the old women "Mother," and the old men "Father." He exchanged quips with most of his patients while he examined them. He turned to me. "It's so awful you might as well be cheerful."

Off and on during those two weeks in Cange, he conducted what he called "a sorcery consult." A woman had decided that one of her sons had "sent" the sickness that killed another son. Farmer was trying to make peace in the family. This would probably take months, because, for one thing, it was useless to try to convince any of the parties that sorcery didn't exist. Farmer said he felt "eighty-six per cent amused." But he saw suffering behind these accusations. Saying that one son had "sold" the other, the mother had used an old Creole word once applied to slaves, and such charges, which often tore friends and families apart, always seemed to spring from the jealousies that great scarcity inspires—in this case, the accused son lived in a better *ti kay* than his mother. Farmer said, "It's not enough that the Haitians get destroyed by everything else, but they also have an exquisite openness to being injured by words."

After office hours, he went on rounds, first to the general hospital and then, with trepidation, to the children's pavilion upstairs, where there always seemed to be a baby with the sticklike limbs, the bloated belly, the reddish hair of kwashiorkor, a form of starvation. Just two weeks earlier, on his first morning back in Cange this year, he'd lost a baby to meningitis, in its ghastly *purpura fulminans* presentation. And, only days later, another baby, from beyond the catchment area—within it, all children are vaccinated free—had died of tetanus. Farmer saved rounds at the TB hospital for last, because just now everyone there was getting better. Most of the patients were sitting on the beds in one of the rooms watching a soccer game on a wavy, snowy TV screen. "Look at you bourgeois people watching TV!" Farmer said.

Everyone laughed. One of the young men looked up at him. "No, Doktè Paul, not bourgeois. If we were bourgeois, we would have an antenna."

"It cheers me up," Farmer said on the way out. "It's not all bad. We're failing on seventy-one levels, but not on one or two." Then it was across the road to his *ti kay,* where he worked with a young American woman from Partners in Health who had been dispatched to help him—on his thank-you notes and upcoming speeches and grant proposals. But on many nights Ti Jean, a handyman, would appear out of the dark, with news that would take Farmer back to the hospital.

A thirteen-year-old girl with meningitis had arrived by donkey ambulance. The young doctors on duty hadn't done a spinal tap, to find out which type of meningitis, and thus which drugs to give her. "Doctors, doctors, what is wrong with you?" Farmer said. Then he did the tap himself. Wild cries from the child: *"Li fe-m mal, mwen grangou."* Farmer looked up from his work and said, "She's crying, 'It hurts, I'm hungry.' Can you believe it? Only in Haiti would a child cry out that she's hungry during a spinal tap."

TWO DAYS BEFORE WE LEFT FOR CUBA, Farmer took a hike to the village of Morne Michel, the most distant of all the settlements in the catchment area. "And beyond the mountains, more mountains" is an old Haitian saying. It appeared to describe the location of Morne Michel. A TB outpatient from the town had missed an appointment. So—this was a rule at Zanmi Lasante—someone had to go and find him. The annals of international health contain many stories of adequately financed projects that failed because "noncompliant" patients didn't take all their medicines. Farmer said, "Only physicians are noncompliant. If the patient doesn't get better, it's your own fault. Fix it." A favorite Doktè Paul story in the village of Kay Epin was of the time, many years earlier, when he chased a man into a field of cane, calling to him plaintively to come out and be treated. He still went after patients occasionally. To inspire the staff, he said. Hence the trip to Morne Michel.

He drove the first leg in a pickup truck, past dirt-floored huts with banana-frond roofs, which leak during the rainy season, so that the dirt floors turn to mud; and little granaries on stilts, which don't prevent rats from taking a third of every farmer's meagre harvest; and yellow dogs so skinny, Haitian peasants say, that they have to lean against trees in order to bark. In a little while, the reservoir that feeds the Péligre Dam came into view, a mountain lake far below the road. The scene looked beautiful: blue waters set among steep, arid mountainsides. But if you saw with peasant eyes, Farmer said, the scene looked violent and ugly—a lake that had buried the good farmland and ravaged the mountainsides.

We parked near the rusted hulk of a small cement factory, beside the concrete dam. In every speech and in all his books, Farmer is at pains to assert the interconnectedness of the rich and poor parts of the world, and here in the dam he had his favorite case study. The dam was planned by engineers from the United States Army during the rather brutal American occupation of Haiti early in the twentieth century, and was built in the mid-fifties, during the reign of one of America's client dictators, by Brown & Root, of Texas, among others, with money from the U.S. Export-Import Bank. The dam had drowned the peasants' farms and driven them into the hills, where farming meant erosion, all in order to improve irrigation for American-owned agribusinesses downstream and, eventually, to supply electricity to Port-au-Prince, especially to the homes of the wealthy élite and the foreign-owned assembly plants, in which peasant girls and boys from Cange still work as servants and laborers, more than a few of them nowadays returning home with AIDS. Most of the peasants didn't get paid for their land. As they liked to say, the project hadn't even brought them electricity or water.

On the other side of the dam, a footpath—loose dirt and stones, slippery-looking—went straight up. Farmer has a slipped disk from eighteen years of travelling the *gwo wout la.* He also suffers from high blood pressure and mild asthma, which developed after he'd recovered from a possible case of tuberculosis. His left leg was surgically repaired after he was hit by a car and turns out at a slight angle—like a kickstand, as one of his brothers says. But when I got to the top of that first hill, sweating and panting, he was sitting on a rock, writing a letter. It was the first of many hills. We passed smiling children carrying water jugs that must have weighed half as much as they did, and the children had no shoes. We passed groups of laughing women washing clothes in the muddy rivulets of gullies. Haitians, Farmer had said, are a fastidious people. "I know. I've been in all their nooks and crannies. But they blow their noses into dresses because they don't have tissues, wipe their asses with leaves, and have to apologize to their children for not having enough to eat."

"Misery," I said.

"And don't think they don't know it," Farmer said. "There's a W.L. line—the 'They're poor but they're happy' line. They do have nice smiles and good senses of humor, but that's entirely different."

We stopped awhile at a cockfight, the national sport, and passed beside many fields of millet, the national dish, which seemed to be growing out of rocks, not soil. We passed small stands of banana trees and, now and then, other tropical species, Farmer pausing to apply the Latin and familiar names: papaw, soursop, mango—a gloomy litany, because there were so many fewer of each variety than there should have been. We paused on hilltops, where the wind was strong and cold on my sweaty skin. Curtains of rain and swaths of sunlight swept across the vast reservoir far below and across the yellow mountains, which, I realized, could never look pretty again to me. I wondered how Haitians avoided hopelessness. I wondered how Farmer did. After about two and a half hours, we arrived at the hut of the noncompliant patient, another shack made of rough-sawn palm wood with a roof of banana fronds and a cooking fire of the kind Haitians call "three rocks." The patient, it turned out, had been given confusing instructions by the staff at Zanmi Lasante, and he hadn't received the money, about ten dollars a month, that all Zanmi Lasante's TB patients get—for extra nutrition, to boost their immune systems. He hadn't missed any doses of his TB drugs, however.

Farmer gave him the money, and we started back through the mountains. I slipped and slid down the paths behind him. "Some people would argue this wasn't worth a five-hour walk," he said over his shoulder. "But you can never invest too much in making sure this stuff works."

"Sure," I said. "But some people would ask, 'How can you expect others to replicate what you're doing here?' What would be your answer to that?"

He turned back and, smiling sweetly, said, "Fuck you."

Then, in a stentorian voice, he corrected himself: "No. I would say, 'The objective is to inculcate in the doctors and nurses the spirit to dedicate themselves to the patients, and especially to having an outcome-oriented view of TB' " He was grinning, his face alight. He looked very young just then. "In other words, 'Fuck you.' "

For a person whose résumé makes one think of Albert Schweitzer—the once popular image of that personage, at least—Farmer has an oddly cheerful and irreverent turn of mind. From time to time, colleagues, and even a close friend or two, have subjected him to moral envy, as if his self-abnegation were meant as a reproach to them. It isn't as though he doesn't preach self-sacrifice, but he practices more than he preaches. He has taken only two vacations in the last twelve years—the first after he was run over, in Cambridge in 1988, the second in 1997 while recovering from hepatitis A, contracted in Peru. Yet he thinks that other people ought to have vacations, and the more luxurious the better. He likes a fine meal, a good bottle of wine, a fancy hotel, and a hot shower. But he doesn't seem to need any of those things, or the money to buy others, in order to be happy.

It was impossible to spend any time with Farmer and not wonder why he'd chosen this life. Maybe some partial explanations can be found in the usual place.

One morning, between airplanes, he and I were standing near the side exit of a crowded bus, and he said, "I feel at home. *Our* bus had doors like these." He added, "Until the bus turned over." He was about twelve. The bus was older. His father had bought it from the State of Alabama, had refashioned it into the Farmer residence, and parked it in a campground in Florida. The Farmers came from western Massachusetts. The whole family—Farmer's father and mother and six children—was heading home from a vacation there when the bus flipped onto its roof. No one was seriously hurt.

"Where did you live after that?" I asked.

"In a tent. Of course. What kind of a question is that?" He was smiling. "This is before it got crazy, before the boat."

His father bought a boat, on which he intended the family to achieve full self-sufficiency. But the one time they went to sea they hauled up only a couple of edible fish, and were buffeted all night in a storm. Then his father got lost heading back for port and grazed a rock. After that, the boat stayed moored in a bayou on the Gulf Coast, north of Tampa.

The bus door opened, and Farmer, returning to the present, looked at the door and said, "My madeleine this morning."

Farmer's father had a profession—schoolteacher, usually. He was a big, vigorous man and a good athlete. He was strict with his children about manners and schoolwork and chores. But in most other respects he shunned convention, stubbornly pursuing his various schemes. "You didn't tell him that he couldn't do something, because then he'd have to prove that he could," Farmer's mother said. "He was a great risktaker, and everything always turned out all right. I mean, no one ever got seriously hurt." He died suddenly, at the age of forty-nine, in apparent good health, while playing basketball.

Farmer, according to his younger sisters, was a scrawny boy, intense in anger and affection, and very smart. He started a herpetology club in fifth grade. No one came to the first meeting, but his father required his siblings to attend the family lectures about plants and animals that Paul delivered at home. He received a scholarship to Duke. There he first discovered wealth.

"How come you put your shirts in plastic?" he asked, watching his roommate unpack. At Duke, he soaked up culture. He was drama critic and art critic for the student paper. The first play he ever saw was one he was sent to review.

But growing up on a bus and a boat, without hot showers, hardly implies a single fate. One of his sisters is a commercial artist, another manages community relations for a hospital's mental-health programs, the third is a motivational speaker. One brother is an electrician, the other a professional wrestler (known to fans as the treacherous New World Order Sting and to his family as the Gentle Giant). A person with Farmer's background might well have yearned for a lucrative suburban practice. He himself doesn't like to make too much of the connections between his present life and his childhood, which, for the most part, he remembers as happy. He did say that it had relieved him of a homing instinct. "I never had a sense of a home town," he said. "It was 'This is my campground.' Then I got to the bottom of the barrel, and it was '*This* is my home town.' " He meant Cange.

Farmer told me, "It stands to reason that a person who lives the way I do is trying to lessen some psychic discomfort." He had wanted to avoid "ambivalence," he said, and had tried to build his life around "areas of moral clarity"—"A.M.C.s," in Partners in Health lingo. These are areas, rare in the world, where what ought to be done seems perfectly clear. But the doing was always complicated, always difficult, in his experience. Thinking of those difficulties, I imagine that most people wouldn't willingly take them on, giving up their comforts. Yet many would like to wake up knowing what they ought to do and that they were doing it. Farmer's life looked hard, but by the time we left Haiti I also thought that it was enviable.

LEAVING HAITI, Farmer didn't stare down through the airplane window at that brown and barren third of an island. "It bothers me even to look at it," he explained, glancing out. "It can't support eight million people, and there they are. There they are, kidnapped from West Africa."

But when we descended toward Havana he gazed out the window intently, making exclamations: "Only ninety miles from Haiti, and look! Trees! Crops! It's all so verdant. At the height of the dry season! The same ecology as Haiti's, and look!"

An American who finds anything good to say about Cuba under Castro runs the risk of being labelled a Communist stooge, and Farmer is fond of Cuba. But not for ideological reasons. He says he distrusts all ideologies, including his own. "It's an 'ology,' after all," he wrote to me once, about liberation theology. "And all ologies fail us at some point." Cuba was a great relief to me. Paved roads and old American cars, instead of litters on the *gwo wout la.* Cuba had food rationing and allotments of coffee adulterated with ground peas, but no starvation, no enforced malnutrition. I noticed groups of prostitutes on one main road, and housing projects in need of repair and paint, like most buildings in the city. But I still had in mind the howling slums of Port-au-Prince, and Cuba looked lovely to me. What looked loveliest to Farmer was its public-health statistics.

Many things affect a public's health, of course—nutrition and transportation, crime and housing, pest control and sanitation, as well as medicine. In Cuba, life expectancies are among the highest in the world. Diseases endemic to Haiti, such as malaria, dengue fever, TB, and AIDS, are rare. Cuba was training medical students gratis from all over Latin America, and exporting doctors gratis—nearly a thousand to Haiti, two en route just now to Zanmi Lasante. In the midst of the hard times that came when the Soviet Union dissolved, the government actually increased its spending on health care. By American standards, Cuban doctors lack equipment, and are very poorly paid, but they are generally well trained. At the moment, Cuba has more doctors per capita than any other country in the world—more than twice as many as the United States. "I can sleep here," Farmer said when we got to our hotel. "Everyone here has a doctor."

Farmer gave two talks at the conference, one on Haiti, the other on "the noxious synergy" between H.I.V. and TB—an active case of one often makes a latent case of the other active, too. He worked on a grant proposal to get anti-retroviral medicines for Cange, and at the conference met a woman who could help. She was in charge of the United Nations' project on AIDS in the Caribbean. He lobbied her over several days. Finally, she said, "O.K., let's make it happen." ("Can I give you a kiss?" Farmer asked. "Can I give you two?") And an old friend, Dr. Jorge Pérez, arranged a private meeting between Farmer and the Secretary of Cuba's Council of State, Dr. José Miyar Barruecos. Farmer asked him if he could send two youths from Cange to Cuban medical school. "Of course," the Secretary replied.

Again and again during our stay, Farmer marvelled at the warmth with which the Cubans received him. What did I think accounted for this?

I said I imagined they liked his connection to Harvard, his published attacks on American foreign policy in Latin America, his admiration of Cuban medicine.

I looked up and found his pale-blue eyes fixed on me. "I think it's because of Haiti," he declared. "I think it's because I serve the poor."

I had the impression that he was angry, disappointed, and a little hurt. An oddly potent combination. And then I felt I was forgiven. Lying in the bed next to Farmer's in the hotel room took me back to late-night talks in college and in the Army. I turned out the light, and he went on talking, his voice growing slurred: "I had a lovely day. I'm lucky. All my days are good. Not all are lovely, but they're good. I wouldn't trade with anyone."

A FEW NIGHTS LATER, we started flying toward Moscow. We stopped off in Paris for eighteen hours, so Farmer could attend his daughter Catherine's second-birthday party. He'd brought short-acting benzodiazepines to get us through the flights. They have left my memories of Paris all wrapped up in gauze. A small apartment in the Marais district, and Farmer in his black suit, dancing with his daughter, holding her to his chest, swaying from side to side in a loopy, long-limbed waltz. And the little girl's dark eyes, which her face hadn't yet grown into, fixed in serious rapture on some invisible object in the ceiling. Later, Farmer sat on the sofa and watched Catherine play with her stuffed animals. His wife, Didi, tall and stately—she probably was, in fact, the most beautiful woman from Cange—called to him from the kitchen. When did he leave for Moscow?

Tomorrow morning, Farmer said.

From the kitchen came the sound of something dropping and a deep-throated exclamation.

Farmer was skipping the first meeting in Moscow to make this stop. He'd said he felt guilty about that. Now I looked over at him. He was clasping his knees with his elbows and covering his mouth with both hands. He seemed to be trying, as Haitians say, to make his body very small. I remember thinking, despite my haze, I'd remember this. It was the first time I'd ever seen him at a loss for words or action.

ABOUT A THIRD OF the world's people have TB bacilli in their bodies, but the bacterium is indolent. It multiplies into lung-consuming, bone-eating illness in only about ten per cent of the infected. The likelihood of getting sick increases greatly, however, for those who suffer from malnutrition or various diseases, such as HIV. People who live in crowded peasant huts and prisons and homeless shelters and slums stand the best chances of inhaling TB bacilli, of having the infection expand into active disease, and also, in some settings, of contracting or generating drug-resistant strains. A person with active disease who receives only one anti-TB drug, or who receives several for too short a time, can become a site of rapid bacterial evolution. The drugs apply the selective pressure. The host winds up sick with bacilli immune to those drugs, then coughs them up for others to share. This occurs infrequently in places of nearly universal poverty like Haiti, where most people don't get treated at all, and most often in places where wealth and poverty are mingled, where the poor receive some therapy but not enough—places like New York City and Peru and post-Soviet Russia.

Farmer was obliged to go to Moscow because five years ago an old friend of his, a priest named Father Jack, had died of drug-resistant tuberculosis. It seemed that he must have caught it while working in Carabayllo, a slum on the outskirts of Lima. Farmer and a close friend and colleague named Jim Yong Kim—a fellow-doctor from Harvard and the executive director of Partners in Health—went to the shantytown and, sure enough, discovered an epidemic of multi-drug-resistant TB.

The patients they found in Carabayllo—about fifty, initially—were mainly young, and all were poor. Most had severely damaged lungs. Partners in Health had already helped finance a small clinic in the shantytown, and a good doctor was on hand, but cures would be difficult. Many of the workers were terrified of inhaling the drug-resistant germs. Farmer and Kim would have to arrange to feed the patients, in order to strengthen their immune systems. They'd have to arrange for laboratory analysis of each patient's TB, so that they'd know which antibiotics to use. Because most patients had TB resistant to all five of the best drugs, they'd have to give them "second-line" antibiotics. In short, they'd have to build a first-rate health-care system out of the shantytown's mediocre one—a system that would administer those drugs reliably and keep the patients' spirits up, because the second-line drugs are weak and have unpleasant side effects, which a patient has to endure for as much as two years. Moreover, the second-line drugs were little used and very expensive, and Farmer and Kim didn't know where they'd find the money to buy them. They couldn't expect help from Peru's medical establishment, which had rather recently established a first-rate TB program and didn't welcome news that the program had a flaw. And they had no chance of getting money from foundations, because the World Health Organization had, in effect, declared projects like theirs too expensive for "resource-poor settings."

The W.H.O. had created a program for worldwide TB control called DOTS—Directly Observed Therapy Short Course. Properly applied, DOTS insures that patients take regular doses of the cheap and powerful first-line antibiotics for six to eight months. DOTS worked well in most places. Farmer had used the program for years in Haiti, even back before it had a name. It was inexpensive. It was all that poor countries could afford. Therefore, the policymakers seem to have reasoned, it had to be sufficient, even in settings where first-line-drug resistance had surfaced. As one expert in international health said later, "The party line had been 'We've got a way of treating TB, which is DOTS, and that's expensive enough. If we were to treat M.D.R. TB, it would be at twenty times the cost.' They said that without thinking through the next step—that if you wipe out drug-sensitive TB and let the other flourish, you've really got a problem."

Farmer and Kim had some allies. The most powerful was Howard Hiatt, the former dean of the Harvard School of Public Health. He watched their project in Peru, a little nervously, from a distance. For a time, he wondered where they were getting the money for the second-line drugs. Then one day the president of the Brigham and Women's Hospital stopped him in a corridor and said, "Your friends Farmer and Kim are in trouble with me. They owe this hospital ninety-two thousand dollars." Hiatt looked into the matter: "Sure enough. They would stop at the pharmacy before they left for Peru and fill their briefcases with drugs. They had sweet-talked various people into letting them walk away with the drugs." Looking back, Hiatt was amused. "That's their Robin Hood attitude."

Actually, they only borrowed the drugs. Tom White, the chief donor to all their causes, soon wrote a check for the entire bill.

It took Farmer about twenty-two hours round trip to travel between Cange and Carabayllo. Over the next three years, he made the journey fifty times. Kim went almost as often. "Peru nearly killed us," Farmer said later—literally true in his case, when he came down with hepatitis A and ignored the symptoms for a time. But the results were very good. Indeed, "astonishing" to Howard Hiatt. He arranged a meeting of the eminent in worldwide TB control, among them some of the policymakers. There Farmer and Kim presented their team's results, and also epidemiological evidence that strains of drug-resistant TB are at least as contagious and virulent as drug-susceptible ones and that in epidemics involving drug resistance DOTS will cure no more than half the victims, will amplify resistance among the rest, and will allow M.D.R. to go on spreading. The atmosphere was heated. Farmer and Kim were upstarts, mere clinicians, and their message was embarrassing to many people there. But they had solid data. And some of the people who received their data at the meeting were, after all, scientists.

"PAUL AND JIM mobilized the world to accept drug-resistant TB as a soluble problem," Hiatt says, looking back. This was no small matter, he believes. "At least two million people a year die of TB, more adults than from any other infectious disease. And when those people who die include predominantly people with drug-resistant strains, as will happen unless a very big and good program gets established, it's not going to be two million. That number could be increased by an order of magnitude."

Many other meetings and arguments followed, but after that one a general strategy for treating M.D.R. officially existed, and was endorsed by the W.H.O.—a strategy like the one Farmer and Kim had used. It even had a name: DOTS-Plus. With help from Hiatt, and later from others as well, Kim worked to reduce the prices of the second-line drugs. (Today, they cost ninety per cent less than when Kim and Farmer started using them in Carabayllo.) But Peru had drained about a million dollars a year from their little treasury—all the money they'd hoped to save as an endowment for Partners in Health.

Farmer asked for help from the Open Society Institute, George Soros's foundation. The O.S.I. turned him down, but sent him a letter of recommendation. It said that the O.S.I. understood the importance of his and Kim's work in Peru, because the O.S.I. had a similar project in Russia. Farmer knew about the TB epidemic in Russia, of course. It had arisen in the turmoil that followed the dissolution of the Soviet Union, the small wars and the thievery and the structural economic readjustments imposed in part by international lenders. The troubles contained ingredients not just for an epidemic but for a drug-resistant epidemic: a failing health-care system had led to many uncompleted therapies; a rising crime rate had led to overcrowded prisons. Farmer knew that Soros had put up twelve and a half million dollars for a pilot project to improve TB control in Russia. But Farmer hadn't known the details until he saw the letter. The O.S.I. would use the DOTS strategy. It would treat all patients as if they had nonresistant, pan-susceptible TB, and for those who didn't get better it would provide hospice care to ease their deaths.

Farmer was appalled. With Hiatt's blessing, he wrote a two-page letter to the O.S.I., explaining that its project was bound to fail.

In the event, Farmer ended up in George Soros's office. Soros shouted over the phone at the director of his project, Alex Goldfarb, for a while, in Farmer's presence. Then he asked Farmer to help fix the pilot project. Farmer wavered. Haiti needed him more. But his basic impulse was to say, "You can't just let a poor person die," and here he thought he saw a chance to apply this creed on every level. "Forgive me for saying this," Kim once remarked. "But the great thing about TB is that it's airborne." TB is only predominantly a disease of the poor. Others get it, too, just from breathing. The affluent world would have to pay attention to the threat of a TB so difficult to treat, to the dire but real possibility that strains resistant to every drug would spread across borders. Conceivably, the affluent nations would decide to protect themselves. Then they'd have to do on a grand scale what Partners in Health had done in Carabayllo and in Cange.

So Farmer took a trip to Siberia with Alex Goldfarb. They returned as friends. What they saw in the overcrowded prisons alarmed Farmer. He and Goldfarb went to see Soros to ask for more money. Soros said that would just delay the international response. Instead, he arranged a meeting at the White House, with Hillary Clinton presiding. Farmer and Goldfarb helped write the talking points for Soros and critiqued the ones prepared for the First Lady. She prevailed upon the World Bank to take action, and the World Bank dispatched something called a mission to Moscow, a group of economists, epidemiologists, and public-health experts who would work out the details of a TB loan to Russia.

This mission to Moscow was a good example of what Farmer meant when he said that in Areas of Moral Clarity only the imperative to take action is clear. He represented the World Bank on the problem of TB in Russian prisons. Alex Goldfarb represented the Russian Ministry of Justice, which ran the prisons. The Ministry of Health, in charge of the civilian sector, felt it should receive the great majority of whatever money was loaned. Some of the World Bank consultants agreed. But nearly half of all cases and most of the drug-resistant ones languished in the prisons, and the prisons were playing the role of what Goldfarb called "an epidemiological pump," spreading TB among prisoners, then sending them back to civilian society. Besides, prisoners were part of Farmer's special constituency—it was in the Gospels, you could look it up. So he and Goldfarb thought the Ministry of Justice should get half the loan. The mixture seemed combustible: World Bank consultants with substantial résumés, some with egos to match, mixed with Russian colonels and generals and former apparatchiks and old TB warriors, members of a defeated empire, on the lookout for condescension.

Farmer had flown to Russia four times already on this business, and he was tired of it, he said. Tired of the meetings and arguments in rooms that grew increasingly airless, where there were no patients, his thoughts straying back to Haiti: when the next meningitis victim came in, would the doctors, in his absence, do a spinal tap? Tired physically, too, just then. Our first morning in Moscow, he said at breakfast, "I'm still biologically deranged." He wore his third and last shirt. One of the buttons was missing. His black suit was rumpled. His face was red, probably because in his mind he was already arguing. One member of the World Bank team had been quoted as saying, "It's ridiculous and too expensive, this proposal for the prisons. It's ridiculous." Now Farmer himself said, over breakfast, "The battle is joined. But this is a ten-year program. This is a very long process. Ten years. I think I should be nonconflictual at least for a day. I'm trying to talk to myself. I'm trying to keep from slugging the guy." He added, "The prisoners are dying. They'll go on dying."

As the days wore on, Farmer's smiles and his vigor returned, and with them, somehow, the illusion of a stylishly dressed man. He seemed to be winning the argument about the percentage of the loan the prisoners would receive. Now he and the other consultants were arguing about details, about whether or not the gaunt, tuberculous prisoners would receive ten cents' worth of extra food a day. "The food fight," Farmer called it. He kept his temper.

Our hotel was situated across from Red Square, and from certain windows you could catch glimpses of the onion-topped towers of St. Basil's Cathedral. Farmer thought that it was one of the most beautiful buildings in the world, but it was marred by the fact that it was built to celebrate Ivan the Terrible's bloody victory over the Tartars. Marred for him. Erasing history, he liked to say, always served the interests of power. He practiced his usual brand of tourism. He visited a prison.

MOSCOW'S CENTRAL PRISON, the largest in the city, is called a *sizo*—a detention center. The building was immense, though I couldn't grasp its actual dimensions, because of the complexity of turnings, through doorways where you had to duck your head, and climbings, up ancient metal staircases, and hikes, down corridors that made me think of abandoned subway tunnels, with some sort of yellow fibreboard slapped haphazardly on the walls. We passed through various climate zones, from warm to cold to warm again, and regions of odor, some of food, others hard to place—it seemed better not to know.

"Don't get lost," a prison official said. "This is not a good place to get lost."

We passed a file of prisoners, all dressed in baggy pants, in ragged coats and caps, gray faces in dim light; one had the crookedest nose I'd ever seen. Then we reached the prison hospital. "Think of Cuba," Farmer whispered to me. "Look at this shitty place." The guides were all doctors or public-health officials, and they deplored the conditions. They opened the door to a cell reserved for patients with AIDS. "There are fewer than in the usual cells," one of the doctors said.

"How many?"

"Only fifty in this room."

Farmer went in first, followed by a translator. A dingy gray room, smaller than many American living rooms, full of double-decker beds, laundry hanging from clotheslines. Most of the men were young. In a moment, Farmer was shaking hands with them, touching arms and shoulders, and in another moment loud voices all around him were competing to air grievances. One prisoner, older than the rest, evidently something like a spokesman, declared that he had merely been a witness to the killing of a man, but because he had AIDS he got five years. The actual killer, who was tried with him, got only three. "And when I get out I will cut his head off," he said. Everyone, prisoners and doctors, laughed, a deafening sound in the cramped cell.

Farmer thanked the prisoners. The spokesman said, "I wish you would come more often."

"I would like to."

Another twisty passage, into the TB department. "The doctors are overworked and have almost no protection," an official said. "The X-ray equipment—it is exhausted." They weren't sure how many patients had drug-resistant strains. "We do not have laboratory support from Moscow. We get no information from the other institutions where the prisoners come from. This is a division of the railway station. Fifty per cent are not from Moscow."

We went into another cell, this one filled with TB patients, the same as the last but a little more crowded and humid—the humidity that comes from many pairs of lungs exhaling. Several men were coughing, each distinctively, I thought—a Chaliapin bass, a baritone, a tenor. Farmer stood beside a bed, his arm resting on the mattress of an upper berth. "You look good," he said to one of the men. "Anybody coughing up blood?"

"No."

"So, pretty much, people are getting better?"

"It's not worse," said a prisoner.

He asked them where they came from.

Grozny, Volga, Baku.

"Tell him I've been to Baku," Farmer said to the translator. "And it's better to be here. Tell him I've been to Colony Three."

A young man sitting on an upper berth said, "I saw you in Colony Three. You were with a woman."

"Yes, I was there with a woman!" Farmer exclaimed. He shook hands with the man. "It's nice to see you again."

It was time to leave. "Good luck," Farmer said through the translator. "Tell them I hope everybody gets better."

We headed back toward the prison office. "I like these prison medical people," Farmer said to me. "They're trying." He turned to the translator. "Tell Ludmilla"—she was one of the doctors—"I've met some extremely dedicated prison doctors." He had singled out Ludmilla because she'd told him a story about an Italian human-rights activist who had accused her of mistreating AIDS patients, by keeping them isolated from the other inmates. Farmer had said, "In a setting where there's a lot of TB? *Not* to isolate them would be a violation of human rights!"

About ten per cent of Russia's one million prisoners had active TB. In many prisons, a majority of them had drug-resistant strains; twenty per cent, it was feared, had M.D.R. On top of that, one of the doctors told Farmer, the incidence of syphilis was rising. Alarming, because rising syphilis announces the imminence of AIDS, and AIDS would grossly magnify the TB epidemic. "It's gonna be a fucking disaster," Farmer said softly to me as we headed back to the central office.

Now the crude conference table there was laid with a feast. Farmer declared, "Oh, thank you! Just what I like!" He murmured to me, "I was afraid of this. I hate vodka." But he knocked it back with expertly feigned pleasure, just as he did in Haiti when eating proffered items of what he called "the fifth food group." Toasts were offered, and counter-toasts. After a while, Farmer's grew lengthy.

"I have been working in Haiti for almost twenty years, ever since I was a young chap, and some years ago I was asked by the State of Massachusetts to be a TB commissioner, and I said, 'What the hell do we do?' I was in Haiti and I had a lot of TB patients and I took sputums and I brought them to Boston. And I took them into the lab and I wrote, 'Paul Farmer, State TB Commissioner.' I wanted them to process my samples from Haiti and they did and never asked any questions, so I did it more and more, and then I did it with sputums from Peru, and, of course, eventually they asked me why. I said,

'Massachusetts is a great state, it has a big TB lab, lots of TB doctors, lots of TB nurses, lots of TB lab specialists. It lacks only one thing: tuberculosis.' "

One of the Russians—a colonel—laughed. A woman doctor said, "We have lots of TB and no labs."

More toasts, more vodka. After a time, the colonel asked Farmer, "Is America a democracy?"

Farmer's face grew serious. "I think whenever a people has enormous resources, it is easy for them to call themselves democratic." He cited the "idiotic" remark that the Italian visitor had made to Ludmilla, and went on, "I think of myself more as a physician than as an American. Ludmilla and I, we belong to the nation of those who care for the sick. Americans are lazy democrats, and it is my belief, as someone who shares the same nationality as Ludmilla—I think that the rich can always call themselves democratic, but the sick people are not among the rich. Look, I'm very proud to be an American. I have many opportunities because I'm American. I can travel freely throughout the world, I can start projects, but that's called privilege, not democracy."

As Farmer talked, the colonel's face had begun showing signs of exertion. Now he let his laughter out. He said, "But I only wanted to know if you would permit me to smoke a cigarette."

IN THE END, Farmer won his skirmishes in Moscow. He'd managed to sneak food into the budget by calling it vitamins. And, for now at least, the prisoners would get about half the loan. The first installment was projected at thirty million dollars. (The figure has since been increased to a hundred and seventy million, paid out over five years.) Farmer believed the world ought to spend the money outright, not lend it, and he figured that stanching the epidemic in Russia and the former republics would cost at least half a billion dollars. Still, he felt happy.

So did Goldfarb. "I am always ambivalent, though," he told Farmer. "It means I have to deal with this thirty million. Keep them from stealing it. And there is a lot at stake with this project. It has to work, or we can forget DOTS-Plus."

Farmer and Goldfarb spent a lot of time together that week. Goldfarb usually appeared in slightly rumpled tweeds and corduroys, threatening to do battle with one or another Russian official or member of the World Bank team. Farmer would argue him out of this. They seemed to have the kind of friendship that thrives on argument. At dinner one night, Goldfarb said, in sonorously accented English, "Prisoners. They are not nice people. They are epeedeemeeologically eempoortant."

"Our big split," said Farmer. He turned to me. "The stench of innocence is what I smell. The stench of guilt is what he smells."

"I should take that back," Goldfarb said. "About half of the people should not be in jail."

"Three-quarters," said Farmer. "Come on, Alex. Those are crimes against property."

"There is twenty-five per cent should be in jail for life," said Goldfarb.

"No. Ten per cent," said Farmer. "You think I'm naïve."

"You are not naïve," said Goldfarb. "You see the whole situation. You just don't accept that . . ."

"People aren't nice."

"No! Bad people. You are not naïve. You can just disregard things which are unpleasant, and that is why you are not scientific. You disregard reality."

"But you still like me," said Farmer.

"Of course I like you!" said Goldfarb.

Farmer had wished for a blizzard in Moscow. He had got just a snowfall. We walked back to the hotel on slippery sidewalks, in the cold, cold night, Farmer with his red scarf over his nose, his glasses fogging up.

I rehearsed his argument with Goldfarb, Farmer's whittling down the number he thought should be in prison. If it had gone on, I thought, he might have got down to one per cent, or zero.

"Do you think I'm crazy?" Farmer asked.

"No. But some of those prisoners have done terrible things."

"I know," he said. "And I believe in historical accuracy."

"But you forgive everyone."

"I guess I do. Do you think that's crazy?"

"No," I said. "But I think it's a fight you can't win."

"That's all right. I'm prepared for defeat."

"But there are the small victories," I said.

"Yes! And I love them!"

Many of Farmer's friends worried about his health, and thought he should cut back his gruelling schedule. One, at least, thought he should retire from his clinical work in Haiti and concentrate on "big issues," such as the Russian TB epidemic. Now, on a street in Moscow, I began to pose a hypothetical question. "But without your clinical practice—"

Farmer interrupted. "I wouldn't be anything," he said.

WE LEFT MOSCOW before dawn, and flew to Zurich, where we boarded a plane to Boston. Farmer carried a large curved sabre made of glass and filled with cognac, a present from the Deputy Justice Minister. The customs agents raised their eyebrows. Other passengers did double takes. He smiled back, but his smile looked wan. Every takeoff and landing nauseated Farmer for a few minutes. When we arrived in Boston the next afternoon, he'd have to go right to another meeting—funding for the women's clinic in Cange was running out.

Most modern descriptions of human behavior give selfishness great explanatory power, even over what look like selfless acts. But after I'd spent a month with Farmer altruism had begun to seem plausible, even normal. On the airplane back to Boston, he offered his own explanation. "There's a social truth and a personal truth," Farmer said, once we'd crammed ourselves into our seats. While he spoke, he traced a finger over his stowed tray table, making evanescent designs, as if of his thoughts. "We live in a time of great ease and bounty. I have complete access to all that ease and bounty. At the same time, I have had the world revealed to me as it really is. It isn't a different world—it's the same world. There is no reason or event. I came back to Catholicism through liberation theology because it's such a powerful rebuke to the hiding away of poverty, but that was after I was already involved in Haiti. I would read stuff from scholarly texts and know they were wrong. Living in Haiti, I realized that a minor error in one setting of power and privilege could have an enormous impact on the poor in another. For me, it was a process, not an event. A slow awakening, as opposed to an epiphany."

He held up a finger and moved it to his left. "I can have this world of privilege, and I like this world of privilege." He moved the finger to his right. "But I'm not willing to erase this world of suffering." He went on, "People don't get up in the morning and say, 'I'm going to erase this world of suffering.' They're just cosseted. They don't have to see the suffering. We live in a country so rich that you can hide away anything in it. What does it mean to be human, as opposed to being American? I believe in signs, kind of jokingly. But here, after a week of haggling over ten cents for a day's worth of food for Russian prisoners, I open the paper and the basic message of Clinton's State of the Union is how are we going to get rid of this huge budget surplus. We were in Moscow central prison two days ago. We were in Morne Michel. What do you have to do to erase the people in those places? How do you say this without sounding like some self-important asshole? I now know the choices I made are the right ones for me. I feel happy. Satisfied. Not self-satisfied. I'm not satisfied about this loan to Russia. It's a *loan*. What does it mean to be human, instead of a cockroach? Solidarity, compassion, sympathy, and love."

He glanced out the airplane window, and began telling me a story. Back before Zanmi Lasante, he said—when he was twenty-three, volunteering as a doctor's assistant in a hospital in Léogâne, in Haiti—he had a long talk with an American doctor, a kindly man who seemed to love the Haitians. The doctor had worked there for a year. Now he was departing.

"Isn't it going to be hard to leave?" Farmer asked him.

"Are you kidding? I can't wait. There's no electricity here. It's just brutal here."

"But aren't you worried about not being able to forget all this? There's so much disease here."

"No," said the doctor. "I'm an American, and I'm going home."

Farmer thought about that conversation all day and into the evening. "What does that mean, 'I'm an American'? How do people classify themselves?" He thought the doctor's answer was sensible, a legitimate answer. But he didn't know his own. He was supposed to start medical school himself in the fall. I'm definitely going to be a doctor, he thought.

Farmer fidgeted in the narrow airplane seat. "So later on that night, a young woman came in. She was pregnant, and she had malaria. . . . It's not as if it hasn't happened since." He stopped, his face turned to the window again.

"She had a very high parasitimia. Bad malaria. She went into a coma and, you know, I didn't know the details then. I do now, because it's my specialty. She needed a transfusion, and her sister was there and—" It was drizzling outside. He stared out at the runway landscape, gray and dull, crying softly.

"It's not about her. It happens all the time in Haiti, but I didn't know that then. So there was no blood at the hospital, and the doctor told her sister to go to Port-au-Prince to get some blood. But she would need some money. I had no money. I ran around the hospital. I rounded up fifteen dollars. I gave her the money and she left, and then she came back, and she didn't have enough money to go to Port-au-Prince. So meanwhile the patient started having respiratory distress. And this pink stuff started coming out of her mouth, and the nurses were saying, 'It's hopeless,' and other people were saying, 'We should do a cesarean delivery.' I said, 'There's got to be some way to get her some blood.' Her sister was beside herself. She was sobbing and crying. The woman had *five kids*. The sister said, 'This is terrible. You can't even get a blood transfusion if you're poor.' And she said, 'We're all human beings.' She said that again and again. 'We're all human beings.' "

The flight crew was preparing the cabin for takeoff. In a moment, Farmer would start feeling sick. "My big struggle is how people can not care, erase, not remember. I'm not a dour person. But I have a terrible message. And I'm not gonna put my seat in an upright position."

He had recovered his normal voice. The death was a memory again. He said, quoting the sister once more, " 'We're all human beings.' As if in answer to my question." He shook his head.

"The other thing about it is, I knew that the physicians and the others focussed on my reaction. The nurses were saying, 'Poor Paul. What a sweet young man.' And the doctors thought, He's new here, he's green, he's naïve." He paused. "Yeah, but I got staying power. That's the thing. I wasn't naïve, in fact."

JACQUES LESLIE

Running Dry

FROM *HARPER'S MAGAZINE*

> In the world there is nothing more submissive and weak than water. Yet for attacking that which is hard and strong nothing can surpass it.
>
> —Lao-tzu (sixth century B.C.)

When I was a war correspondent twenty-five years ago, I paid more attention to blood than to water. Carnage transfixed and terrified me; water seemed to flow inconsequentially through the embattled landscape before me. The Mekong River, in its multifingered brownness, and the reverse-flowing, hugely contracting and expanding great Cambodian lung-lake, the Tonle Sap, barely registered on my psyche, except as obstacles: my thoughts on water consisted chiefly of military observations, such as that the onset of the rainy season slowed down Khmer Rouge advances more effectively than the regime's hapless human enemies did. I can't say when the pendulum began to swing—probably about the time I stopped thinking about the war, a decade or two later. More recently, I started to notice how many news stories involved water. The subjects weren't just hurricanes, droughts, and floods but less predictable phenomena, such as the accelerating destruction of U.S. watersheds caused by urban sprawl, violent protests in Bolivia prompted by a water-utility-rate increase, and a death sentence handed down to a Chinese administrator who embezzled nearly $2 million set aside for the resettlement of people displaced by Three Gorges Dam. I began reading every water text I could find, and in March 2000 I attended an international water conference in The Hague at which water ministers from 115 countries declined to agree on how to address the problem of water scarcity. Now when I envision the globe, I try to see beyond political boundaries to the world as it really is: a collection of watersheds, lakes, rivers, and aquifers that together maintain the earth's biota—which is to say, us. Now the world's quotidian skirmishes and conflagrations are mere background noise. Now it is water that scares me.

WE FACE AN unassailable fact: we are running out of freshwater. In the last century we humans have so vastly expanded our use of water to meet the needs of industry, agriculture, and a burgeoning population that now, after thousands of years in which water has been plentiful and virtually free, its scarcity threatens the supply of food, human health, and global ecosystems. With global population hurtling toward roughly 9 billion people by 2050, projections suggest that if we continue consuming water with our habitual disregard all those needs cannot be met at once.

The world's supply of freshwater remains roughly constant, at about 2½ percent of all water, and of that, almost two thirds is stored in ice caps and glaciers, inaccessible to humans; what must change is how we use the available supply. Humans have grown so numerous that the usual response to anticipated water scarcity—to increase supply with dams, aqueducts, canals, and wells—is beginning to push against an absolute limit.

In the developed world widespread water shortages are projected but not yet broadly experienced. In the developing world the crisis has already arrived. As many as 1.2 billion people—one out of five on the globe—lack access to clean drinking water. Nearly 3 billion live without sanitation: no underground sewage, toilets, or even latrines. More than 5 million people a year die of easily preventable waterborne diseases such as diarrhea, dysentery, and cholera; in fact, most disease in the developing world is water-related. As Peter Gleick writes in *The World's Water 1998–1999*, "For nearly three billion people, access to a sanitation system comparable to that of ancient Rome would be a significant improvement in their quality of life."

To be sure, the water shortages that give rise to these conditions so far are regional, not global, and often involve inequality of distribution and high pollution levels as much as absolute scarcity. Thus one water basin experiences a shortage while neighboring basins enjoy ample supplies. Water doesn't ship well, except in unusual circumstances, such as the provisioning of some Greek and Caribbean islands by tanker or barge: water is far too cheap and unwieldy to justify long-distance transport. This tends to keep shortages confined to specific areas, but it also means that they can't easily be alleviated with water from another region.

In one way, however, the impact of water shortages has already registered globally, thanks to water's role in agricultural production. Indeed, water experts refer to grain as "virtual water," since many countries facing water shortages respond by importing grain. It takes roughly a thousand tons of water to produce a ton of grain, so importing grain has an obvious shipping advantage. As a result, stockpiling grain is one way to counter water shortages. The billion-dollar question among water and agriculture experts, in fact, is whether, owing in part to water scarcity, the human race in the twenty-first century will lose the capacity to feed itself. For now, the answer is unknowable, since predictions inevitably rest on highly speculative assumptions. One forecaster, Lester Brown of the Washington, D.C.-based Worldwatch Institute, has advanced a dire scenario in which China's water shortage forces it to import so much grain that poorer nations are priced out of the international grain market, inducing widespread starvation. More plausibly, others argue that in many developing countries water scarcity is the most significant component of environmental degradation, which in turn is an underlying cause of mass migration, peasant revolt, and urban insurrection. Such notions have not escaped the attention of U.S. policymakers, as evidenced by meetings of officials from the Department of Defense, the CIA, the State Department, and the White House last September to consider the global implications of water conflicts. At a time when the First World is obsessed with computer technology, genetics, and the froth of media entertainments, we would be well advised to remember our relationship to the two atoms of hydrogen and one of oxygen that, bound in nature, support all life.

Irrigation and Its Discontents

IN THE TIGRIS-EUPHRATES, Indus, and Yellow river basins, ancient civilizations flourished when they devised ways to grow crops with irrigated water and foundered when the systems collapsed, either because sediment clogged their canals or waterborne salt poisoned their soils. We like to think that we've mastered irrigation—indeed, in the last two centuries humans have increased land under irrigation thirtyfold. Yet the daunting obstacles we face in maintaining irrigation systems are not so different from those that brought down the Sumerian and Indus civilizations. "The overriding lesson from history is that most irrigation-based civilizations fail," writes Sandra Postel in her compelling survey of the global water crisis, *Pillar of Sand: Can the Irrigation Miracle Last?* "As we enter the third millennium A.D., the question is: Will ours be any different?"

Now, more than ever, humans depend on irrigation: less than a fifth of the world's cropland is irrigated, but because irrigation typically enables higher yields and two or three crops a year, irrigated land produces two fifths of the world's food. Even so, the planet's reliance on irrigated crops undoubtedly will intensify in the coming decades. The world's food supply comes from three major sources: cropland, rangeland, and fisheries. Livestock have already grown so numerous that 20 percent of the earth's rangeland has lost productivity because of overgrazing, and most of the world's fisheries have been decimated by overfishing. By default, the likely source of food for the roughly 3 billion additional humans expected in the next fifty years will be cropland. Yet the amount of cropland is not likely to grow much: newly cultivated land probably will barely surpass the amount of land lost to agriculture because of erosion, urbanization, and salination. Moreover, the best cropland is already in use; much of the land still awaiting cultivation has the potential to be only marginally productive. The result is that population growth is already outstripping growth of irrigated land. In fact, the area of global per capita irrigated land peaked in 1978 and has dropped 5 percent since then. Projections by international agencies suggest that by 2020, per capita irrigated land will have dropped 17–28 percent from the 1978 peak. Success in feeding all the people who will populate the earth in the mid-twenty-first century therefore depends largely on increasing the productivity of existing cropland. "The difference between the Malthusian pessimists and the cornucopian optimists," says Postel, "comes down to little more than an assumption about grainland productivity over the next several decades—specifically, whether yields will grow at closer to the 1 percent rate of the 1990s or the 2 percent rate of the previous four decades."

There's reason to worry. The 2 percent rate occurred as farmers applied Green Revolution techniques to land irrigated by groundwater or reservoirs, but those techniques have largely fulfilled their promise, and yields in recent years have either stagnated or declined. One reason may be irrigation itself:

some scientists believe that soils become depleted when repeatedly subjected to the two or three annual crops that irrigation enables. In addition, Green Revolution agriculture depends on copious applications not just of pesticides and fertilizer but of water: between 1950 and 1995 grainland productivity increased 240 percent while water use for irrigation increased 220 percent. With global depletion of groundwater and increasing diversions of agricultural water for industrial, urban, and environmental needs, the scarcity of water is likely to become the most important factor in limiting agricultural production. That means that more people may hunger for relatively less food.

Unseen Lakes, Pumped Dry

COMPARED WITH THE earth's visible freshwater—in lakes, ponds, and rivers—the amount of water stored in underground aquifers is sixty times as large. A stock that immense might seem beyond our capacity to exhaust, yet in many parts of the world groundwater is being depleted at an unsustainable rate. The Ogallala Aquifer, one of the world's largest stores of groundwater, covers 225,000 square miles beneath parts of eight U.S. states, from Texas to South Dakota, and feeds a fifth of the nation's irrigated lands. Although its stock is "fossil water"—water locked underground for thousands of years, with few sources of replenishment—it is being depleted so rapidly that many farmers who once depended on it now must rely on rainwater, significantly lowering yield. The amount of acreage supported by the Ogallala in six states fell from its peak in 1978 by nearly 20 percent in less than a decade; despite efforts to limit use of Ogallala water, substantial withdrawals continue.

Of course, unlike the Ogallala, most aquifers are naturally "recharged"—replenished by rain and runoff—but even these are being depleted dramatically, as the rate of withdrawal easily surpasses the recharged amount. India's volume of annual groundwater overdraft is higher than any other nation's. Almost everywhere in the country, water withdrawals are proceeding at double the rate of recharge, causing a drop in aquifers of three to ten feet per year; in the state of Tamil Nadu, groundwater levels have dropped as much as ninety-nine feet since the 1970s, and some aquifers there have become useless. The cost of land subsidence caused by aquifer depletion in the United States is about $400 million per year, with incidents occurring in Houston, New Orleans, and California's Santa Clara County and San Joaquin valley; Beijing is sinking at an annual rate of about four inches a year; and certain Mexico City barrios sink as much as a foot a year. In both Florida and the Indian state of

Gujarat, the water table has dropped so low that seawater has invaded the aquifers, limiting their usefulness for drinking or irrigation. In Palestine's Gaza Strip, which relies almost entirely on groundwater, saltwater intrusion from the Mediterranean has been detected as far as a mile inland, and some experts predict that the aquifer will become totally salinized. Groundwater depletion, says the International Water Management Institute, a World Bank-supported group in Sri Lanka, is "the single most serious problem in the entire field of water resources management. . . . Many of the most populous countries of the world—China, India, Pakistan, Mexico and nearly all of the countries of the Middle East and North Africa—have literally been having a free ride over the past two or three decades by depleting their groundwater resources. The penalty of mismanagement of this valuable resource is now coming due, and it is no exaggeration to say that the results could be catastrophic for these countries, and, given their importance, for the world as a whole."

Humans alone cannot deplete aquifers: we lack the strength to draw that much water or dig wells that deep. Rather, groundwater depletion is a phenomenon of the late twentieth century, made possible by the availability of electricity and cheap pumps. IWMI calls the spread of small pump sets throughout the world "one of the most dramatic yet generally unappreciated revolutions in water resource technology." In some ways pump irrigation is ideal: the water is stored underground and shielded from evaporation, so it can be used during the dry season, when crops need water most. In many Asian countries pump irrigation alone deserves much of the credit for high Green Revolution yields.

Yet in many countries the new technology shattered traditional water-sharing arrangements that had worked for centuries. John Briscoe, a senior water adviser at the World Bank, cites the example of Yemen, which once had "very sophisticated ancient water management techniques" that handled everything from floods to water allocation. "Then you come to the post-Second World War, with deep wells and electricity and diesel pumps for groundwater, and people pump like there's no tomorrow. You have a lot of food production as a result of this, but in the basin around San'a, the capital, for instance, four times more water is being pumped out than is being recharged, and the aquifer is dropping three meters a year."

Often the new technology combines catastrophically with government policies.[1] Until the early 1990s, individual farmers in Mexico used powerful pumps, concluding, "If I don't pump fast, my neighbor will, so I might as well pump faster than he does." On top of this, the government subsidized everything from fertilizer to energy costs and imposed tariffs on competing foreign crops, accelerating the waste of water. "The whole thing was a total disaster," Briscoe says. The government finally phased out tariffs and created subsidies to encourage sustainable water use. The reforms forced thousands of Mexican farmers off the land, yet, Briscoe says, there are "very clear signs" that the remaining farmers have begun to manage their water use.

In places such as Punjab and Haryana, India's breadbasket states, the new technology also widened the gap between rich and poor. As water tables dropped, farmers had to drill deeper wells and buy more powerful pumps, but only rich farmers could afford the new equipment. Poor farmers, whose shallow pumps became useless, were forced to rent their land to richer farmers, for whom they became laborers. The IWMI report lists the consequences if this trend is not reversed: "Lakes and rivers dry up as the aquifer recedes. . . . The costs of pumping become so high that the pumps are shut down and the whole house of cards collapses. It is not difficult to believe that India could lose 25% or more of its total crop production under such a scenario."

From the earth's surface, groundwater is invisible: farmers don't realize they've used up an aquifer until it's too late. Even in countries where limits on withdrawals exist, enforcement is virtually impossible, so no governments have established regulations for sustainable groundwater use. Yet reliance on groundwater in agriculture causes food to be grossly undervalued. Postel estimates the global annual groundwater overdraft in the mid-1990s at about 163 million acre-feet, or roughly enough water to grow about 198 million tons of grain, a tenth of the global harvest. Agricultural prices are now at their lowest point in two decades and have forced some American farmers out of work, but if overpumping were to cease, grain prices probably would rise significantly. Instead, the mounting cost of pumping groundwater from deeper and deeper levels may eventually produce the same result.

Forever Dammed

WHAT AQUIFERS ARE below ground, dams create above ground. Many environmentalists will tell you, however, that the very concept is faulty, that anything as destructive as a dam cannot be an uncomplicated good. Even by their reckoning, however, the best dam—the one that is the closest to the ideal—surely is Hoover Dam, the first of the modern water era. Hoover is America's Great Pyramid, whose face was designed without adornment to emphasize its power, to focus the eye on its smooth, arcing, awe-inspiring bulk. Yet the dam nods to beauty with a grace that seems more precious year by year: its suave Art Deco railings, fluted brass fixtures, and three miles of polished terrazzo granite walkways are the sort of features missing from the purely utilitarian public-works projects of more recent decades. Hoover is a miraculous giant thumbnail that happens to have transformed the American West. Take it away, and you take away water and power from more than 20 million people. Take it away, and you remove a slice of American history, including a piece of the recovery from the Depression, when news of each step in the dam's construction—the drilling of the diversion tunnels, the building of the earth-and-rock cofferdams, the digging to bedrock, the first pouring of foundation, the accretion of five-foot-high cement terraces that eventually formed the face—heartened hungry and dejected people across the land.

The dam and Las Vegas more or less vivified each other; if Hoover evokes glory, Las Vegas, only thirty miles away, is its malignant twin. Even now, Hoover provides 85 percent of Las Vegas's water, turning a desert outpost into the fastest-growing metropolis in the country—so, by all means, take away Las Vegas. Take away Hoover, and you might also have to take away the Allied victory in World War II, which partly depended on warplanes and ships built in southern California with Hoover's hydroelectric current. And take away modern Los Angeles, San Diego, and Phoenix: you reverse the twentieth-century shift of American economic power from East Coast to West. Take away Hoover and the dams it spawned on the Colorado—Glen Canyon, Davis, Parker, Headgate Rock, Palo Verde, all the way to Morelos across the Mexican border—and you restore much of the American Southwest's landscape, including a portion of its abundant agricultural land, to shrub and cactus desert. Above all, take away Hoover, and you take away the American belief in technology, now on a millennial crest of enthusiasm. At Hoover's September 30, 1935, dedication, Interior Secretary Harold Ickes reflected the common understanding when he declared: "Pridefully, man acclaims his conquest of nature." After Hoover every country wanted dams, and every major country, regardless of ideology, built them. (Even now, the ubiquitousness of dams is one of their most striking features: the world's highest dam is in Tadzhikistan, the largest reservoir is in Uganda, and the dam with the biggest hydroelectric capacity is on the Brazil-Paraguay border.) At its completion Hoover towered 280 feet above the world's second-highest dam, the Arrowrock in Idaho, and was the planet's largest source of electricity, but its current ranking, sixteenth in height and lower than twentieth in hydroelectric capacity, reflects the momentum that the dam movement eventually gathered. Take away Hoover Dam, and you take away a bearing, a confidence, a sense of what nations are for.

Yet in a sense that's what's happening. Even if Hoover lasts another 1,100 years (when Bureau of Reclamation officials say Lake Mead will be filled with sediment, turning the dam into an expensive waterfall), its teleological edifice is crumbling. In sixty-five years we have learned that if you take away Hoover, you also take away millions of tons of salt that the Colorado once carried to the sea but which have instead been strewn across the irrigated landscape, slowly poisoning the soil. Take away the Colorado River dams, and you return the silt gathering behind them to a free-flowing river, allowing it again to enrich the wetlands downstream and the once fantastically abundant, now often caked, arid, and refuse-fouled delta. Take away the dams, and the Cocopa Indians, whose ancestors fished and farmed the delta for more than a millennium, might again have a chance of avoiding cultural extinction. Take away the dams, and the Colorado would again bring its nutrients to the Gulf of California, helping that depleted fishery to recover the status it held a half-century ago as an unparalleled repository of marine life. Take away the dams, finally, and the Colorado River returns to its virgin state: tempestuous, fickle, in some stretches astonishing.

What we have learned is that we have overestimated dams and underestimated the water that runs through them. In the era of big dams that has at last peaked and started to decline, river water that reached the sea was considered wasted because it had not been turned to human ends. Only recently have we noticed that the human good is not served by the depleted rivers and wetlands that the diversions create. We would have been wise to listen to Aldo Leopold, the celebrated naturalist, who wrote in 1933, two years before Hoover Dam's dedication: "We build storage reservoirs or power dams to store water, and mortgage our irrigated valleys and our industries to pay for them, but every year they store a little less water and a little more mud. Reclamation, which should be for all time, thus becomes in part the source of a merely temporary prosperity."

The prosperity is evident, but so, increasingly, is its transience. Dams have lifetimes as surely as any natural thing. The rate at which a reservoir fills depends on its size and the amount of sediment flowing into it. Sediment has filled more than half the storage capacity of some dams within a decade. Other dams, like Hoover, have a projected lifetime of more than a thousand years—though Hoover is deceptive because the Glen Canyon Dam upstream traps most of the sediment that would otherwise reach it. On average, sediment annually reduces by 1 percent the storage capacity of the world's reservoirs. In China, where soil erodes easily, reservoirs fill at a rate of 2.3 percent a year. One dam on the silty Yellow River, the Yangouxia, lost almost a third of its storage capacity even before it was commissioned.

Radiating outward from any dam, irrigated water slowly poisons the land with salt. Salinity has affected a fifth of the world's agricultural land; each year it forces farmers to abandon a million hectares and affects an additional 2 million hectares. If in the course of a year a farmer applies the unremarkable sum of 10,000 tons of water to a single hectare, the land will collect two to five tons of salt. It's precisely the process by which ancient Mesopotamia turned into the barren desert of contemporary southern Iraq. Salt problems are severe in China, India, Pakistan, Central Asia, and the Colorado and San Joaquin river basins of the American West. In many arid areas the soil is naturally saline. As rainwater and snowmelt flow through a saline watershed to a river, they collect salt throughout their path. A few billion years ago the oceans were full of freshwater, then were gradually turned saline by riverborne salt. Now, in the modern water era, dams divert both the water and the salt. Because reservoirs expose so much water to the sun, those in hot climates lose a huge quantity to evaporation: for example, a full third of the Colorado's flow evaporates from reservoirs. In the remaining water, salt concentrations increase. Some water is distributed to surrounding croplands, where the salt collects. As the water permeates the soil, it accumulates more salt, then returns to the river with a more concentrated share; on a single trip down the Colorado, the same water may be used for irrigation eighteen times. Human use of the Colorado has approximately doubled its salinity. Neither the environment nor urban areas are spared salt's effects: it kills aquatic organisms in the lower river and corrodes pipes in Los Angeles, San Diego, and Phoenix.

The world's most spectacular saline catastrophe is Central Asia's Aral Sea. Decades ago Soviet planners diverted two major rivers that feed the Aral in order to turn the surrounding desert into a cotton cornucopia. As cotton bloomed, however, the sea wilted: it now contains a third of its former volume and may disappear.

All twenty-four native fish species in the Aral have already vanished, and the fish catch has dropped from 48,000 tons to none. The regional climate has declined, producing less rainfall and greater temperature extremes. Each year windstorms pick up 44 million tons of salt and dust from the dried seabed and scatter them over the river basin. Cotton output is dropping. The drinking water is contaminated with high concentrations of salt and agricultural chemicals. Inhabitants suffer plagues of cancer, respiratory illnesses, and waterborne diseases such as hepatitis and typhoid fever.

All dams cause environmental damage: they fragment the riverine ecosystem, isolating upstream and downstream populations, and, by preventing floods, cut off the river from its floodplain. Within the reservoir lake, water temperature changes dramatically. Deep reservoir water is usually colder in summer and warmer in winter than river water. Thus water leaving Glen Canyon Dam never varies more than a few degrees from its 46 degree average. For 240 miles below the dam the water is too cold for native fish to reproduce.[2]

The reservoir lake traps not just sediment but nutrients. Algae thrive on the nutrients and end up consuming the lake's oxygen, turning the water acidic. It comes out of the dam "hungry," more energetic after shedding its sediment load, ready to capture new sediment from the riverbed and -bank. As it scours the downstream river, the bed deepens, losing its gravel habitats for spawning fish and the tiny invertebrates they feed on. Within nine years after Hoover Dam was sealed, hungry water took 89,000 acre-feet of material from the first 87-mile stretch of riverbed beneath. In places the riverbed dropped by more than thirteen feet, and it sometimes took floodplain water tables down with it. In addition, riverbank erosion has undermined some embankments and flood-control levees.

"A dammed river," Wallace Stegner wrote, "is not only stoppered like a bathtub, but it is turned on and off like a tap." Instead of varying with snowmelt and rainfall, its flow is regulated to meet the requirements of power generation and human recreation. Most fluctuations reflect electricity demand: the river level changes hour by hour and is lower on Sundays and holidays. These quick fluctuations intensify erosion, eventually washing away riverbank trees, shrubs, and grasses as well as riverine nesting areas. Riverside creatures lose needed food and shelter.

The changes are registered all the way to the river's mouth and beyond. Because of dams, many major rivers—including the Colorado, the Yellow, and the Nile—flow to the sea only intermittently. Without its customary allotment of sediment, the coastline is subject to erosion. By one estimate, dams have reduced by four fifths the sediment reaching the southern California coast, causing once wide beaches to disappear and cliffs to fall into the ocean. Estuaries, where riverine freshwater mixes with ocean saltwater, are crucial in the development of plankton, which in turn supports a huge abundance of marine life; deprived of large portions of freshwater and nutrients, the estuaries decline, and with them so do fisheries. Migrating fish such as salmon and steelhead trout find their paths obstructed, both as juveniles swimming downstream to mature and as adults going upstream to spawn. For this reason, the Columbia River, where 2 million fish returned annually to spawn just before the dam era began, has hosted half that number in recent years, and most remaining stocks in the upper river are in danger of extinction.

Only by multiplying all these effects by the number of the world's river basins studded with dams—an overwhelming majority—can the full environmental impact of dams be appreciated. The numbers are stunning. The planet accommodates 40,000 large dams—dams more than four stories high—and some 800,000 small ones. They have shifted so much weight that geophysicists believe they have slightly altered the speed of the earth's rotation, the tilt of its axis, and the shape of its gravitational field. Together they blot out a terrain bigger than California.

The most obvious beneficiaries of dams are politicians, bureaucrats, and builders, all of whom profit from the dams' huge price tags. Think of the towering political leaders of the twentieth century—Roosevelt, Stalin, Mao, Nehru. They all loved dams. Dams provide jobs and a generous amount of money to constituents, some of whom don't mind donating a portion back to the politicians. Bureaucrats like dams because that's where the action is: the expense of dams ensures power to its overseers. The constituents include dam builders, road builders, engineers, electricians, carpenters, cooks, plus every sort of professional boomtowns attract, from developers to prostitutes. In fact, dams, which provide nearly a fifth of the world's electricity, are also among the world's costliest public-works projects; by the time China's Three Gorges Dam is completed (in about 2009), it will have become the world's largest and most expensive, with an estimated cost of up to $75 billion.

The attraction of dams to farmers is obvious. Supported by funding from central governments and international agencies, farmers rarely pay more than 20 percent of the real cost of the irrigated water. The subsidies distort the farmers' economic outlook: instead of planting crops that match the hydrology of their fields, farmers take advantage of abundant cheap water to plant crops that guzzle water, even if the crops bring a low return. In the San Joaquin valley of California, the richest irrigated land in the world, some farmers grow water-guzzling cotton, or, worse (because it is fed to cows, the most notorious guzzlers of all), alfalfa. It takes at least 15,000 tons of water to produce a ton of beef and nearly that much to produce a ton of cotton; comparatively, a ton of grain requires 1,000 tons of water.

Still, many farmers founder. For one thing, canal maintenance is often underfunded and neglected, particularly in developing countries. Planners often overestimate the amount of water available to the system and underestimate leakage, evaporation, and waste. Farmers near a canal head—the "head-enders"—almost invariably receive much more water than those far down the canal—the "tail-enders." The head-enders may have bought their position with bribes; they are often wealthy enough to afford the new equipment, seeds, fertilizers, and pesticides that irrigation farming promotes. At the other end, the tail-enders may be forced to borrow money at high rates; deeply indebted, they often end up as tenants on their own land.

The biggest losers are people displaced by dams. They're usually minorities, often uneducated and powerless, and therefore hard to count or even notice, particularly by a government's ruling elite. If the government bothers to relocate them, it's usually to inferior land, where settled residents resent them. Rates of illness and death usually increase after relocation. One estimate puts the worldwide total of people displaced by dams at 30 to 60 million. As startling as that sum is, it omits another huge group, the floodplain residents living downstream from dams whose livelihoods are jeopardized by the sudden loss of regular nutrient-bearing floods or other hydrological changes.

If dams are so destructive in so many ways, why don't we tear them down? The most obvious answer is that we can't afford to; dismantling dams is nearly as expensive as building them. Some dams may be decommissioned and drained, but in the foreseeable future even those will be few, for the world's reliance on dams for electric power and irrigation has grown too great to do without them: a world abruptly deprived of a fifth of its electricity and a significant portion of its food supply would not remain tranquil for long. Boxed in by the size of our population, we have approached a natural limit, damned if we do dam and damned if we don't dam.

The result is a kind of standoff. While dam building has largely stopped in the United States and northern Europe, companies based in North America, Europe, and Japan continue to lead construction efforts in developing nations. But even Third World governments increasingly must finance dams themselves or look for support from private investors. The World Bank once enthusiastically financed dams throughout the Third World, until a series of embarrassments, culminating in militant opposition to a project to build 30 large dams, 135 medium-sized ones, and 3,000 small ones in the Narmada valley of India caused it to reconsider. "We now build very few dams," says Briscoe, the World Bank water specialist.

Although most water experts appreciate the destructive impact of dams, few oppose them entirely. IWMI, the World Bank-supported water agency, concluded a gloomy survey of global water needs in 2025 by noting that "medium and small dams will almost certainly . . . be needed." Postel, whose book enumerates dams' many liabilities, nevertheless told me, "I think there's no way we could be supporting a population of 6 billion today without dams.

Water comes at uneven times of the year, and we've got to have a way to store it. The question is how."

Rain, Rain, Go Away

GLOBAL WARMING, we know, is here. Some people think the change chiefly involves temperature, but the phrase is misleading—it leaves out water. Nearly every significant indicator of hydrologic activity—rainfall, snowmelt, glacial melt, evaporation, transpiration, soil moisture, sedimentation, salinity, and sea level—is changing at an accelerating pace. Alaskan and Siberian permafrost is beginning to thaw; in Antarctica scientists are finding beaches and islands exposed after being covered by ice for thousands of years. The sea level has risen between four and ten inches in the last century. Precipitation is increasing, but so are evaporation, floods, and droughts.

Pick any point of the hydrologic cycle and note the disruption. One analysis of 1900–1998 data pegged the increase in precipitation at 2 percent over the century. In water terms this sounds like good news, promising increased supply, but the changing timing and composition of the precipitation more than neutralizes the advantage. For one thing, it is likely that more of the precipitation will fall in intense episodes, with flooding a reasonable prospect. In addition, while rainfall will increase, snowfall will decrease. This means that in watersheds that depend on snowmelt, like the Indus, Ganges, Colorado, and San Joaquin river basins, less water will be stored as snow, and more of it will flow in the winter, when it plays no agricultural role; conversely, less of it will flow in the summer, when it is most needed. One computer model showed that on the Animas River at Durango, Colorado, an increase in temperature of 3.6 degrees Fahrenheit—the global change predicted from now to 2100—would cause runoff to rise by 85 percent from January to March but drop by 40 percent from July to September. The rise in temperature increases the probability and intensity of spring floods and threatens dam safety, which is predicated on lower runoff projections. Dams in arid areas also may face increased sedimentation, since a 10 percent annual increase in precipitation can double the volume of sediment washed into rivers.

The consequences multiply. Soil moisture will intensify at the highest northern latitudes, where precipitation will grow far more than evaporation and plant transpiration but where agriculture is nonexistent. At the same time, precipitation will drop over northern mid-latitude continents in summer months, when ample soil moisture is an agricultural necessity.

* DOES THE WHOLE ARTICLE LACK CREDIBLITY?

Meanwhile the sea level will continue to rise as temperatures warm, accelerating saline contamination of freshwater aquifers and river deltas. This already has occurred in Florida, Gaza, and the Nile River delta. The temperature rise will cause increased evaporation, which in turn will lead to a greater incidence of drought. In fact, extreme water-related events such as storms, floods, and droughts will become more frequent and intense.

Perhaps most disturbing of all, the hydrologic cycle is becoming increasingly unpredictable. This means that the last century's hydrologic record—the set of assumptions about water on which modern irrigation is based—is becoming unreliable. Build a dam too large, and it may not generate its designed power; build it too small, and it may collapse or flood. Release too little dam runoff in the spring and risk flood, as the snowmelt cascades downstream with unexpected volume; release too much and the water won't be available for farmers when they need it. At a time when water scarcity calls out for intensified planning, planning itself may be stymied.

Water Wars of the Future

In the modern era we fight wars over oil and take water for granted, yet of the two liquids water is far more capricious and confounding. Oil induces fear because we sense it can make or break empires; water has already made and broken quite a few. Think of oil, and you conjure up gushers, cartels, and economic dominance; think of water and you contemplate the elixir of life. Oil belongs to whoever owns the land above it; water, with its sprawling underground aquifers and long sinuous rivers, complicates ownership and intertwines nations' fates. Oil promotes grandiosity; water teaches humility.

The handy cliché is that sooner or later water will cause war. In a quote that caroms ceaselessly from one water publication to another, World Bank vice president Ismail Serageldin declared in 1995, "The wars of the next century will be over water." Many foreign leaders have expressed similar sentiments. In the late 1980s, Egyptian foreign minister and soon-to-be U.N. secretary general Boutros Boutros-Ghali said that the next Middle East war "will be over the waters of the Nile, not politics." Jordan's King Hussein said in 1990 that water was the only issue that could prompt a war between Jordan and Israel.

Yet such wars haven't quite happened. Aaron Wolf, an Oregon State University specialist in water conflicts, maintains that the last war over water was fought between the Mesopotamian city states of Lagash and Umma 4,500 years ago. Wolf has found that during the twentieth century only 7 minor skirmishes were fought over water while 145 water-related treaties were signed. He argues that one reason is strategic: in a conflict involving river water, the aggressor would have to be both downstream (since the upstream nation enjoys unhampered access to the river) and militarily superior. As Wolf puts it, "An upstream riparian would have no cause to launch an attack, and a weaker state would be foolhardy to do so." And if a powerful downstream nation retaliates against a water diversion by, say, destroying its weak upstream neighbor's dam, it still risks the consequences, in the form of flood or pollution or poison from upstream.

So, until now, water conflicts have simmered but rarely boiled, perhaps because of the universality of the need for water. Almost two fifths of the world's people live in the 214 river basins shared by two or more countries; the Nile links ten countries, whose leaders are profoundly aware of one another's hydrologic behavior. Countries usually manage to cooperate about water, even in unlikely circumstances. In 1957, Cambodia, Laos, Thailand, and South Vietnam formed the Mekong Committee, which exchanged information throughout the Vietnam War. Through the 1980s and into the 1990s, Israeli and Jordanian officials secretly met once or twice a year at a picnic table on the banks of the Yarmūk River to allocate the river's water supply; these so-called picnic-table summits occurred while the two nations disavowed formal diplomatic contact. Jerome Delli Priscoli, editor of a thoughtful trade journal called *Water Policy* and a social scientist at the U.S. Army Corps of Engineers, believes the whole notion of water conflict is overemphasized: "Water irrigation helped build early communities and bring those communities together in larger-functional arrangements. Such community networking was a primary impetus to the growth of civilization. Indeed, water may actually be one of humanity's great learning grounds for building community. . . . The thirst for water may be more persuasive than the impulse toward conflict."

On the other hand, water has often been the goal, tool, or target of conflicts that fall just short of war or that contain non-water-related dimensions. Recent history is full of examples. In 1965, Syria tried to divert the Jordan River from Israel, provoking Israeli airstrikes that forced Syria to abandon the effort. Colin Powell, chairman of the U.S. Joint Chiefs of Staff during the 1991 Gulf War, said in 1996 that the United States considered bombing dams on the Euphrates and Tigris rivers north of Baghdad but desisted, apparently because of the likelihood of high civilian casualties. The allies also discussed asking

Turkey to reduce the Euphrates flow at the Atatürk Dam upstream from Iraq. As it was, the allies targeted Baghdad's water-supply system while the Iraqis destroyed Kuwait's desalination plants.

Postel believes water hostilities are most likely to occur when a river's water is insufficient to meet projected demand, water allocation is considered inequitable, and involved nations have made no water-sharing agreement. In five of the world's most contentious water basins—the Aral Sea region, the Ganges, the Jordan, the Nile, and the Tigris-Euphrates—rapid projected population growth—up to 75 percent by 2025—threatens to turn the basins into cauldrons of hostility. For instance, in the Tigris-Euphrates basin, Turkey's position upstream gives it enormous leverage over its downstream neighbors, Syria and Iraq. It's likely that Syria's longtime support of the separatist Kurdistan Workers' Party in Turkey was at least partly a way of countering Turkey's control over 80 percent or more of Syria's water supply. But once Turkey captured Abdullah Ocalan, a Kurdish guerrilla leader who had lived in Syria for nearly two decades, Syria's leverage against Turkey declined. Turkey is now in the midst of a huge dam-building program that will further diminish the Euphrates's flow into Syria, increasing Syria's grievances.

In the Nile basin the situation is more volatile, because the downstream nation, Egypt, dominates the region. Egypt already diverts so much Nile water that the river barely flows to its mouth in the Mediterranean Sea and the Nile delta is subsiding because sediment no longer reaches it. Nevertheless, Egypt is launching vast new irrigation projects that will divert even more Nile water. One project will irrigate about 500,000 acres of Egypt's southwestern desert; another will divert water beneath the Suez Canal to irrigate 625,000 acres of the Sinai Desert. At the same time, Ethiopia, the source of 86 percent of the Nile's flow, intends to launch its own irrigation and hydroelectric projects, which could dramatically reduce downstream water. Steve Lonergan, a specialist in water and security issues at the University of Victoria, British Columbia, told me, "I don't doubt that if Ethiopia starts building water projects that restrict the flow of the Nile, Egypt will bomb them."

Even if water wars remain rare, other sorts of water-related violence already occur frequently and are certain to increase. Thomas F. Homer-Dixon, a pioneer in the emerging field of environmental security, cites the Israeli-Palestinian conflict as an example of how environmental scarcity affects politics. In his 1999 book, *Environment, Scarcity, and Violence,* he argues that the capture by a dominant group of such resources as water, cropland, and forest occurs most often just at the point when the resource is becoming scarce and its price is rising, enabling speculation and increased profits. In the case of water, that time is now. In Israel water is growing increasingly scarce; although the nation has been a trailblazer in the development of water-conservation technologies, it continues to extract groundwater at an unsustainable rate.

Soon after the occupation of the West Bank in 1967, Israeli authorities instituted a rationing program that by the early 1990s gave four times as much water per capita to Israeli settlers as to Arabs. Israelis also required Arabs to seek permission to drill wells. When Arabs sought approval to drill over the West Bank's "Mountain" aquifer, the biggest aquifer in Israeli-controlled territory, they were invariably turned down; in other areas permission was given to Arabs infrequently. In addition, because Israelis had access to more sophisticated technology, their wells went deeper, often sucking Arab wells dry or exposing them to saltwater intrusion. Partly as a result, irrigated Arab farmland dropped from 27 percent to as low as 3.5 percent of the area of all West Bank cropland. Many Arab farmers abandoned their fields for towns, where they worked as day laborers, if at all. When the Palestinians revolted in 1987, the disenfranchised farmers were presumably primed to participate. "It is reasonable to conclude," Homer-Dixon writes, "that water scarcity and its economic effects contributed to the grievances behind the *intifadah.*"

In this manner, water scarcity encourages insurgencies. It reduces economic productivity and forces migration from depleted countryside to ill-prepared city. Social institutions may break down. Division into ethnic, religious, or linguistic groups increases. "Water scarcity rarely causes interstate wars," Homer-Dixon writes. "Rather its impacts are more insidious and indirect: it constrains economic development and contributes to a host of corrosive social processes that can, in turn, produce violence within societies."

Over the Horizon, China

CHINA IS NOT ONLY the world's most populous nation, with 1.3 billion people now and 1.5 billion projected by 2050; it also embodies the planet's extremes of water management and water disaster. China suffers from both severe droughts and severe floods. It is building what will be the world's largest dam, and the number of people that the dam will displace—at least 1.2 million—also will be a record. The Yellow River is the world's siltiest river by a factor of nine. At least 50 million rural Chinese live with an extreme scarcity of drinking water, never mind water for less immediate uses, such as bathing and sanitation.

The essence of China's water problem is that while the nation possesses 21 percent of the world's population, it has access to only 7 percent of the globe's freshwater. More specifically, densely populated northern China includes one third of China's territory, two fifths of its population, and produces 45 percent of its industrial output but receives only a quarter of the country's precipitation. One result is a profound reliance on irrigation: 70 percent of China's grain crop grows on irrigated land (compared with 15 percent in the United States). This hydrologic riddle is nothing new: failure to resolve it has ended dynasties and may yet again. With its ideological claims to legitimacy nullified by its abandonment of Communism, the current government draws what strength it can from the nation's huge economic expansion. But the expansion, volatile and vastly uneven, already has created enormous waves of social change, such as tens of millions of destitute rural migrants to the cities and the rising expectation among city dwellers of running water, indoor toilets, and diets rich in water-intensive beef and pork. China's recent displays of wealth are deceptive, since the nation's leaders are forever trying merely to hold a course amid the country's turbulent demographic currents. The government consequently takes food-related issues seriously. Most officials lived through the country's 1959–61 famine, which killed 30 million people, and have no desire to repeat the experience. That famine only increased Chinese leaders' desire for grain self-sufficiency. Through the 1990s, that policy required that at least 95 percent of China's grain be produced domestically.

The irrigation system, unfortunately, is a mess. Of China's 30,000 miles of major rivers, 80 percent are too polluted to support fish. Every year since 1985, the Yellow River, which flows through the heart of northern China's farmlands, has failed to reach its mouth for weeks, and the number of dry days each year has grown progressively, all the way to 226 days in the drought year of 1997. For long stretches, the Yellow hasn't even flowed to Shandong Province, the last province it waters before reaching the sea. Shandong farmers grow a fifth of China's wheat and an eighth of its corn; these days many of them are contemplating a return to rain-fed agriculture, which means they must drop back to one crop a year instead of two or three. More serious still, farmers all over northern China have been depleting aquifers to grow food. One Chinese survey reported that the water table beneath the North China Plain had dropped roughly five feet a year over a recent five-year period.

Rapid industrialization has intensified water scarcity. Water used in Chinese industry produces seventy times as much economic value as water used in agriculture, so industry's needs routinely take precedence over farmers'. Indeed, one reason Shandong farmers get so little Yellow River water is that it is being diverted to factories upstream. Moreover, as cities grow, farmland is taken out of production and turned into industrial and residential areas. Chinese officials are so desperate to develop new sources of water for northern China that they are planning a mammoth diversion that would dwarf Three Gorges Dam in cost and scope. One route would siphon water from a Yangtze River tributary, pump it under the Yellow River, and deliver it to the Beijing region, more than 600 miles away. A U.S. intelligence study places its cost in the hundreds of billions of dollars, enough to dampen government expenditures for other projects for many years into the future. Even that project would probably serve industrial and residential needs before agricultural ones, and its environmental damage would be enormous.

In 1994 the extremity of China's water situation produced an argument across the globe, in Washington. Lester Brown, the environmental researcher and founder of the Worldwatch Institute, claimed that within four decades China's water scarcity would compel a huge drop in the country's grain production, forcing China onto the world market in such volume that it would price out poorer countries and induce widespread famine. Most American China experts thought Brown egregiously overstated China's predicament, but the U.S. National Intelligence Council took him seriously enough to sponsor an expensive study. At Sandia National Laboratories intelligence officers gathered satellite photos to determine precisely the extent of Chinese agricultural acreage. They found that Brown greatly underestimated the acreage, which seemed to discredit his theory, but then, having deflated Brown, the NIC concluded that China still would need to import 193 million tons of grain by 2025—an estimate falling only slightly short of Brown's low-range forecast of 238 million tons by 2030. Meanwhile, Brown's critics were tossing around numbers far below 100 million tons.

Chinese officials considered Brown's claims an affront: instead of commending them for their remarkable economic gains, a Westerner was accusing them of being poised to starve the world. Brown's motive was to persuade the Chinese to begin conserving water, but they drew nearly the opposite conclusion: instead of switching to more lucrative and water-thrifty export crops, as even Brown's critics (and the NIC) advocated, Chinese officials tried to prove Brown wrong by *increasing* grain production. This meant that Chinese rivers and aquifers would be depleted at an even faster rate. Although Chinese officials lately have shown signs of revising this policy, the argument over Brown's claims still festers. I found this out when I mentioned his name to

Vaclav Smil, author of the estimable 1993 book *China's Environmental Crisis.* Smil went off like a firecracker: "Stay off Brown! He's a nut! He's a guy who predicts the end of the world and massive food shortages and high prices, yet year after year we have the biggest surpluses of food and the lowest food prices in history. . . . Come on, get serious." And Smil was just getting started.

Of course, all these estimates are certain to be wrong, or, at best, right for the wrong reasons. The volume of Chinese grain imports thirty years from now is unknowable, because so many variables will influence the outcome. Every estimate takes into account only a fraction of all the variables and makes different assumptions about the variables used. Brown downplays the impact of prices, which the optimists believe will limit meat, grain, and water consumption; on the other hand, Brown thinks the optimists underestimate the impact of water scarcity. Will China's agricultural yields increase, as the optimists assume, or will they stagnate, as Brown believes? How much money will China invest in agricultural research? At what rate will China's land erode, its dams fill with sediment, its water become fouled by industrial waste and raw sewage? The questions go on and on, and suggest the folly of predicting production levels decades into the future.

In fact, the China debate is a microcosm of the larger argument over the impact of water scarcity on global food production half a century from now. In this dispute the optimists and pessimists are more evenly divided. Among the optimists are the U.N. Food and Agriculture Organization and the International Food Policy Research Institute, a World Bank-supported nonprofit; the pessimists include Brown, Postel, and IWMI, another World Bank-supported nonprofit. In preparation for a chapter of *The World's Water 2000–2001,* Peter Gleick found himself getting into arguments with people on both sides. "I realized that you can't answer the question without understanding what they're assuming, whether they tell you that they're assuming it or not," Gleick said over the phone. "My conclusion is that people in both camps don't know the answer—they're making a whole bunch of assumptions that they aren't telling us. And their crystal ball is no better than anybody else's."

Gleick focused on elucidating the disputed range of each of the key variables that will determine whether the world will be able to feed itself. Among his variables: Will the world's population in 2050 be closer to 10.7 billion, the U.N.'s high projection, or 7.3 billion, the U.N.'s low projection? Will most people eat 2,300 calories a day, the minimum level for health set by the FAO, or will they eat 3,300 calories a day, as people in the wealthier industrialized nations do? What portion of those calories will come from meat? This is a significant statistic, since meat consumption requires that grain be fed to livestock instead of humans. By one estimate, all the grain fed to U.S. livestock is equivalent to the amount needed to feed 400 million people. What will crop yields be? What fraction of crops will be lost to plant disease, pests, storm damage, harvesting inefficiency, spoilage, and waste? These sums now are enormous. Diseases, insects, and weeds destroy about a third of all crops, and a 1997 study estimated that in the United States, 27 percent of all edible food for humans was lost at the retail, consumer, and food-service levels.

Water scarcity dictates another set of questions: What will be the water requirements of the crops grown? A ton of potatoes, for instance, needs 500 to 1,500 tons of water, while a ton of chicken needs 3,500 to 5,700 tons of water, and a ton of beef needs 15,000 to 70,000 tons of water. What percentage of cropland will be irrigated? How efficiently will irrigation water be used? At one extreme, flood irrigation requires a low capital expenditure but wastes a vast amount of water. At the other extreme, drip irrigation requires expensive technology but uses water with high efficiency. And will the water come from rainfall, rivers and streams, lakes and reservoirs, groundwater, or reclaimed wastewater? As groundwater is depleted, will other sources be available?

What, finally, will be the impact of climate change? Gleick calls this "the down card in the poker game—you can't see it, but you know it's going to be a factor in all the other answers." If we ignore these questions, he says, the likelihood increases that food and water shortages will be a pivotal feature of twenty-first-century life. "The bottom line," he notes, "is that a lot of things have to go right to avoid a severe crisis."

The Mirage of Big Technology

IT IS INDICATIVE of the bind we're in that even though technology helped get us into it, technology also will have to help get us out. Of course, "technology" includes a wide range of tools, from five-dollar drip-irrigation bucket kits to the Three Gorges Dam. The optimists' preference is for big technology, which, as always, seems to promise a painless way out. The gleam in their eye now is desalination, the process of turning saltwater into freshwater. "As soon as desalination technology gets water below, say, thirty cents per cubic meter, you really run out of a problem," says Aaron Wolf, in what the pessimists would call an overstatement. Desalination is useful chiefly as a source for industrial and municipal water in coastal areas, but the plants are usually too far from farmland to justify the ample pumping expense—and agriculture consumes 70 percent of all water used by humans. Of the 11,000 desalination plants that now exist, 60 percent are in the Middle East, where fuel is cheap and state budgets are relatively flush. The price of desalinated water has dropped in recent years, but it still typically costs $1 to $2 per cubic meter. Tantalizingly, a new desalination plant planned for Tampa, Florida, will sell water at 55 cents per cubic meter, but the Gulf water it treats is less saline than ocean water, and the plant enjoys financing and energy advantages that may make it unique. As it stands, desalination accounts for less than 1 percent of human water needs.

If desalination can't help us dramatically expand the supply of water, we have no choice but to reduce our demand for it. Here again, some optimists look to a high-tech, big-money solution—genetic engineering, which could produce crops with lower water requirements and higher resistance to insects, disease, and toxic substances. But the future of genetic engineering is uncertain because of safety concerns and political opposition in both Europe and the United States. "I absolutely believe we need to work on crop genetics," Gleick says. "But do we bet the house on it? That's dangerous. I think you have to address all the food and water questions."

Inevitably, this means increasing water productivity, getting more "crop per drop." The potential here is vast, since by some estimates the worldwide efficiency of agricultural water is 40 percent, which means that most water diverted for agriculture never even contributes to food production. Instead, it's lost to evaporation, leaky pipes, unlined canals, and wasteful irrigation practices. But whereas the Green Revolution offered a single strategy for increasing crop yield, no single equivalent exists for increasing water yield. In place of one approach, many have emerged. They usually use fewer resources, cause less environmental disruption, and cost less than their twentieth-century predecessors. In contrast to big projects such as dams, many of these approaches give local farmers a stake in the outcome and catalyze them to improve management techniques. "There is considerable evidence that farmer-controlled small-scale irrigation has a better record of performance than government-controlled large- or small-scale systems," writes Mark Rosegrant, an IFPRI analyst. The list of potentially useful small-scale methods is long and encompasses technical, managerial, institutional, and agronomic realms. In some places the best technique is a traditional one, such as rooftop or mountain-slope water-harvesting that was ill-advisedly superseded by a big but ultimately wasteful project.

At the top of most lists of appropriate water technology is drip irrigation, which was developed in Israel in the 1960s after cheap plastic tubing became available. Drip systems deliver water directly to individual plant roots, eliminating evaporation and saving water and energy. Drip irrigation not only produces water efficiency as high as 95 percent but also increases yields, since plants receive water on a regular basis instead of the boom-and-bust cycle of flood irrigation. Studies in many countries show that drip irrigation reduces water use by 30 to 70 percent and increases yields by 20 to 90 percent. Since only 1 percent of the world's irrigated lands now use drip and other high-efficiency methods, the potential for water conservation is huge. In India, for instance, 20 percent of irrigated land may be suitable for the technology.

Drip irrigation's major liability is its expense, typically $500 to $1,000 per acre, which has meant that only large farmers growing high-value crops use it. This is one facet of a huge income gap that irrigation technology has helped foster: a large majority of the farmers in developing countries can't afford the tools of irrigation and so are left out of the global economy. Among them are most of the world's 790 million undernourished people. To Paul Polak, president of a Lakewood, Colorado, nonprofit called International Development Enterprises, this is "a market chasm instead of a market niche." IDE has tried to fill it by working with small businesses in developing countries to design, field-test, manufacture, and market irrigation technology for poor farmers. At the low end of its product line is an easy-to-maintain $5 drip bucket kit, which can irrigate a 10- by 16-foot kitchen garden with two buckets of water a day. If a farmer grows income-generating crops, he can make enough money to move up to the next product in the line, a 55-gallon-drum kit for $26 that can water a 1,300-square-foot field. In China, where most farms are smaller than an acre, a poor farmer eventually could irrigate his entire field with a $300 system. In Nepal and India, IDE-assisted businesses have sold 10,000 drip kits in two years, enabling farmers to double yields without increasing water consumption.[3]

The Dream of the Oasis

LAS VEGAS IS America's city of fantasy, and water, not wealth, is its greatest fantasy of all. The city that Hoover Dam made possible is the nation's fastest-growing metropolis in the country's driest state, the perfect manifestation of the notion that water will never run out. Las Vegas and the desert don't match: the city looks as if it didn't so much emerge from its surroundings as get deposited on them. In this desert of ostentation, water is displayed more lasciviously than sex. Among the city's hotel casinos, Caesars Palace laid down the archetype, festooning its property with fountains and aqueducts in 1966. Now the Mirage sports a one-acre outcropping of terraced waterfalls, and a rain forest has been installed beneath a glass canopy at the entrance. At Treasure Island the main feature is a naval battle between British and pirate ships that employs live actors and a large supply of fireworks; this event attracts a few hundred sidewalk onlookers five times a night. The Mandalay Bay's grounds include a sandy beach with three- to four-foot waves. In pursuit of an impressive water display, I recently chose to stay at the Venetian, which features—can you guess?—canals, but unfortunately they resemble nothing so much as brightly lit, elongated bathtubs. The Venetian even contains a Grand Canal and a Basilica di San Marco, whose dissimilarities from the originals include being miniature and plasterboard and on the second floor. Bewildered tourists wait in line for the chance to pay money to stripe-shirted "gondoliers," who pole them down the hall, singing into the air-conditioning ducts.

For all its hydraulic glory, the Venetian has been upstaged by the Bellagio half a mile away. There, hotelier Steve Winn spent $40 million on choreographed spigots that dance to "Singin' in the Rain" and other tunes. Created by "water feature" specialist WET Design of Universal City, California, the installation is set within an 11-acre artificial lake for which the hotel serves as backdrop. Inside the lake are 27 million gallons of water (which a WET Design press release points out are equivalent to 3,000 swimming pools), 4,500 lights, 798 "MiniShooters," 213 "Oarsmen," 192 "SuperShooters," 350 miles of electrical wires, 120 miles of electrical cables, and 5 miles of pipe. The electrical load of this assemblage is 7.5 megawatts, enough for 7,500 homes. Every half hour, speakers all around the lake introduce a melody, drawing from an eclectic repertoire that gives equal billing to Aaron Copland, Luciano Pavarotti, Lionel Richie, and Marvin Hamlisch. Then, as the music plays, the nozzles rhythmically spew water in sinuous, synchronous arcs or in pulsed skyward streams as high as 250 feet. Mist rises lubriciously from the lake. If the sweating spectators are lucky, some of it wafts their way. When I asked Carolyn Nott, WET Design's vice president for business development, why so many Las Vegas hotels feature water, she had a quick answer: "People in the desert have always been fascinated by water. It's the idea of the oasis."

I mentioned this notion to the voluble Pat Mulroy, who as general manager of the Southern Nevada Water Authority is one of the state's most prominent officials. Mulroy, whose poufed and bejeweled appearance belies a canny grasp of western water issues, understands that the "oasis" is a construct created for Las Vegas's 30 million tourists a year. The real Las Vegas is so short of water that even if it adheres to its current conservation plan, it will probably run out of Colorado River water by 2007; then, Mulroy says, "other mechanisms have to come into play." The "mechanisms," however, are uncertain bets. In the short term, Mulroy is trying to persuade reluctant Arizonans to sell part of their allotment of Colorado River water. In the long run, she is pinning her hopes on California water, which she thinks could become available if desalination plants start supplying California. "That's the only logical place to go," she says.

Since its institution in 1995, Las Vegas's conservation plan has already pared 16 percent off the city's projected water use and is calibrated to reach 25 percent by 2010. The biggest problem, Mulroy notes, is the insistence by so many residents on growing lawns: two thirds of the city's water is used outdoors. The conservation plan has instituted tiered water rates that force profligate residential users to pay $900 a month or more for water, and the city limits the size of front lawns. The hotels, on the other hand, are forgiven their conspicuous use of water because they are central to Nevada's economy. "The hotels generate somewhere around 70 percent of the state's gross product, and they use 8 percent of all the water we deliver," Mulroy says. "That's not a bad investment." Even so, the hotels pay top-tier rates for their water, and most use treated gray water for their displays: the Mirage and Treasure Island share one underground water-treatment plant, while the Bellagio houses another. Of course, these facilities are hidden from the hotels' guests, for whom the illusion of bountiful water is carefully preserved. I found this out when I asked Mulroy why the hotels don't advise their guests to reuse towels and stint on water use, as other desert resorts do. She said, "Las Vegas sells fantasy. Anything that jars people back to reality is viewed by those who run the hotels as a disincentive." It was the next sentence that clicked inside my brain, as I realized that it summed up the human approach to water at the end of the twentieth century. "People don't want to live in reality," she said.

But reality has a way of forcing its way into human consciousness, and sooner or later we must acknowledge that our relationship to water is intimate, complex, and primal: if we abuse it, we inevitably suffer the consequences. Remove trees from the watershed, and the river below floods; deplete aquifers, and the land above subsides; pollute or obstruct the river, and the effects flow all the way to the sea. We must accommodate ourselves to water, not the other way around. Neither the pollution of our air and soil nor the destruction of wilderness nor even the probable extinction of a majority of the earth's creatures and the threat of catastrophic climate change has prompted us to change our behavior. Now it is the turn of water, the very foundation of life, to teach us to be good animals.

Notes

1. Saudi Arabia provides perhaps the world's best example of extravagant groundwater depletion. After helping to launch the OPEC oil embargo of 1973, Saudis feared other countries might retaliate with a grain embargo, so they embarked on a program to make the country self-sufficient in grain. As Postel explains, the nation subsidized farmers' land, equipment, and water, and paid them several times the world market price for grain. The result was that for a short time Saudi Arabia managed to become a grain exporter. Because of the country's hot, arid climate, each ton of grain required 3,000 tons of water, triple the usual ratio. When Saudi Arabia was forced to make budget cuts in the mid-1990s, the effort could not be sustained. The curtailment of subsidies caused grain production to fall by 60 percent. That may be just as well, since Saudi Arabia otherwise would have run out of groundwater by 2040. Even now, with a more modest agricultural program, the Saudis continue to run a significant groundwater deficit.

2. In the new lakes, sport fish stocked for humans' recreation—catfish, bass, and sunfish, and minnows for all of them to feed on—arrive previously adapted to stable lake environments and thrive. They prey on the native fish, which are now disadvantaged by being suited to river conditions. The humpback chub, native to the Colorado, has an odd-looking hump behind its head that contains extra muscles connecting to its tail; before the dam era, it used those muscles to survive in the Colorado's occasionally torrential waters. Now the chub, like virtually all other native Lower Colorado fish, courts extinction. In little more than half a century a foreign fish population has essentially replaced the Colorado's native one.

3. The path of appropriate water management often isn't smooth. IDE's biggest success is in Bangladesh, where it has overseen the sale and installation of 1.3 million treadle pumps since 1984. Treadle pumps are useful in areas like Bangladesh's Ganges delta, where aquifers are replenished during summer monsoons and the major problem is finding water during the scorching dry season. Farmers peddle the treadles for two to six hours a day; the difficulty of pumping water this way assures its judicious use. Treadle pumps cost $35 and enable farmers to earn at least $100 a year in increased crop production. In recent years, however, scientists have discovered that much of the underground water in the Ganges delta is contaminated with naturally occurring arsenic. The result has been what the World Bank calls "perhaps the largest mass poisoning in history"; 20 million people may be poisoning themselves, and several hundred thousand already display symptoms. The difficulty arose after officials promoted wells to counter a more immediate health problem, the spread of waterborne diseases as a result of drinking dirty pond water. Although IDE's treadle pumps are used chiefly for agriculture, not human consumption, Polak says he assumes that even crops grown with contaminated groundwater are affected, and IDE has joined a massive effort to replace contaminated wells. The larger lesson, of course, is that testing should occur when wells are dug.

ROBERT L. PARK

Welcome to Planet Earth

FROM *THE SCIENCES*

In the summer of 1954, when I was a young Air Force lieutenant, I was sent on temporary assignment to Walker Air Force Base in Roswell, New Mexico, to oversee the installation of a new radar system. Late one night I was returning to the base after a weekend visit with my family in Texas. I was driving on a totally deserted stretch of highway. The sky was moonless but very clear, and I could make out a range of ragged hills off to my left, silhouetted against the background of stars. Suddenly the entire countryside was lit up by a dazzling blue-green light, streaking across the sky just above the horizon.

The light flashed on and off as it passed behind the hills, then vanished without a sound. It was all over in perhaps two seconds. At the time, reported sightings of unidentified flying objects—UFOs—made the news almost daily. Indeed, the town where I was stationed, Roswell, was the hub of many such speculations. But I prided myself on being a skeptical thinker, and I had little patience for wacky ideas about flying saucers invading the earth.

In fact, I had a perfectly plausible explanation for the spectacular event I had just witnessed. Pale blue-green is the characteristic color of the light emitted by certain frozen free radicals as they warm up. A free radical is a fragment of a molecule, and one well-known variety of free radical is the so-called hydroxide radical—a water molecule that is missing one of its hydrogen atoms.

Free radicals are energetically predisposed to reconnect with their missing parts, and for that reason they are highly reactive: ordinarily they do not stick around very long.

But if molecules are broken up into free radicals by radiation at low temperatures, the radicals can be frozen in place. Then, when the severed parts of the molecule are warmed up, they readily recombine to form the same kinds of stable molecules from which they originated. The energy that is liberated when hydroxide radicals recombine with hydrogen atoms to form water appears as blue-green fluorescence. It occurred to me that an ice meteoroid would gradually accumulate hydroxide radicals as a result of cosmic-ray bombardment. What I had had the good fortune to see just then, I reasoned, was a meteor plunging into the earth's upper atmosphere, where it warmed, setting off the recombination reaction.

As I continued driving down the empty highway and crossed into New Mexico, I felt rather smug. The UFO hysteria that was sweeping the country, I told myself, was for people who don't understand science. Then I saw the flying saucer.

It was off to my left, between the highway and the distant hills, racing along just above the range-land. It appeared to be a shiny metal disk, thicker in the center than at the edges, and it was traveling at almost the same speed I was. Was it following me? I stepped hard on the gas pedal of the Oldsmobile—and the saucer accelerated. I slammed on the brakes—and it stopped. Only then could I see that it was my own headlights, reflecting off a telephone line strung parallel to the highway. The apparition no longer looked like a flying saucer at all.

It was a humbling experience. My cerebral cortex might have sneered at stories of flying saucers, but the part of my brain where those stories were stored had been activated by the powerful experience of the icy meteorite. At an unconscious level, my mind was busy making connections and associations. I was primed to see a flying saucer—and my brain filled in the details.

WHO HAS NOT "SEEN" an animal in dusky twilight that turns into a bush as one takes a closer look? But something more than the mind playing tricks with patterns of light is needed to explain why hundreds—by some accounts thousands—of people claim to have been abducted by aliens, whisked aboard a spaceship and subjected to some kind of physical examination, usually focusing on their erogenous zones. After the examination, the aliens are frequently said to insert a miniature implant into the abductee's body. Often the memory of an abduction has a dreamlike quality, and subjects can recall the details only under hypnosis.

Scientists themselves are not immune to such beliefs. In 1992 a five-day conference was held at the Massachusetts Institute of Technology to assess the similarities among various accounts of alien abduction. The conference was organized by John E. Mack, a Harvard psychiatrist, and David E. Pritchard, a prize-winning MIT physicist. Mack had been treating patients who thought they had been kidnapped by aliens. His treatment was to reassure them that they were not hallucinating but really had been abducted.

Pritchard, an experimentalist, was more interested in the physical evidence of the kidnappings, particularly the minuscule implants. The most promising candidate seemed to be an implant that abductee Richard Price said had been inserted midshaft into his penis. The implant, amber in color and the size of a grain of rice, was clearly visible. Under a microscope, what appeared to be fine wires could be seen protruding from it. What wonders of alien technology might be revealed by a sophisticated analysis of that diminutive device? Amid high expectations, the "implant" was removed and examined. The conclusion? It was not from Andromeda. Its origins were distinctly terrestrial: human tissue that had accreted fibers of cotton from Price's underwear.

It is hardly surprising that there are similarities in the accounts of people who claim to have been abducted by aliens. All of us have been exposed to the same images and stories in the popular media. My local bookstore stocks three times as many books about UFOs as it carries about science. Aliens stare at us from the covers of magazines and make cameo appearances in television commercials. As time goes by, the depictions become increasingly uniform. Any six-year-old can now sketch what an alien looks like. Popular culture is, in fact, undergoing a kind of alien evolution: each new creation by a filmmaker or sci-fi writer acts as a mutation, and the selection mechanism is audience approval. Aliens subtly evolve to satisfy public expectations.

THE WIDESPREAD BELIEF in alien abductions is just one example of the growing influence of pseudoscience. Two hundred years ago, educated people imagined that the greatest contribution of science would be to free the world from superstition and humbug. It has not happened. Ancient beliefs in demons and magic still sweep across the modern landscape, but they are now dressed in the language and symbols of science: A best-selling health guru asserts that cancer can be banished from the body by the power of the mind. If anyone should doubt it, he explains that it's all firmly grounded in quantum theory. Inventors claim to have built perpetual-motion machines that circumvent the laws of physics. Educated people wear magnets in their shoes to draw "energy" from the earth.

Voodoo science is everywhere. But why? Perhaps the most endearing characteristic of Americans is their sympathy for the underdog. They resent arrogant scientists who talk down to them in unfamiliar language, and government bureaucrats who hide behind rules. Scientists, meanwhile, often look the other way when science is being abused, expecting bogus claims to self-destruct. But members of the public are often not in a position to distinguish between fabulous but verifiable phenomena, such as hermaphrodites and antimatter, and fanciful ones, such as touch therapy and astrology. It's up to the scientists to inform the nonscientists—and to remember how easy it can be to subscribe to erroneous ideas. Whenever I become impatient with UFO enthusiasts, as I often do, I try to remember that night in New Mexico when, for a few seconds, I, too, believed in flying saucers.

The current fascination with aliens can be traced back to the strange events that took place near Roswell, New Mexico, in the summer of 1947. On June 14 of that year, William Brazel, the foreman of the Foster Ranch, seventy-five miles northwest of Roswell, spotted a large area of wreckage about seven miles from the ranch house. The debris included neoprene strips, tape, metal foil, cardboard and sticks. Brazel didn't bother to examine it closely at the time, but a few weeks later he heard about reports of flying saucers and wondered if what he had seen might be related. He went back with his wife and gathered up some of the pieces. The next day he drove to the little town of Corona, New Mexico, to sell wool, and while he was there he "whispered kinda confidential like" to the Lincoln County sheriff, George Wilcox, that he might have found pieces of one of those "flying discs" people were talking about. The sheriff reported the matter to the nearby army air base—the same base, in fact, where I would be stationed seven years later (before my time, though the Air Corps was still part of the army, and the base was known as Roswell Army Air Field).

The army sent an intelligence officer, Major Jesse Marcel, to check out the report. Marcel thought the debris looked like pieces of a weather balloon or a radar reflector; in any event, all of it fit easily into the trunk of his car. There the incident might have ended—except for the garbled account the public-information office at the base issued to the press the next day. The army, the press office noted, had "gained possession of a flying disc through the cooperation of a local rancher and the sheriff's office." The army quickly issued a correction describing the debris as a standard radar target, but it was too late. The Roswell incident had been launched. With the passage of years, the retraction of that original press release would come to look more and more like a cover-up.

BY 1978, THIRTY YEARS AFTER Brazel spotted wreckage on his ranch, actual alien bodies had begun to show up in accounts of the "crash." Major Marcel's story about loading sticks, cardboard and metal foil into the trunk of his car had mutated into the saga of a major military operation, which allegedly recovered an entire alien spaceship and secretly transported it to Wright-Patterson Air Force Base in Ohio. Even as the number of people who might recall the original events dwindled, incredible new details were added by second- and third-hand sources: There was not one crash but two or three. The aliens were small, with large heads and suction cups on their fingers. One alien survived for a time but was kept hidden by the government—and on and on.

Like a giant vacuum cleaner, the story had sucked in and mingled together snippets from reports of unrelated plane crashes and high-altitude parachute experiments involving anthropomorphic dummies, even though some of those events took place years later and miles away. And, with years' worth of imaginative energy to drive their basic beliefs, various UFO "investigators" managed to stitch those snippets into a full-scale myth of an encounter with extraterrestrials—an encounter that had been covered up by the government. The truth, according to the believers, was simply too frightening to share with the public.

Roswell became a gold mine. The unverified accounts spawned a string of profitable books, and were shamelessly exploited for their entertainment value on television programs and talk shows—even serious ones, such as CBS's *48 Hours,* then hosted by the eminent anchorman Dan Rather, and CNN's *Larry King Live.* The low point was reached by Fox TV. In 1995 the network began showing grainy black-and-white footage of what was purported to be a government autopsy of one of the aliens—a broadcast that garnered such exceptional ratings (and such exceptional advertising revenues) that it was rerun repeatedly for three years. Then, when ratings finally began to wane, Fox dramatically "exposed" the entire thing as a hoax.

In 1994, to the astonishment of believers and skeptics alike, a search of military records for information about the Roswell incident uncovered a still-secret government program from the 1940s called Project Mogul. There really had been a cover-up, it turned out—but not of an alien spaceship.

IN THE SUMMER OF 1947 the U.S.S.R. had not yet detonated its first atomic bomb, but it had become clear by then that it was only a matter of time. It was imperative that the United States know about the event when it happened. A variety of ways to detect that first Soviet nuclear test were being explored. Project Mogul was an attempt to "listen" for the explosion with low-frequency acoustic microphones flown to high altitudes in the upper atmosphere. The idea was not entirely harebrained: the interface between the troposphere and the stratosphere creates an acoustic channel through which sound waves can propagate around the globe. Acoustic sensors, radar tracking reflectors and other equipment were sent aloft on long trains of weather balloons, in the hope that they would be able to pick up the sound of an atomic explosion.

The balloon trains were launched from Alamogordo, New Mexico, about a hundred miles west of Roswell. One of the surviving scientists from Project Mogul, the physicist Charles B. Moore, professor emeritus at the New Mexico Institute of Mining and Technology in Socorro, recalls that Flight 4, launched on June 4, 1947, was tracked to within seventeen miles of the spot where Brazel found wreckage ten days later. Then, Moore says, contact was lost. The debris found on the Foster Ranch closely matched the materials used in the balloon trains. The Air Force now concludes that it was, beyond any reasonable doubt, the crash of Flight 4 that set off the bizarre series of events known as the Roswell incident. Had Project Mogul not been highly secret, unknown even to the military authorities in Roswell, the entire episode would probably have ended in July 1947.

It is hard to understand why Project Mogul was secret at all. Even before the Soviets tested their first atomic bomb, the project was abandoned, pushed aside by more promising detection technologies. There was nothing in Project Mogul that could have provided the Soviets with anything but amusement, yet it was a tightly kept secret for nearly half a century; even its code name was secret. The project would still be a secret if not for an investigation initiated in 1994 by Steven H. Schiff, a Republican congressman from New Mexico. Schiff insisted that an all-out search for records and witnesses was needed to reassure the public that there had been no government cover-up in Roswell.

By 1997 the Air Force had collected every scrap of information dealing with the Roswell incident into a massive report, in hopes of bringing the story to an end. In fact, the enormous task of locating and sifting through old files and tracking down surviving witnesses had actually begun even before Schiff's call for full disclosure. Responding to requests from self-appointed UFO investigators acting under the Freedom of Information Act had become a heavy burden on the Air Force staff at the Pentagon, and they were eager to get ahead of the Roswell incident. The release of *The Roswell Report: Case Closed* drew one of the largest crowds on record for a Pentagon press conference.

Although the people involved insist that it was mere coincidence, the Air Force report was completed just in time for the fiftieth anniversary of Brazel's discovery of the Project Mogul wreckage. Thousands of UFO enthusiasts descended on Roswell, now a popular tourist destination, in July 1997 for a golden-anniversary celebration. They bought alien dolls and commemorative T-shirts, and snatched up every book they could find on UFOs and aliens. The only book that sold poorly was the Air Force report.

If there is any mystery still surrounding the Roswell incident, it is why uncovering Project Mogul in 1994 failed to put an end to the UFO myth. Several reasons seem plausible, and they are all related to the fact that the truth came out almost half a century too late. The disclosures about Project Mogul were pounced on by UFO believers as proof that everything the government had said before was a lie. What reason was there to think that Project Mogul was not just another one?

Furthermore, Project Mogul was not the only secret government program that bolstered belief in UFOs. During the cold war, U-2 spy planes often flew over the Soviet Union. At first, U-2s were silver-colored, and their shiny skins strongly reflected sunlight, making them highly visible—particularly in the morning and evening, when the surface below was dark. In fact, the CIA estimates that more than half of all the UFO reports from the late 1950s and throughout the 1960s were actually sightings of secret U-2 reconnaissance flights. To allay public concerns at the time, the Air Force concocted far-fetched explanations involving natural phenomena. Keeping secrets, as most people learn early in life, inevitably leads to telling lies.

But secrecy, it seems, is an integral part of military culture, and it has generated a mountain of classified material. No one really knows the size of that mountain, and despite periodic efforts at reform, more classified documents exist today than there were at the height of the cold war. The government estimates that the direct cost of maintaining those records is about $3.4 billion per year, but the true cost—in loss of credibility for the government—is immeasurable. In a desperate attempt to bring the system under control, in 1995 President Clinton issued an executive order that will automatically declassify documents that are more than twenty-five years old—estimated at well in excess of a billion pages—beginning in 2000.

RECENT POLLS INDICATE that a growing number of people think the government is covering up information about UFOs. Nevertheless, it is easy to read too much significance into reports of widespread public belief in alien visits to earth. The late astronomer and science popularizer Carl Sagan saw in the myth of the space alien the modern equivalent of the demons that haunted medieval society, and for a susceptible few they are a frightening reality. But for most people, UFOs and aliens merely add a touch of excitement and mystery to uneventful lives. They also provide a handy way for people to thumb their noses at the government.

The real cost of the Roswell incident must be measured in terms of the erosion of public trust. In the interests of security, people in every society must grant their governments a license to keep secrets, and in times of perceived national danger, that license is broadened. It is a perilous bargain. A curtain of official secrecy can conceal waste, corruption and foolishness, and information can be selectively leaked for political advantage. That is a convenient arrangement for government officials, but in the long run, as the Roswell episode teaches, it often backfires. Secrets and lies leave the government powerless to reassure its citizens in the face of far-fetched conspiracy theories. Concealment is the soil in which pseudoscience flourishes.

ALAN LIGHTMAN

A Portrait of the Novelist as a Young Scientist

FROM THE *NEW YORK TIMES*

When I turned 35, I wrote an essay for *The New York Times Magazine* about my distressing awareness that I would soon be an old man in my field.

That field was theoretical physics, where people do their best work at a famously young age. Now, 16 years later, having long since given up physics for a profession in which I am still young, I find myself looking back on my life as a scientist and what I so miss.

I miss the purity. Theoretical physicists, and many other kinds of scientists, work in a world of the mind. It is a mathematical world without bodies, without people, without the vagaries of human emotion. A physicist can imagine a weight hung from a spring, bouncing up and down, and fix this mental image with an equation. If friction with air becomes an unwanted nuisance, just imagine the weight in a vacuum.

Much of science, in fact, is built on these pure pictures of the mind. And the equations themselves are beautiful. The equations have a precision and elegance, a magnificent serenity, an indisputable rightness.

I remember so often finding a sweet comfort in my equations after arguing with my wife about this or that domestic concern or fretting over some difficult decision in my life or feeling confused by a person I'd met. I miss that purity, that calm.

Of course, other occupations also deal with ideas. But the ideas are often complicated with the ambiguity of human nature. The exquisite contradictions and uncertainties of the human heart do indeed make life interesting; they are why God held the apple in front of Eve and then forbade her to eat it, they inspire artists and art, they are why the poet Rainer Maria Rilke wrote that "We should try to love the questions themselves, like locked rooms and like books that are written in a very foreign tongue."

All that is necessary and good. But I miss the answers. I miss the rooms I could enter, the language that sounded clear as a struck bell.

I miss the exhilaration of seeing brilliant people at work, watching their minds leap right in front of me, not the brooding intelligence of writers, but an immediate mental agility, pole vaults and somersaults and triple axels on the ice. Richard Feynman once walked into my tiny office at Caltech and, in 20 minutes at the blackboard, outlined the basic equations for the quantum evaporation of spinning black holes, an ingenious idea that had just occurred to him on the spot.

When I was stymied by a tough astrophysics problem at Cornell, the great theoretician Edwin Salpeter, while lying on the floor of his living room with back pain, instantly drew an analogy between the slow drift of stars orbiting a disruptive mass and the random motion of marbles bumping around on a table with a hole in its center.

Others I watched from the front row: the British astrophysicist and astronomer royal, Martin Rees, the Nobel Prize winning particle physicist, Steven Weinberg. In the presence of these minds I felt humbled as well as excited. I miss the humility; it made me crouch down and observe. I listened more than I talked. I took in.

Most of all, I miss the intensity. I miss being grabbed by a science problem so that I could think of nothing else, consumed by it during the day and then through the night, hunched over the kitchen table with my pencil and pad of white paper while the dark world slept, tireless, electrified, working on until daylight and beyond.

Every creative field has its moment of inspiration, the struggle to that moment and then the surge of insight. In most occupations, the aftermath is a slow working out of the idea. As a writer, even when I am writing well, I cannot write more than six hours at a time. After that I am exhausted and my vision has become clouded by the inherent subtleties and uncertainties of the work. Then I must wait for the words to shift and settle on the page and my own strength to return.

But as a scientist, I could be gripped for days at a time. I could go for days without stopping. Because I wanted to know the answer. I wanted to know that telltale behavior of matter spiraling into a black hole, or the maximum temperature of a gas of electrons and positrons, or what was left after a cluster of stars had slowly lost mass and drawn in on itself and collapsed.

When in the throes of a new problem, I was driven night and day, compelled because I knew there was a definite answer, I knew that the equations inexorably led to an answer, an answer that had never been known before, an answer waiting for me.

That certainty and power, and the intensity of effort it causes, I dearly miss. It cannot be found in most other professions.

Sometimes, I wonder if what I really miss is my youth. Purity, exhilaration, intensity—these are aspects of the young.

In a way, it is not possible at age 50 for me to look back on myself in my 20's and early 30's and understand anything more than the delicious feeling of immortality, the clarity of youth, the feeling that everything was possible.

I do miss my youth. And now, as a writer still finding my stride, while I can reasonably expect another couple of decades to thrive, I know that this second profession too will come to an end, that I will inevitably dwindle down to the physical as well as mental and artistic end, the final end. Of course, I want to be young again.

Yet, if given a chance to start over, I would do just what I did, to be not only a young man in the shimmering of youth but a scientist. I would want again to be driven day and night by my research. I would want the beauty and power of the equations. I would want to hear that call of certain truth.

FREEMAN J. DYSON

Science, Guided by Ethics, Can Lift Up the Poor

FROM THE *INTERNATIONAL HERALD TRIBUNE*

Throughout history, people have used technology to change the world. Our technology has been of two kinds, green and gray. Green technology is seeds and plants, gardens and vineyards and orchards, domesticated horses and cows and pigs, milk and cheese, leather and wool. Gray technology is bronze and steel, spears and guns, coal and oil and electricity, automobiles and airplanes and rockets, telephones and computers. Civilization began with green technology, with agriculture and animal-breeding, 10,000 years ago. Then, beginning about 3,000 years ago, gray technology became dominant, with mining and metallurgy and machinery. For the last 500 years, gray technology has been racing ahead and has given birth to the modern world of cities and factories and supermarkets.

The dominance of gray technology is coming to an end. During the last 50 years, we have achieved a fundamental understanding of the processes in living cells. With understanding comes the ability to exploit and control. Out of the knowledge acquired by modern biology, modern biotechnology is growing. The new green technology will give us the power, using only sunlight as a source of energy, and air and water and soil as materials, to manufacture and recycle chemicals of all kinds. Our gray technology of machines and computers will not disappear, but green technology will be moving ahead even faster.

Green technology can be cleaner, more flexible and less wasteful than our existing chemical industries. A great variety of manufactured objects could be grown instead of made. Green technology could supply human needs with far less damage to the natural environment. And green technology could be a great equalizer, bringing wealth to the tropical areas, of the planet, which have most of the world's sunshine, people and poverty. I am saying that green technology could do all these good things, not that green technology will do all these good things.

To make these good things happen, we need not only the new technology but the political and economic conditions that will give people all over the world a chance to use it. To make these things happen, we need a powerful push from ethics. We need a consensus of public opinion around the world that the existing gross inequalities in the distribution of wealth are intolerable. In reaching such a consensus, religions must play an essential role. Neither technology alone nor religion alone is powerful enough to bring social justice to human societies, but technology and religion working together might do the job.

We all know that green technology has a dark side, just as gray technology has a dark side. Gray technology brought us hydrogen bombs as well as telephones. Green technology brought us anthrax bombs as well as antibiotics. Besides the dangers of biological weapons, green technology brings other dangers having nothing to do with weapons. The ultimate danger of green technology comes from its power to change the nature of human beings by the application of genetic engineering to human embryos. If we allow a free market in human genes, wealthy parents will be able to buy what they consider superior genes for their babies. This could cause a splitting of humanity into hereditary castes. Within a few generations, the children of rich and poor could become separate species. Humanity would then have regressed all the way back to a society of masters and slaves. No matter how strongly we believe in the virtues of a free market economy, the free market must not extend to human genes.

I see two tremendous goods coming from biotechnology: first, the alleviation of human misery through progress in medicine, and second, the transformation of the global economy through green technology spreading wealth more equitably around the world. The two great evils to be avoided are the use of biological weapons and the corruption of human nature by buying and selling genes. I see no scientific reason why we should not achieve the good and avoid the evil. The obstacles to achieving the good are political rather than technical. Unfortunately a large number of people in many countries are strongly opposed to green technology, for reasons having little to do with the real dangers. It is important to treat the opponents with respect, to pay attention to their fears, to go gently into the new world of green technology so that neither human dignity nor religious conviction is violated. If we can go gently, we have a good chance of achieving within a hundred years the goals of ecological sustainability and social justice that green technology brings within our reach.

The great question for our time is how to make sure that the continuing scientific revolution brings benefits to everybody rather than widening the gap between rich and poor. To lift up poor countries, and poor people in rich countries, from poverty, technology is not enough. Technology must be guided and driven by ethics if it is to do more than provide new toys for the rich. Scientists and business leaders who care about social justice should join forces with environmental and religious organizations to give political clout to ethics. Science and religion should work together to abolish the gross inequalities that prevail in the modern world. That is my vision, and it is the same vision that inspired Francis Bacon 400 years ago, when he prayed that through science God would "endow the human family with new mercies."

About the Contributors

JOEL ACHENBACH is a reporter and columnist for the *Washington Post.* He is the author of five books, including *It Looks Like a President Only Smaller,* based on his political columns for washingtonpost.com; *Captured by Aliens,* a rumination on the idea of extraterrestrial life; and three books based on his syndicated column, "Why Things Are." He is a regular commentator on National Public Radio's *Morning Edition* and has written for such publications as *Slate, Smithsonian,* and *GQ.* A native of Gainesville, Florida, Achenbach graduated from Princeton University and worked for eight years at the *Miami Herald* before joining the *Post.* He lives in Washington with his wife and three daughters.

"Like many people, I assumed that it would be arrogant of us to claim that we were the only intelligent beings in the universe," he says. "Now, I'm not so sure. There are so many unknowns. How easily does life originate? How inevitably does it evolve into complex, multicellular organisms? How likely is it that an animal will become technological, a creature that dramatically reengineers its environment (and builds spaceships)? No one knows. We just make guesses. My guess is that intelligence is pretty darn rare in any given corner of any given galaxy . . . and this article gives some of the reasons why I say that. There are constraints that limit the abundance of intelligent creatures. For one thing, it apparently takes billions of years for simple life to evolve to the point where it can build a fire, much less a spaceship. How many places in the universe are habitable for such vast stretches of time? No one yet knows. My basic message: It's a good thing to be alive, to be a human being, to have a brain capable of interesting thoughts, in a universe that's mostly rocks and gas and dust and emptiness."

NATALIE ANGIER, whose science writing for the *New York Times* won her the 1991 Pulitzer Prize, started her career as a founding staff member of *Discover* magazine, where her beat was biology. In 1990 she joined the *Times*, where she has covered genetics, evolutionary biology, medicine, and other subjects. Her work has appeared in a number of major publications, and she is the author of three books: *Natural Obsessions,* about the world of cancer research (recently reissued in a new paperback edition); *The Beauty of the Beastly;* and the national bestseller *Woman: An Intimate Geography,* published originally in 1999 and now available in paperback. She is also the recipient of the American Association for the Advancement of Science–Westinghouse Award for excellence in science journalism and the Lewis Thomas Award for distinguished writing in the life sciences. She lives in Takoma Park, Maryland, with her husband, Rick Weiss, a science reporter for the *Washington Post,* and their daughter, Katherine Ida Weiss Angier.

"How amusing it is," Natalie Angier writes, "to go from the traditional view of female mandrills as meek little gals in a harem, beholden to their alpha male and even relying on his harlequin nose to guide them through the forest, to our new understanding of them as simian Amazons, living in all-female societies and consorting with males only when necessary, during mating season. It's also extraordinary that we have only just discovered the dimensions of these mandrill sororities: Whereas most primate social groups consist of a few dozen individuals, mandrills congregate in societies numbering five hundred, eight hundred, sometimes thirteen hundred females in all.

"Male mandrills offer surprises of their own: most of the year, it seems, they live in solitude, away from each other and from the female hordes. Only the orangutan matches the he-mandrill in its monkish habits. And now that we're finally learning about the mandrill behind the myth, our next task is to keep them from the paws, jaws, and chainsaws of the most populous primate of all."

ERIK ASPHAUG was born in Norway and grew up there and in the United States. His current position, Assistant Professor in the Earth Sciences Department at the University of California, Santa Cruz, is either destroying him or making him stronger, but he hesitates to complain, since there are few things so fun and rewarding as belonging to a vibrant academic department. His research involves modeling the evolution of asteroids, the formation of impact craters, and giant collisions such as that which formed the Moon. An additional interest is the exploration of near-Earth objects and the development of scientific missions that could guide our strategic or technological response to hazardous asteroids. His hobbies include going for walks with his young kids and gardening with his wife, Tracy, since, he believes, a moment spent outdoors is never wasted.

"I have long been fascinated by these objects that are neither boulders nor planets," he writes. "My doctoral research began with the fact that nobody knew, to within five orders of magnitude, what size asteroid can hold onto its pieces during a collision. We still remain ignorant of how asteroids evolve, but there is some hope. George Wetherill once commented that we can learn about star formation by understanding planets, since if we get the planets right, we will have constrained the mass of the nebula and its method of condensation. But getting the planets right is itself a formidable task. Maybe by studying lowly asteroids, so near and so common, we can learn about planets, and from there uncover a few of the broader strokes of the universe—'infinity in a grain of sand,' as Blake wrote."

DEBBIE BOOKCHIN has been a journalist for more than two decades. As a daily reporter she covered politics and health issues in Vermont, winning several awards for her reporting. Her work has appeared in numerous publications, including *The Nation* and the *New York Times.* JIM SCHUMACHER, who is married to Bookchin and who is a lawyer and management consultant in Vermont, once seriously considered a career in medicine. He and Bookchin have written stories about SV40 and polio vaccine that have appeared in the *Boston Globe, Boston Magazine, Newsday,* and *The Atlantic Monthly.* They are currently working on a book on the subject.

"I was working on a case for a client whose unwillingness to vaccinate her children could have led to her losing custody of them," Jim Schumacher writes. "My job was to read every respectable scientific paper that could support her belief that vaccines can sometimes be dangerous. One of them mentioned that polio vaccine had once been widely contaminated by a monkey virus that might cause cancer. I spent the next several weeks in a medical library until I found some of the original papers from almost forty years ago that describe the discovery of SV40 in the vaccine. I was stunned. Half the

American population was exposed, yet there were almost no references to this fact in the scientific literature or anywhere else. To me, the story of SV40 speaks directly to an issue that isn't often addressed in scientific writing—that the unforeseen consequences of a new technology may be with us far longer than the anticipated benefits."

"Jim suggested we try to research and write about this topic," Bookchin recalls, "and shortly before we began, the first modern-day evidence that SV40 was causing disease had been published by Michele Carbone, the Italian scientist whose work we feature in 'The Virus and the Vaccine.' We contacted him and for the past six years have had an inside look at the expansion of his discoveries and the development of his theories. When we started, he was the only scientist talking about SV40 and cancer. Now scientists from around the world are seriously researching this topic. Observing the response to Carbone's research and its challenges to older paradigms, you see firsthand how science, which we think of as a disinterested pursuit, is as much influenced as any other endeavor by prejudice, predilection, and personality."

PETER J. BOYER worked as a correspondent for the *Los Angeles Times*, the *New York Times*, and CBS News, before becoming a staff writer at *The New Yorker* in 1992. His *New Yorker* stories cover a range of subjects, from politics to a memoir about a faith-healing evangelist. As a correspondent for public television's *Frontline* documentary series, Boyer has won several awards, including an Emmy for a film about the politics afflicting the U.S. Navy and a Peabody for a film about the Waco disaster. He lives in Ossining, New York, with his wife, Kari, and his children, Samuel and Eleanor Jane.

"When I researched and wrote this story, in the summer of 1999," he writes, "the great DNA awakening within the criminal justice system was in full flush. At last, some of the fundamental questions in the abiding debates over crime and punishment—whether some innocent people were incarcerated or even executed because of racial prejudice, misidentification, or otherwise flawed prosecutions—seemed answerable, thanks to science. Advances in forensic DNA testing had made it possible to reopen old cases, sometimes vindicating wrongly convicted defendants on the basis of biological evidence once deemed too small or too old to be of any forensic use.

"But there were problems. Despite the apparent immutability of the science, it turned out that, in the adversarial context of a criminal case, DNA testing was as subject to dispute as old-fashioned blood-typing, fingerprinting, or any other type of forensic evidence. Prosecutors, performing contortions of legal theory, could make the most convincing DNA test result seem meaningless. If, for example, DNA taken from a rape victim did not match that of the accused, it did not necessarily exonerate him; prosecutors could (and do) argue that the semen in question came from another, separate sexual encounter and, thus, didn't prove anything. Peter Neufeld has derided such reasoning as the 'unindicted co-ejaculator theory.' Yet, it turned out that even the fiercest proponents of DNA-based exoneration can, in the context of a legal battle, also find wildly imaginative ways to assail the science when it undermines their legal opinion.

"In a courtroom, the 'irrefutable' science of DNA typing is often only as compelling as the legal argument it is meant to prove, or to refute."

GREG CRITSER was born in Steubenville, Ohio, and was educated at Occidental College in Los Angeles. He is a freelance writer specializing in nutrition, health, and medical issues. Critser's work on these subjects appears regularly in *USA Today,* where he is a member of the paper's board of contributors. His essays and features have also appeared as cover stories in *Harper's Magazine, Worth,* and the *Washington Post Magazine.* His writing on obesity earned him a James Beard nomination for best feature writing in 1999, and he is frequently interviewed by PBS and other media on the subject of food and food politics. Critser's historical writings have appeared in the American Historical Association's *Pacific Historical Review* and the *UCLA Journal of History.* He is the author of *National Geographic California* (May 2000), a comprehensive guidebook to the state's history, culture, and principal destinations. His next book, *Supersize* (Houghton Mifflin), will appear in fall of 2002.

" 'Let Them Eat Fat' represented the culmination of two years of research on the subject of weight loss and obesity," he writes. "The study began when my physician suggested that I try the diet drug Meridia in order to lose weight. In two lengthy personal essays, one in the *LA Weekly,* the other in *Worth,* I described the psychological and physical changes brought on by the drug while interweaving investigative work about how Meridia gained FDA approval in the first place. These fueled my curiosity and moved me into an exploration of how poor Americans, only a generation ago feared to be undernourished, became the fattest people in the world. Upon its publication, 'Let Them Eat Fat' generated an avalanche of news and media coverage, from PBS to Reuters. The controversy came from liberals who accused me of 'blaming the victim'—I had the audacity to suggest that the abandonment of cultural boundaries had something to do with the problem—and from freemarket conservatives who hated my attack on the fast-food industry, and who claimed that I was somehow calling for more government regulation (I wasn't). This response told me that I may actually have gotten things right. Apparently, so did a few others. I plan to expand on the themes struck in the essay in my book, tentatively titled *Supersize*, which I am working on now. Now all I have to do is stay away from the donuts."

FREEMAN J. DYSON was born in 1923 in Crowthorne, England. He received a B.A. in mathematics from the University of Cambridge in 1945 and came to the United States in 1947 as a Commonwealth Fellow at Cornell University. He settled in the United States permanently in 1951, became a professor of physics at the Institute for Advanced Study at Princeton in 1953, and retired as a professor emeritus in 1994.

Professor Dyson began his career as a mathematician but then turned to the exciting new developments in physics in the 1940s, particularly the theory of quantized fields. He wrote two papers on the foundations of quantum electrodynamics that have had a lasting influence on many branches of modern physics. He went on to work in condensed-matter physics, statistical mechanics, nuclear engineering, climate studies, astrophysics, and biology.

Beyond his professional work in physics, Dyson has a keen awareness of the human side of science and of the human consequences of technology. His books for the general public include *Disturbing the Universe, Weapons and Hope, Infinite in All Directions, Origins of Life and the Sun,* and *The Genome and the Internet.* In 2000 he was awarded the Templeton Prize for Progress in Religion.

"Science, Guided by Ethics, Can Lift Up the Poor" is an extract from Professor Dyson's Templeton Prize acceptance speech, given at the Washington National Cathedral in May 2000. The rest of the speech was concerned more with religion and less with science. Summarizing his remarks in one sentence at the end, he said, "Science and religion should work together to abolish the gross inequalities that prevail in the modern world."

HELEN EPSTEIN holds a Ph.D. in molecular biology and has taught biochemistry at Makerere University in Uganda and conducted AIDS research there. She has written for *Granta, New Scientist, The New York Review of Books,* and other magazines and has worked as a consultant for organizations such as the Panos Institute, the London School of Hygiene and Tropical Medicine, and the Rockefeller Foundation.

She writes: "Most pregnant HIV-positive South African women still lack access to anti-retroviral treatment that might spare their babies infection with the virus that causes AIDS. However, in the winter of 2001, the government of Thabo Mbeki finally agreed to offer the drug nevirapine free in around twenty public maternity clinics around the country. This move was welcome, but Mbeki's delay in establishing these programs remains mysterious. Meanwhile, HIV continues to spread throughout the subcontinent. This is not only because government leaders like Thabo Mbeki have been slow to address the epidemic. HIV spreads fastest where gradients of inequality are steepest, and where great wealth meets extreme poverty. In places, for example, like Atteridgeville.

"In April 2001, South Africa won a court case against a consortium of pharmaceutical companies, giving it the right to circumvent patent rules and import generic drugs, including cheap versions of anti-retroviral AIDS medications made in India and other countries. Shortly after the ruling was announced, South Africa's health minister, Manto Tshabalala-Msimang, announced her intention to import the technology to manufacture anti-retrovirals so that South Africa could make and sell its own AIDS drugs. Was this the point all along? For years, South Africa delayed offering anti-retroviral drugs to poor pregnant HIV-positive women, causing thousands of babies, who might have been spared, to be born with HIV. Was this delay an attempt to keep Western pharmaceutical companies out of the local AIDS market so that South African firms, and the South African government, could eventually take control of it?"

MALCOLM GLADWELL is a staff writer for *The New Yorker* and the author of *The Tipping Point,* published in 2000 by Little, Brown. He grew up in Canada, attended the University of Toronto, and began his career as a reporter at the *Washington Post.*

He writes: " 'John Rock's Error' grew out of a conversation I had years ago with a friend of mine who is an OB/GYN and who loved to hold forth on the 'unnaturalness' of monthly menstruation. I was fascinated, but despaired of ever finding a way of making that particular topic appeal to a wider audience. Then, for some reason (and I cannot remember how) I stumbled across the story of John Rock, and the result is 'John Rock's Error'—my roundabout way of talking about a very arcane subject."

STEPHEN JAY GOULD is currently Alexander Agassiz Professor of Zoology at Harvard University, where he also holds the positions of Professor of

Geology in the Department of Earth and Planetary Science and Curator of Invertebrate Paleontology of the Museum of Comparative Zoology. He is also the Vincent Astor Visiting Research Professor of Biology at New York University. The recipient of numerous honors and awards, including a John D. and Catherine T. MacArthur Foundation Fellowship, he is well known as the author of the column "This View of Life," whose twenty-seven-year run in *Natural History* magazine came to an end in 2001, and several popular books, among them: *The Panda's Thumb,* which won the American and National Book Awards; *The Mismeasure of Man,* winner of the National Book Critics Circle Award; *Wonderful Life,* which won the Rhone-Poulenc Prize; and, most recently, *Full House* and *Questioning the Millennium.*

STEPHEN S. HALL has been a journalist since the age of sixteen. Currently he is a contributing writer for *The New York Times Magazine,* where he covers biomedicine and the impact of science on the culture at large. His work has appeared in several magazines, and he is the author of three books on science: *Invisible Frontiers,* about cloning and the birth of the biotech industry; *Mapping the Next Millennium,* a survey of recent scientific work in the fields of geophysics, biology, mathematics, and astronomy within the historical context of mapmaking; and *A Commotion in the Blood,* about the immune system and how it can be enlisted to fight cancer and other diseases. He is currently working on a book about the molecular biology of life extension and the rise of "regenerative medicine."

"I remember first hearing about stem cells in the mid-1980s," Hall says, "but at that time they were a kind of unicorn cell: mythically powerful, a wonder to behold, but almost impossible even to glimpse in the lab, much less tame. By late 1998, however, several groups of scientists had succeeded in isolating human embryonic stem cells, and I was curious to find out how this triumph occurred, what promise these cells might hold as a futuristic kind of medicine, and what thorny ethical and political terrain biologists would have to traverse before realizing that medical promise. It turned out to be a fascinating, exciting, and, of course, early chapter in what is now being hailed as 'regenerative medicine.' "

TRACY KIDDER graduated from Harvard, studied at the University of Iowa, and served as an army officer in Vietnam. His books include *The Soul of a New Machine, House, Among Schoolchildren, Old Friends,* and, most recently, *Home Town.* He has won the Pulitzer Prize, the National Book Award, the

Robert F. Kennedy Award, and many other literary prizes. He is currently working on a book about Paul Farmer and the organization he founded.

"Paul Farmer is obviously an extraordinary person," Kidder says, "worthy of a *New Yorker* profile. But the reason I am attempting a book about him and his colleagues is the example they have created. They have shown that supposedly intractable problems, such as treating AIDS in places like Haiti, are not insoluble at all. The story of how they have done this is full of drama, and for me it represents an antidote to all the usual forms of despair."

JACQUES LESLIE began his career as a *Los Angeles Times* foreign correspondent. He was stationed successively in Saigon, Phnom Penh, Washington, D.C., New Delhi, Madrid, and Hong Kong, usually moving to a new assignment just as a crisis was erupting. He was the first American journalist to enter Viet Cong territory, and won a Sigma Delta Chi Journalism Society Distinguished Service Award and an Overseas Press Club citation for his reporting from Vietnam. Project Censored designated his book about his war experiences, *The Mark: A War Correspondent's Memoir of Vietnam and Cambodia,* one of the top "censored" books of 1995. He has written personal essays, reportage, profiles, and humor for a broad range of publications, including *Harper's Magazine, The Atlantic Monthly, The New York Times Magazine, Newsweek,* and *Wired,* where is a contributing writer. "Running Dry" was named a finalist for the John B. Oakes Award in Environmental Reporting.

Leslie writes: "It was my friend Robert Dawson, an honored photographer of the American West, who alerted me to the growing significance of water scarcity around the world. In particular, he referred me to a report called 'Blue Gold: The Global Water Crisis and the Commodification of the World's Water Supply,' written by Maude Barlow and published by the International Forum on Globalization. While I ultimately found myself disagreeing with the report's most sensational conclusions, it served to fix my attention on water. Many months later, without any prompting from me, the *Harper's* design department felicitously closed the circle by illustrating 'Running Dry' with seven photographs, two of which were Dawson's."

ALAN LIGHTMAN is a novelist, essayist, physicist, and educator. He was educated at Princeton University and at the California Institute of Technology, where he received a Ph.D. in theoretical physics. He has taught on the faculties of Harvard and M.I.T. His research in astrophysics has concerned black holes, general relativity, stellar dynamics, and radiative processes. His short fiction, essays, and reviews have appeared in *The Atlantic Monthly, Harper's, Granta, The New Yorker, The New York Review of Books,* and other publications. His two most recent books are a collection of essays and fables, *Dance for Two,* and the novels *Einstein's Dreams, Good Benito,* and *The Diagnosis,* which was a finalist for the National Book Award in fiction for 2000.

"I wrote 'Portrait of the Novelist as a Young Scientist,' " he explains, "to convey some of the pleasures of scientific research from the perspective of someone no longer active in the field, just as an aging ex-athlete might look back on his career as, say, a gymnast."

ERNST MAYR, one of the most influential evolutionary theorists of our time, is Alexander Agassiz Professor of Zoology, Emeritus, at Harvard University, whose faculty he joined in 1953. Born in Germany in 1904, Mayr received his doctoral degree in zoology from the University of Berlin in 1926; five years later he emigrated to the United States, taking up a position at the American Museum of Natural History in New York City. His work has focused on the evolution of new species by sharpening the definition of the species concept through the introduction of population thinking and by establishing the principle that genetic and, most often, geographic isolation are the key elements in speciation. He has put forward these views in several important books, including *Systematics and the Origins of Species* (1942) and *Animal Species and Evolution* (1963). He is also the author of books on the history and philosophy of evolution and biology, including *Growth of Biological Thought, One Long Argument: Charles Darwin and the Genesis of Modern Evolutionary Thought,* and *This Is Biology: The Science of the Living World.* He is the recipient of numerous honorary doctorates and awards, including the Darwin-Wallace Medal, the Darwin Medal of the Royal Society, the Benjamin Franklin Medal, and the Balzan Prize.

"Darwin's Influence on Modern Thought" is based on Mayr's speech in acceptance of the Crafoord Prize of the Royal Swedish Academy of Science, which was awarded in 1999. He wanted to point out another aspect of Darwin's significance, he says: "I noticed in recent years that as Darwin's thought was coming more and more to the foreground, the emphasis was always on evolution and natural selection, but I felt the breakthrough he made was broader in scope. One of Darwin's great achievements is that he, more than anyone else, even Newton, established secular science. All of Darwin's teachers at Cambridge were ordained ministers of the Anglican church and science largely at that time was entwined with religion. In the refutation of essentialism and finalism—the idea that the natural world reveals God's plan—Darwin made major contributions, and yet this is virtually never mentioned in the literature. So I wanted to stress that Darwin had a major impact on our current way of thinking, an impact that has not been properly recognized. These are achievements that are outside biology."

ROBERT L. PARK is professor of physics at the University of Maryland and Director of the Washington Office of the American Physical Society. His preparation for law school at the University of Texas was interrupted in 1950 by the Korean War. Ignoring his legal talent, the U.S. Air Force insisted he should be an electronics officer. On his return to the University of Texas in 1956, Park decided the Air Force might have been onto something. He switched to physics and graduated Phi Betta Kappa with High Honors two years later. In 1960 he became the Edgar Lewis Marston Fellow at Brown University, where he studied surface physics under the late Harry Farnsworth, one of the pioneers of the field. Park received his Ph.D. in 1964.

In 1965 he joined Sandia Laboratories in Albuquerque, and in 1969 became head of the Surface Physics Division at Sandia. He was appointed Professor of Physics and Director of the Center of Materials Research at the University of Maryland in College Park in 1974. Four years later, he became chair of the Department of Physics and Astronomy. He is also the founding editor of *Applications of Surface Science* and he is a Fellow of the American Vacuum Society, the American Association for the Advancement of Science, and the American Physical Society.

On his sabbatical year in 1982, he was asked by the American Physical Society to open an Office of Public Affairs in Washington, D.C. His sabbatical still seems to be going on; he divides his time between the APS and the University of Maryland. Park is the author of *What's New,* a controversial weekly electronic commentary on science policy issues. He is also a regular contributor of opinion articles in major newspapers, and a frequent guest on radio and television news programs. In 1998, he received the Joseph A. Burton Award of the American Physical Society for his contributions to the public understanding of issues involving the interface of physics and society. He is the author of the book *Voodoo Science,* which was published in 2000 by Oxford University Press.

He writes: "Most people, of course, having little or no background in science, must try to find their way through the forest of false claims and preposterous myths that masquerade as science. A recent special on Fox TV, for example, took the absurd position that the Apollo Moon landing was actually an elaborate government hoax. Five years earlier, Fox had shown a grainy, black-and-white film that purported to show the army's secret autopsy of a Roswell space alien. Both programs attracted their share of believers. *The Sciences* excerpted a portion of my book *Voodoo Science* that dealt with the 'Roswell Incident' as 'Welcome to Planet Earth.' In writing about such popular delusions from the perspective of a skeptical scientist, the task may not be to teach people science so much as to share with them a scientific world-view—an understanding that we live in an orderly universe governed by natural laws."

RICHARD PRESTON, forty-six, is the bestselling author of four books, including *The Hot Zone* and *The Cobra Event,* both of which were *New York Times* bestsellers. He has been a regular contributor to *The New Yorker* since 1985. He holds a Ph.D. in English from Princeton University, and he currently serves on the Academic Advisory Councils of both the Princeton English Department and the Princeton Astrophysical Sciences Department. He has won numerous awards for his writing, including the American Institute of Physics Award, the McDermott Award in the Arts from M.I.T., and the Centers for Disease Control's Champion of Prevention Award. Asteroid 3686 is named "Preston" in honor of his book about astronomy, *First Light.* (Preston is a ball of rock the size of the Matterhorn that orbits near Mars, and could collide with Mars or the Earth, causing an explosion not unlike the one that killed the dinosaurs.) Preston's recent *New Yorker* article on smallpox, "The Demon in the Freezer," won the National Magazine Award. He lives near Princeton with his wife and three children. He is currently at work on a new book.

He writes: "Genomics—the decipherment and study of the genetic codes of living organisms—may be the most powerful science of our time. It unlocks the code of life, and will someday allow us to reshape life. J. Craig Venter, the profilee in my piece, is a dramatic, larger-than-life character who will surely go down as a major figure in scientific history, whether you like him or hate him. (Nobody is indifferent to Craig Venter.) Venter is the chief scientist and a founder of Celera Genomics, a publicly traded company that has produced one of the two first, nearly complete sequences of the human genome.

"I got to know Craig Venter when he called me up one day to tell me that he'd had a conversation with President Bill Clinton about *The Cobra Event,* my novel about biological weapons. Clinton said the book had kept him awake all night—and Clinton asked Venter whether advanced biological weapons would really work. Venter said, 'Preston got it right,' and then called me. Thus began an acquaintance and ultimately a friendship—though in 'The Genome

Warrior' I relished quoting Craig's enemies, who are articulate, motivated, and numerous.

"Celera's human genome was published in *Science* magazine on February 16, 2001, simultaneously with the rival publication in *Nature* of the human genome done by the publicly funded Human Genome Project. While there were several attempts to merge the work and publish jointly, the two competing projects remained bitter rivals to the end."

JAMES SCHWARTZ received an A.B. degree in English Literature from Harvard College and a Ph.D. in mathematics from M.I.T. After completing his Ph.D., he wrote a novel. Currently he is working on a four-part series of articles on evolutionary biology for *Lingua Franca.* The first article in the series, "Oh My Darwin!" was published in November 1999. "Death of an Altruist" is the second. He has recently published a parody of literary life in *Poets & Writers* magazine. He lives in Brookline, Massachusetts, with his wife and two sons.

Discussing the enigmatic nature of his subject, Schwartz writes, "I was drawn to George Price because he was an unorthodox and relatively unknown figure who played a significant role in the development of evolutionary theory. Blessed and hampered by his multiple gifts, Price was both a highly original research scientist, who made important contributions in several different fields, and a remarkably lucid popular science writer. Perhaps because he came to evolutionary biology late in life and was unencumbered by a long period of education in the accepted ways of seeing population genetics, he was able to solve several long-standing puzzles in the field.

"His story is also interesting because he personally embodied the tension between rational scientific atheist and devout fundamentalist Christian. Price provided an elegant algebraic derivation of Kin Selection Theory, the idea that one behaves altruistically toward genetic relatives. In the same period, he began to live the life of the Good Samaritan, devoting himself to the welfare of genetic strangers. The common thread in these two contradictory impulses was the belief in a supreme and deterministic power, natural selection on the one hand, and God on the other."

ANDREW SULLIVAN became the editor of *The New Republic* in 1991 at the age of twenty-seven, and he held the position for 250 issues and five years. He was educated at Magdalen College, Oxford, and has a Ph.D. from the John F. Kennedy School of Government at Harvard. He is the author of *Virtually Normal: An Argument About Homosexuality* and *Love Undetectable: Notes on Friendship, Sex and Survival.* In the last four years, he has been a contributing writer and columnist for *The New York Times Magazine,* a regular contributor to *The New York Times Book Review,* and a weekly columnist for *The Sunday Times* of London. A frequent guest on television and radio talk shows, he contributes the TRB column to *The New Republic,* and he is at work on a new book on what science is saying about human difference, provisionally titled *Nature,* to be published by HarperCollins in 2003. He also maintains his own Web site, at www.andrewsullivan.com.

JOHN TERBORGH is a James B. Duke Professor of Environmental Science and is Co-Director of the Center for Tropical Conservation at Duke University. He is a member of the National Academy of Science, and, for the past thirty-five years, he has been actively involved in tropical ecology and conservation issues. An authority on avian and mammalian ecology in neotropical forests, Dr. Terborgh has published numerous articles and books on conservation themes. Since 1973 he has operated the Cocha Cashu biological field station in Peru's Manu National Park, where he has overseen the research of more than a hundred investigators. Dr. Terborgh earlier served on the faculties of the University of Maryland and Princeton University. In June 1992 he was awarded a John D. and Catherine T. MacArthur Fellowship in recognition of his distinguished work in tropical ecology, and in April 1996 he was awarded the National Academy of Science Daniel Giraud Elliot Medal for his research, and for his book *Diversity and the Tropical Rainforest.* He serves on several boards and advisory committees related to conservation, including the Wildlands Project, Cultural Survival, The Nature Conservancy, The World Wildlife Fund, and both the Primate and Ecology Specialist Groups of the International Union for the Conservation of Nature.

Of "In the Company of Humans," he writes: "All my life I have sought to live close to nature, best of all, immersed in it. I have thus had unusual opportunities to observe animals, especially at the research station in Peru where I go on 'retreat' for several months each year. There, humans and animals live intermingled lives that bring us into close contact every day. The monkeys, guans, peccaries, and other creatures mentioned in the article are our neighbors, and they, I'm sure, feel a certain curiosity about us, as we do about them. In these close quarters, familiarity has not bred contempt, but rather mutual accommodation, to the apparent satisfaction of all."

MICHAEL S. TURNER is the Bruce V. and Diana M. Rauner Distinguished Service Professor and Chair of the Department of Astronomy and Astrophysics at the University of Chicago. He also holds appointments in the Department of Physics and Enrico Fermi Institute at Chicago and is member of the scientific staff at the Fermi National Accelerator Laboratory. Turner received his B.S. in Physics from the California Institute of Technology (1971) and his Ph.D. in Physics from Stanford University (1978). His association with the University of Chicago began in 1978 as an Enrico Fermi Fellow, and in 1980 he joined the faculty. Turner is a Fellow of the American Physical Society and of the American Academy of Arts and Sciences and is a member of the National Academy of Sciences. He has been honored with the Helen B. Warner Prize of the American Astronomical Society, the Julius Edgar Lilienfeld Prize of the American Physical Society, the Halley Lectureship at Oxford University, and the Quantrell Award for Excellence in Undergraduate Teaching at Chicago.

A cosmologist whose research focuses on the earliest moments of the universe, Turner has made important contributions to inflationary universe theory, the understanding of dark matter and the origin of structure. He and Edward Kolb helped to establish the Theoretical Astrophysics Group at Fermilab and wrote the monograph "The Early Universe." Eleven of Turner's former students and postdocs hold faculty positions at universities in Canada and the United States.

Turner writes: "Cosmology, the scientific study of the origin and evolution of the universe, is a bold and fragile enterprise. The universe is very big and very old; our ability to explore it directly, both in space and in time, is limited. Progress is erratic, and at times haltingly slow. Because of recent conceptual and technological breakthroughs, cosmology is now in the midst of the most exciting period of discovery ever. I wrote this article to describe a few of the forefront issues and to convey the sense of excitement that those of us involved in this great adventure feel every day. While progress is being made toward answering many profound questions about our universe, I choose to focus on a single question: What is the universe made of?

"The answer to this question that is emerging—from the mysterious dark matter that provides the bulk of the gravitational glue that holds the universe together to the even more mysterious dark energy whose repulsive gravity is causing it to speed up—illustrates the dramatic progress that is being made in understanding how the universe is put together, the puzzles that remain to be solved, and the deep (and unexpected) connections that exist between the inner space of the elementary particles and the outer space of the cosmos."

JOHN UPDIKE was born in 1932 in Shillington, Pennsylvania. He graduated from Harvard in 1954 and has lived in Massachusetts as a freelance writer since

1957. His novels have won the Pulitzer Prize, the National Book Award, the American Book Award, the National Book Critics Circle Award, and the Howells Medal. "Transparent Stratagems" appears in his seventh collection of poems, *Americana,* published by Knopf this year. He was inspired, of course, by Sönke Johnsen's striking article in *Scientific American* ("Transparent Animals") but also by a lifelong fear of the ocean and the weird creatures that inhabit it. In his early days as a writer of light verse, he more than once versified scientific truths, with their sometimes abundant poetic, humorous, and philosophical content. His poem here was a throwback to this old literary subspecies, and took much verbal tweaking to arrive at its present form.

JOHN ARCHIBALD WHEELER, in a career spanning more than sixty-five years, has left his fingerprints just about everywhere that physicists have wandered in the last century—from deep within the atom to the vastness of the whole universe. Early in 1939, he was the first American to learn that the nucleus had been split, and in short order he and Niels Bohr provided a theory of nuclear fission. After contributing to the development of fission and fusion weapons during and after World War II, he turned his attention to Einstein's theory of general relativity and what it had to say about the way the world is put together. Almost single-handedly, Wheeler transformed this field from the playground for mathematicians that it had become to the vital field of physics—both theoretical and observational—that it is today. Wheeler has been showered with honors for his research (including memberships in Benjamin Franklin's American Philosophical Society, Isaac Newton's Royal Society of London, and Galileo Galilei's Lincean Academy of Rome). He is known, too, for his inspirational teaching, and holds the Oersted Medal of the American Association of Physics Teachers. He is famous for his coinages, which include "black hole" and "quantum foam." The book *Gravitation* that Wheeler co-authored in 1973 with two of his former students, Charles Misner and Kip Thorne, is a publishing phenomenon, a bestselling 1,200-page advanced monograph, still in print and still moving off the shelves. Among a dozen other books that Wheeler has authored or co-authored are *A Journey into Gravity and Spacetime* (1990), *At Home in the Universe* (1994), and *Geons, Black Holes, and Quantum Foam: A Life in Physics* (1998).

"Relativity is exciting almost beyond measure," he writes, "yet there is nothing so mysterious about it. Quantum mechanics is a different story—an incredibly successful theory that has steered much of twentieth-century science yet remains, at its core, entirely mysterious. When Dennis Overbye of the

New York Times asked me to contribute an essay in honor of the hundredth anniversary of Max Planck's great December 1900 discovery of the quantum principle (the principle that things in nature change in jumps or pulses, not smoothly), I was happy to accept the assignment. In my later years, I come back again and again to the question, 'Why the quantum?' Throughout my career, I have tried to look beyond the immediacies of this or that calculation to ask how it all hangs together. In my vision of the world there is a reason, a simple reason, not only for every individual phenomenon, but for every general theory. This magnificent edifice of quantum mechanics is sitting there with, so far, no clear reason for its being.

"I may not live to see that reason unearthed, but I try, in my small way, to encourage the young to pursue that vision and find the reason. It could make the twenty-first century as dramatically exciting for physicists as the twentieth has been."

Reprinted with permission from *The New York Review of Books.* Copyright © 2000 by NYREV, Inc. "The Virus and the Vaccine" by Debbie Bookchin and Jim Schumacher. Copyright © 2000 by Debbie Bookchin and Jim Schumacher. Reprinted by permission of the authors. Originally published in *The Atlantic Monthly.* "Syphilis and the Shepherd of Atlantis" by Stephen Jay Gould. With permission from *Natural History,* October 2000. Copyright © 2000 by the American Museum of Natural History. "The Good Doctor" by Tracy Kidder. Copyright © 2000 by John Tracy Kidder. Originally published in *The New Yorker.* All rights reserved. "Running Dry" by Jacques Leslie. Copyright © 2000 by *Harper's Magazine.* All rights reserved. Reproduced from the July issue by special permission. "Welcome to Planet Earth" by Robert L. Park. Adapted from "Only Mushrooms Grow in the Dark," from *Voodoo Science: The Road from Foolishness to Fraud* by Robert L. Park. Copyright © 2000 by Robert L. Park. Used by permission of Oxford University Press, Inc. Originally appeared in *The Sciences.* "A Portrait of the Novelist as a Young Scientist" by Alan Lightman. Copyright © 2000 by Alan Lightman. Originally published in the *New York Times* under the title of "Looking Back at the Pure World of Theoretical Physics." "Science, Guided by Ethics, Can Lift Up the Poor" by Freeman J. Dyson. © 2000 by Freeman J. Dyson. Originally published in the *International Herald-Tribune.* Reprinted by permission.

Some slight textual changes, for the sake of clarifying references to dates, have been made, with the authors' approval, in the following essays: "The Small Planets," "The Recycled Generation," "The Genome Warrior," "DNA on Trial," "Death of an Altruist," "Let Them Eat Fat," "The He Hormone," "John Rock's Error," "The Mystery of AIDS in South Africa," "The Virus and the Vaccine," "The Good Doctor," "Running Dry," and "Welcome to Planet Earth."